FLORA ZAMBESIACA

Flora terrarum Zambesii aquis conjunctarum

VOLUME FIVE: PART TWO

FLORA ZAMBESIACA

MOZAMBIQUE

MALAWI, ZAMBIA, ZIMBABWE

BOTSWANA

VOLUME FIVE: PART TWO

Edited by
G. V. POPE

on behalf of the Editorial Board:

S. J. OWENS
Royal Botanic Gardens, Kew

I. MOREIRA
*Centro de Botânica, Instituto de Investigação
Científica Tropical, Lisboa*

G. V. POPE
Royal Botanic Gardens, Kew

Published by the Royal Botanic Gardens, Kew,
for the Flora Zambesiaca Managing Committee
1998

Typeset at the Royal Botanic Gardens, Kew, by Christine Beard

Printed in Great Britain by
Whitstable Litho Printers Ltd., Whitstable, Kent

ISBN 1 900347 41 5

CONTENTS

FAMILY INCLUDED IN VOLUME V, PART 2

94. Rubiaceae (Tribe Vanguerieae)

LIST OF NEW NAMES PUBLISHED IN THIS WORK

Acknowledgements

The Flora Zambesiaca Managing Committee thanks M.A. Diniz and E. Martins of the Centro de Botânica, Lisbon, for their valuable help in reading and commenting on the text.

vi

94. RUBIACEAE

By D. Bridson*

Subfamily CINCHONOIDEAE Raf.
Subfamily **Cinchonoideae** Raf. in Ann. Gén. Sci. Phys. **6**: 81 (1820), as "*Cinchonaria*".
—Verdcourt in Bull. Jard. Bot. État **28**: 280 (1958).
    Ixoroideae Raf. in Ann. Gén. Sci. Phys. **6**: 84 (1820), as "*Ixorinia*".

Rhaphides absent from the tissues. Trees, shrubs or rarely herbs. Indumentum of stem, foliage and inflorescence, if present, of thick-walled non-septate or incompletely septate hairs. Stipules free or partly to fully connate above the axils, interpetiolar**, entire, often topped by an awn, or occasionally divided. Complete heterostyly is rare, but limited heterostyly is present in a few tribes; secondary pollen presentation (see Ixoroid pollination mechanism in Flora Zambesiaca **5**, 1: 2 (1989)) is frequently present. Corolla lobes valvate, contorted or imbricate. Style present; if secondary pollen presentation occurs then the pollen presenter (receptaculum pollinis) clavate to variously capitate with a limited stigmatic surface area, or stigma bifid with receptive surface restricted to inner face; or if secondary pollen presentation is absent then stigma divided into lobes or fusiform, less often capitate. Ovules solitary, or numerous in 1–many locules, attached to the septum or pendulous, or embedded in 1-several placentas, rarely from the base, or occasionally parietal. Fruit dehiscent or indehiscent, dry or succulent, berry-like or drupe-like with woody to cartilaginous pyrenes, occasionally syncarpic. Seeds with endosperm; testa cells in some tribes with very conspicuous pits.

The tribal classification adopted in this Flora treatment is based on Verdcourt in Bull. Jard. Bot. État **28** (1958), with modifications from Bremekamp in Acta Bot. Neerl. **31**: 1–35 (1966) and Robbrecht in Opera Bot. Belg. **1** (1988). However, for practical reasons the tribe Vanguerieae is here published out of sequence — the flora account of this tribe being ready ahead of the others. The position of this tribe in this Flora does not therefore reflect the arrangement accepted by Bridson and Verdcourt.

Where appropriate, comment is provided on Robbrecht's 'Supplement to the 1988 outline of the classification of the Rubiaceae, Index to Genera' (in Opera Bot. Belg. **6**: 173–196 (1993)) and other recent works. The tribal synopses below are diagnostic and do not necessarily account for all characters or all character states outside Africa.

Synopsis of the tribes and subtribes in subfamily Cinchonoideae

Tribe **Vanguerieae** Dumort., Anal. Fam. Pl.: 32 (1829), as "*Vaugnerieae*".
Shrubs, climbers, trees, or more rarely woody-based herbs. Stipules shortly connate, often bearing a lobe or awn. Inflorescences always axillary. Flowers hermaphrodite or unisexual, with secondary pollen presentation present. Corolla lobes valvate and thickened (i.e. with contact zone). Stamens inserted at the throat. Pollen presenter cylindrical, coroniform or mitriform, mostly hollow and attached within or sometimes attached at the base. Ovary 2–10(12)-locular, the ovules solitary, attached to the upper part of the septum, and pendulous. Fruit fleshy, containing 1–10 pyrenes. Pyrenes cartilaginous to strongly woody with apical preformed germination slits. Seeds with oily endosperm and relatively large embryos; radicles superior. Pollen grains porate to colporate, the exine finely to more openly porate or sometimes reticulate. *Genera 38–54.*

---

* Genera 38–48 were prepared from a draft written by B. Verdcourt
** intrapetiolar stipules do occur in New World genera

Tribe **Naucleeae** Miq., Fl. Ned. Ind. 2: 130, 132 (1856). —Ridsdale in Blumea **24**: 307–366 (1978).

Large trees, shrubs or sometimes climbers. Stipules usually free, triangular or ovate. Inflorescence positions various, individual flowers aggregated into spherical heads. Flowers with secondary pollen presentation present. Calyx tubes fused or free, if free then interfloral bracteoles usually present. Stamens inserted in corolla throat. Ovary 2-locular; ovules 2–numerous in each locule. Fruit a succulent syncarp or collection of 2-locular capsules. Testa cells pitted, at least in the capsular-fruited species. Pollen grains generally 3-colporate.

Subtribe **Naucleinae** DC., Prodr. **4**: 343 (1830). —Ridsdale in Blumea **24**: 320 (1978).

Corolla lobes imbricate. Pollen presenter spindle-shaped, with stigmatic region in lower part. Placenta shape and attachment various (but not as in **Adininae**). Ovules spreading in all directions but predominantly pendulous. Fruits connate and fleshy, or dry and free. Seeds variously shaped, sometimes with a narrow wing. *Genera 55–56.*

Subtribe **Adininae** Ridsdale in Blumea **24**: 319 (1978).

Corolla lobes imbricate, rarely valvate. Pollen presenter ellipsoid to capitate. Placenta shortly obovoid, attached to upper third of septum. Ovules few to several, pendulous. Fruits dry and free. Seeds flattened, ± winged or not. *Genus 57.*

Tribe **Cephalantheae** Kunth ex Ridsdale in Blumea **23**: 179 (1976).

Shrubs, trees or sometimes woody climbers. Stipules not connate. Inflorescences of terminal or axillary compact heads; flowers quite separate; interfloral bracteoles present. Flowers with secondary pollen presentation present. Corolla lobes imbricate. Stamens inserted in the corolla throat. Pollen presenter clavate; the stigmatic region across the apex. Ovary 2-locular with an apically attached pendulous ovule in each locule. Fruit fleshy, comprised of 2, 1-seeded pyrenes; pyrenes with apical preformed germination slits. Seeds oblong with an apical stony aril-like structure; endosperm oily. Embryo large with a superior radicle. Exotestal cells with delicate thickenings along inner tangential wall. Pollen grains 3–4-colporate. *Genus 58.*

Robbrecht, in Opera Bot. Belg. **1**: 183 (1988), includes this monotypic tribe in the *Antirheoideae.*

Tribe **Cinchoneae** DC. in Ann. Mus. Natl. Hist. Nat. **9**: 217 (1807).

Subtribe **Mitragyninae** Havil. in J. Linn. Soc., Bot. **33**: 1–97 (1897). —Ridsdale in Blumea **24**: 43–100 (1978).

Coptosapelteae sensu Andersson & Persson in Pl. Syst. Evol. **178**: 89 (1991), non Bremek. ex Darwin (1976).

Trees, shrubs or sometimes hooked climbers. Stipules triangular, ovate or bifid. Inflorescences terminal or axillary, sometimes spicate or contracted spherical heads. Flowers with secondary pollen presentation usually present. Corolla lobes valvate, imbricate or contorted. Stamens inserted in throat. Ovary 2-locular; placentas mostly pendulous with numerous upwardly overlapping, basally attached ovules, or rarely with only 2 or 5–6 ovules per placenta. Fruit mostly capsular, loculicidally and/or septicidally dehiscent. Seeds winged; exotestal cells with thickenings along radial and inner tangential walls. Pollen grains 3-colporate. *Genera 59–61.*

Andersson & Persson (op. cit., 1991) recently restricted *Cinchoneae* to selected New World representatives, transferring the remaining New World genera to the new tribe *Calycophylleae* and placing the Old World taxa in *Coptosapelteae*. Although Robbrecht in Opera Bot. Belg. **6**: 101–141 (1993) accepted this circumscription he expressed well-justified doubts over the inclusion of both the Asiatic genus *Coptosapelta* and the African genus *Crossopteryx*. Since we share his opinion concerning the nominating genus we have adopted a broadened concept of *Mitragyninae* as a provisional solution.

Tribe **Virectarieae** Verdc. in Kew Bull. **30**: 366 (1975).

Herbs or subshrubby herbs. Stipules triangular or fimbriate. Flowers not heterostylous. Corolla lobes valvate. Ovary 2-locular; the placentas attached to the central septum; ovules numerous. Stamens inserted at throat, well exserted. Stigma capitate, much exserted. Fruit capsular, ovoid or subglobose, splitting in a plane at right-angles to the central septum, very often 1 valve falling off and the other persisting (but this dehiscence is not perfect in all species). Seeds small. *Genus 62.*

Bremekamp, in Acta Bot. Neerl. **15**: 1–33 (1966), includes this in the otherwise monotypic tribe *Ophiorrhizeae*, while Robbrecht (op. cit., 1988) includes it in the *Hedyotideae*.

Tribe **Sabiceeae** Bremek. in Rec. Trav. Bot. Néerl. **31**: 253 (1934), as "*Sabiceae*". —Andersson in Opera Bot. Belg. **7**: 157 (1996).

Cinchonaceae subtribe *Sabiceinae* Griseb., Fl. Brit. W. Ind. Is.: 322 (1861), as "*Sabicieae*".

Herbs or more often shrubs or climbers. Leaves very often pubescent, velvety or felted beneath; stipules not fused at the base, entire or sometimes laciniate but not bipartite. Inflorescences axillary. Flowers heterostylous or homostylous. Corolla lobes valvate and thickened (i.e. with contact zone) or sometimes reduplicate-valvate. Ovary 2–5-locular; placentas elongate attached throughout their length, each with numerous erect ovules. Fruit ± fleshy, berry-like. Seeds very small, angular or round; endosperm copious, oily; testa cells pitted. Pollen grains 3(4)-zonoaperturate, usually porate or colporate with a short and distinct colpus. *Genera 63–64.*

Andersson (cit. sup.) tentatively included the monotypic Asiatic tribe *Acranthereae* Bremek. ex Darwin in synonymy of *Sabiceeae*. Previously Robbrecht (op. cit., 1988) placed it in *Isertieae*, later (op. cit., 1993) excluding it.

Tribe **Mussaendeae** Hook.f. in Gen. Pl. **2**: 8 (1873), pro parte excl. *Isertia*.
   Isertiae sensu auctt. pro parte, non A. Rich. ex DC.

Mostly trees, shrubs or woody climbers. Stipules entire or bifid, sometimes divided almost to the base. Inflorescences usually terminal. Flowers heterostylous, rarely the anthers and stigma both included in the corolla tube, or homostylous. Often a few calyx lobes in each inflorescence developing into coloured leaf-like structures. Corolla lobes induplicate-valvate or reduplicate-valvate or less often imbricate; throat very often with a brightly coloured often pubescent eye. Ovary 2-locular, each locule with numerous ovules. Fruit fleshy or dry, indehiscent or rarely capsular (*Pseudomussaenda*). Seeds very small, not winged, angular or round; endosperm oily; testa cells pitted. Pollen grains 3–4-zonocolporate, colpus often short.

The *Mussaendeae* as circumscribed by Hook.f. was illegitimate because it included the type genus of the earlier published tribe *Isertieae*. The *Mussaendeae* was therefore included in synonymy under *Isertieae* in the F.T.E.A. treatment (1988), and by Andersson in Opera Bot. Belg. **7**: 157 (1996) who at the same time recognised *Sabiceeae* as a distinct tribe. We here accept tribal status for *Sabiceeae* but prefer to recognise *Mussaendeae* as distinct from the *Isertieae*.

In the strict sense the tribe *Isertieae* occurs only in the New World. However, it was very broadly defined by Robbrecht (op. cit., 1988) who included in it the *Acranthereae*, the *Heinsenieae* and *Sabiceeae* as well as many raphide-bearing genera. To accommodate the latter he emended his description (op. cit., 1993) to include "raphides present or absent". All raphide-bearing genera should be excluded from the *Isertieae* and returned to the *Hedyotideae*.

Subtribe **Mussaendinae**
Stipules entire to deeply 2-lobed. Corolla lobes induplicate-valvate or reduplicate-valvate. *Genera 65–66.*

Pseudomussaenda was tentatively included in *Condamineae* by Robbrecht (op. cit., 1988). However, recent studies by Puff, Igersheim and Rohrhofer in Bull. Jard. Bot. Nat. Belg. **62**: 35–68 (1993) have demonstrated that it is closely allied to *Mussaenda*, as alleged by Verdcourt (in F.T.E.A., Rubiaceae: 467 (1988)).

Subtribe **Heinsiinae** Verdc. in Bull. Jard. Bot. État **28**: 281 (1958) (as subtribe of "*Mussaendeae*"), pro parte, excluding *Bertiera*.
   Heinsieae (Verdc.) Verdc. in Kew Bull. **31**: 183 (1976), pro parte, excluding *Bertiera*.
Stipules divided almost to the base. Corolla lobes imbricate, quincuncial. *Genus 67.*

F. Hallé in Adansonia **1**: 266 (1961), N. Hallé, Fl. Gabon **12**: 131 (1966), Robbrecht (op. cit., 1988) and Andersson in Opera Bot. Belg. **7**: 157 (1996) all put *Heinsia* in the *Isertieae* (incl. *Mussaendeae*). This relationship has been confirmed by Robbrecht, Rohrhofer & Puff in Opera Bot. Belg. **6**: 101–141 (1993).

Tribe **Aulacocalyceae** Robbr. & Puff in Bot. Jahrb. Syst. **108**: 126 (1986).
Shrubs, rarely trees. Stipules triangular, caducous. Inflorescences terminal on much abbreviated branches, often above a single leaf (TAB. **46**/3, 4). Flowers presumably with secondary pollen presentation. Corolla lobes contorted to the left. Stamens inserted in the corolla throat exserted or included. Pollen presenter filiform to club-shaped, 10-ribbed, the two arms appressed. Ovary 2-locular, each pendulous placenta bearing 1–2(3) embedded ovules. Fruit fleshy, containing (1)2–4(5) seeds embedded in placental tissue which may be difficult to distinguish from the fruit wall. Seed lacking seed coat. Embryo with superior radicle. Pollen grains 3-colp(oid)orate. *Genera 68–69.*

Tribe **Gardenieae** DC., Prodr. **4**: 342, 347 (1830), as "*Gardeniaceae*".
Trees, shrubs or climbers. Inflorescences terminal, axillary or lateral. Flowers with secondary pollen presentation present. Corolla lobes contorted to the left or occasionally to the right.

Although this tribe is still not fully understood the division into two subtribes as suggested by Robbrecht & Puff (cit. sup., 1986) has been followed here. We recognise a third subtribe to accommodate *Bertiera*.

Subtribe **Bertierinae** K. Schum. in Engl. & Prantl, Nat. Pflanzenfam., Div. IV, **4**: 73 (1891), as "*Bertiereae*".

Shrubs or small trees, or less often dwarf shrubs, climbers or stoloniferous herbs. Stipules entire, shortly connate above the axils, not awned. Inflorescences terminal or less often axillary, much branched, spike-like or capitate. Corolla lobes contorted to the left. Stamens inserted in upper part of corolla tube. Pollen presenter club-shaped (the two limbs appressed), 10-winged, only the tip exserted from corolla tube. Ovary 2-locular; placentas peltate, covered with ovules on both sides. Fruit berry-like, fleshy or coriaceous. Seeds numerous, not embedded in placental tissue, 1–2 mm across, angular; seed coat reticulate, areolate. Pollen grains single, 3(5)-colporate. *Genus 70*.

Verdcourt (1958) included *Bertiera* in his *Heinsieae*. However, N. Hallé, Fl. Gabon **17**: 32 (1970) treated it as an aberrant member of the *Gardenieae*. This position has been confirmed by Robbrecht, Rohrhofer & Puff in Opera Bot. Belg. **6**: 101–141 (1993), who regard it as occupying an isolated position in the very diverse *Gardeniinae*. Recognising it as a subtribe is considered here to be a more practical solution.

Subtribe **Gardeniinae**

Shrubs or small trees, sometimes climbers; branching often sympodial. Inflorescences terminal, pseudo-axillary or leaf-opposed. Flowers often large, (4)5(12)-merous; corolla contorted to the left, or sometimes to the right. Ovary 1–9-locular (or incompletely so); placentas axile or parietal; ovules numerous or rarely few. Fruits mostly large, woody, leathery or sometimes fleshy. Seeds usually embedded in fleshy or pulpy placental tissue, occasionally free, mostly lenticular but occasionally angular; endosperm entire; testa cells with thickenings mainly along the radial and inner (rarely also outer) tangential walls. Pollen grains single or in tetrads, 3-porate or 3-colporate. *Genera 71–81*.

Subtribe **Diplosporinae** Miq., Fl. Ned. Ind. **2**: 237 (1857), as "*Diplosporeae*". —Robbrecht & Puff in Bot. Jahrb. Syst. **108**: 114 (1986), excl. *Cremasporeae* in syn.

Shrubs, or occasionally pyrophytic subshrubs; branching usually monopodial. Stipules mostly partly to fully connate, frequently awned. Inflorescences axillary or rarely terminal (*Argocoffeopsis*). Flowers with secondary pollen presentation, but rarely ± heterostylous (*Tricalysia* subgen. *Ephedranthera* and *Belonophora*), mostly small, 4–5(8)-merous. Corolla contorted to the left. Ovary 2-locular; placentas axile; ovules 1–many. Fruit mostly small. Seeds usually not embedded in placental tissue, or wholly or partly enveloped by arilloid placental tissue; endosperm entire, or rarely ruminate; testa cells isodiametric or elongated, thickenings mostly absent or ± weak, smooth along outer tangential walls; embryo with inferior radicle. Pollen grains single, 3-colporate. *Genera 82–86*.

Robbrecht & Puff, in Bot. Jahrb. Syst. **108**: 126 (1986), tentatively place *Belonophora* with *Aulacocalyceae* on account of its superior embryo radicles. Because of its axillary inflorescences and well developed seed coat we prefer to provisionally maintain it in the *Diplosporinae*. Robbrecht (op. cit., 1988) included *Cremaspora* in this subtribe, but it is maintained as a separate monotypic tribe in this work, see below.

Tribe **Cremasporeae** (Verdc.) S.P. Darwin in Taxon **25**: 601 (1976) (excluding *Scyphiphora*). *Cremasporinae* Verdc. in Bull. Jard. Bot. État **28**: 281 (1958).

Shrubs, small trees or scramblers. Inflorescences axillary. Stipules not awned, soon caducous. Flowers hermaphrodite with secondary pollen presentation present. Corollas small, the lobes contorted to the left. Style filiform, exserted, hairy, bilobed only at the extreme apex. Ovary 2-locular; ovules solitary in each locule, descending. Fruit small, indehiscent, 1–2-seeded; pericarp leathery; placenta elongate, lying along the whole side of the seed; radicle appearing inferior; testa characteristically finely transversely wrinkled-striate; exotestal cells with strongly thickened radial walls; endosperm not ruminate. Pollen 3-colpate. *Genus 87*.

Robbrecht & Puff in Bot. Jahrb. Syst. **108**: 115 (1986) include this in *Gardenieae* subtribe *Diplosporinae*.

Tribe **Octotropideae** Beddome, Fl. Sylv. S. India [Forester's Manual]: cxxxiv-12 (1872). *Hypobathreae* (Miq.) Robbr. in Bull. Jard. Bot. État **50**: 75 (1980).

Trees or shrubs. Stipules mostly triangular, not awned, entire. Inflorescences axillary or supra-axillary. Flowers hermaphrodite or rarely functionally unisexual; secondary pollen presentation present. Corolla lobes contorted to the left. Ovary 2-locular, rarely 2-locular above and 1-locular below, (or 1-locular outside the Flora Zambesiaca area); the ovules 1–many, on axile placentas, (pendulous from placenta attached to apex of the septum or 2-seriate along the septum), or occasionally on 2 parietal placentas (but not in Flora Zambesiaca area). Fruits mostly small, with fleshy or rarely leathery walls, with 1–many pendulous or horizontal seeds. Seeds with endosperm ruminate or entire; testa often appearing fibrous with a fingerprint-like pattern at low magnifications; exotesta mostly folded with regard to endotesta, composed of elongated cells with thickenings along the radial wall or less often ± isodiametric, parenchyma-like; radicle superior or orientated towards the septum in horizontal seeds. Pollen grains single, 3-colporate. *Genera 88–91*.

Tribe **Pavetteae** Dumort., Anal. Fam. Pl.: 33 (1829). —Robbrecht in Pl. Syst. Evol. **145**: 105–118 (1984). —Bridson & Robbrecht in Bull. Jard. Bot. Belg. **55**: 83–115 (1985).

Trees, shrubs, sometimes scandent, or occasionally pyrophytic subshrubs; branching usually monopodial, often sympodial in *Pavetta*. Stipules partly to fully connate, often awned, or sometimes ± triangular. Inflorescences terminal on main or lateral branches, pseudo-axillary or occasionally cauliflorous. Flowers with secondary pollen presentation present. Corolla small, sometimes quite long; lobes 4–5(6), contorted to the left. Anthers inserted at the corolla throat, usually exserted, locellate in a few species. Style exserted, rarely included; pollen presenter fusiform to clavate, not lobed or shortly to distinctly lobed. Ovary 2(7)-locular; ovules 1–numerous on fleshy placentas attached near the middle of the septum, or less often the placenta basal with single ascending ovule. Fruit mostly small, sometimes quite large, berry-like or drupe-like with relatively thin-walled pyrenes, or occasionally ± woody, 1–2(7)-locular, 1–many-seeded. Seeds not embedded in placental tissue with or without an excavation on ventral face; endosperm ruminate in several genera; radicle inferior; often with a characteristic annular zone of thickened exotestal cells around the excavation. Pollen grains colporate with perforate to reticulate exine. *Genera 92–96.*

Tribe **Coffeeae** DC. in Ann. Mus. Natl. Hist. Nat. **9**: 217 (1807), as "*Coffeaceae*".

Shrubs or small trees; branching monopodial or sometimes sympodial. Stipules shortly connate, shortly awned. Flowers hermaphrodite, moderately small, (4)5–12-merous; secondary pollen presentation lacking. Style exserted or included; stigma bilobed. Ovary 2-locular; ovules solitary. Fruit small to medium-sized, drupaceous; endocarp crustaceous. Seeds with a deep ventral groove; endosperm entire; embryo with inferior radicle; testa crushed during development of the endosperm, comprised of thin elongate parenchymatic cells usually containing many ± isolated fibres. Pollen grains single, 3–4-colporate, with reticulate exine and without endoapertures. *Genus 97.*

Robbrecht & Puff in Bot. Jahrb. Syst. **108**: 122–123 (1986) restrict this tribe to the genera *Coffea* and *Psilanthus*, and exclude the Asiatic genus *Nostolachma* T. Durand (*Lachnostoma* Korth.) which Leroy (ASIC 9ᵉ Colloque Londres: 473–477 (1980)) associated with *Coffeeae.*

Key to the tribes and subtribes in subfamily Cinchonoideae

1. Herbs or subshrubs; corolla lobes valvate, very narrowly lanceolate, ± equalling the tube in length, suberect; style and stamens exserted; capsule often with 1 persistent and 1 falling valve; seeds numerous, unwinged · · · · · · · · · · · · · · · · · · · **17. Virectarieae** (vol. 5,3)
– Small shrubs to trees, climbers or pyrophytic plants; corolla lobes if valvate then not as above; fruit not as above; seeds 1–numerous, winged or unwinged · · · · · · · · · · · · · · 2
2. Corolla lobes valvate; pollen presenter knob-like, globose, cylindrical or mitriform, often ridged, 2–several-lobed at apex when mature; pyrenes woody or sometimes crustaceous; inflorescence always axillary · **13. Vanguerieae**
– Corolla lobes contorted or imbricate, if valvate then stigmas 2-lobed or pollen presenter not as above; occasionally heterostylous; if pyrenes present then cartilaginous to crustaceous; inflorescences various · 3
3. Fruit a dehiscent capsule; seeds numerous, winged; inflorescences terminal, perfectly spherical heads (if heads solitary see couplet 4), spike-like (with corollas glabrous), or corymbose · · · · · · · · · · · · · · · · · · · **16. Cinchoneae** subtribe **Mitragyninae** (vol. 5,3)
– Fruit leathery to succulent, very rarely dehiscent; seeds 2–numerous, unwinged; inflorescences axillary or terminal, only occasionally spherical or spike-like · · · · · · · · · 4
4. Inflorescences/infructescences terminal, spherical, with calyx tubes (ovaries) joined together or free · 5
– Inflorescences/infructescences sometimes tightly capitate but not as above · · · · · · · · · 7
5. Ovules solitary and pendulous in each locule; calyx tubes free; fruits free and fleshy · **15. Cephalantheae** (vol. 5,3)
– Ovules few to numerous in each locule; calyx tubes free or fused together; fruits forming a fleshy syncarp, or if free then dry · 6
6. Inflorescence heads 2–5 cm in diameter; pollen presenter spindle-shaped; ovules numerous; fruits dry and free or fused to form a fleshy syncarp · **14. Naucleeae** subtribe **Naucleinae** (vol. 5,3)
– Inflorescence heads 1.5–2.5 cm in diameter; pollen presenter ± capitate; ovules few (2–5); fruits sepicidal capsules · · · · · · · · · · · · · · · **14. Naucleeae** subtribe **Adininae** (vol. 5,3)
7. Corolla lobes valvate (simple, induplicate or reduplicate) or imbricate; flowers heterostylous or homostylous; ovules numerous in each locule; shrubs, trees or climbers, rarely herbs · 8

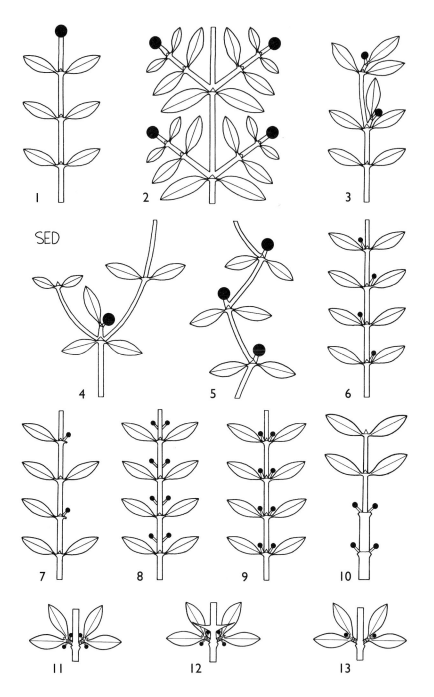

Tab. 46. Arrangement of inflorescences in Cinchonoideae. 1–2, terminal-types. 3–7, *Gardeniinae*-types (3 & 4, *Rothmannia*-types; 5, *Sherbournia*-type; 6, *Oxyanthus*-type; 7, *Aidia*-type). 8, supra-axillary-type. 9–10, axillary-types (10, axillary on mature stems). 11–13, inflorescences borne on brachyblasts (11, axillary; 12, below spines, axillary; 13, terminal). Drawn by Sally Dawson.

- Corolla lobes contorted; flowers with secondary pollen presentation usually but not always present; ovules solitary, or few to numerous in each locule · 10
8. Stipules entire, sometimes emarginate; inflorescences axillary; never with showy calyx lobes; corolla lobes simply valvate; ovary (2)3–5-locular · · · · · · · · **18. Sabiceeae** (vol. 5,3)
- Stipules 1 or 2-lobed; inflorescences terminal; calyx often with one–few expanded showy lobes in each inflorescence; corolla often with showy eye at throat; corolla lobes induplicate-valvate, reduplicate-valvate or imbricate; ovary 2-locular · · · · · · · · · · · · · · 9
9. Stipules entire to deeply 2-lobed; corolla lobes induplicate-valvate or reduplicate-valvate · **19. Mussaendeae** subtribe **Mussaendinae** (vol. 5,3)
- Stipules 2-lobed virtually to base; corolla lobes imbricate · **19. Mussaendeae** subtribe **Heinsiinae** (vol. 5,3)
10. Inflorescences axillary (in both axils) (TAB. **46**/9, 10) or rarely terminal on short lateral spurs (*Argocoffeopsis*); flowers not showy ; fruit not large · 11
- Inflorescences terminal, lateral (in only one axil), pseudo-axillary (as a result of extreme reduction of lateral branches), supra-axillary (TAB. **46**/1–8) or if truly axillary then usually with showy flowers and large fruit · 14
11. Calyx limb mostly reduced to a rim; seed 1 in each locule, grooved on the ventral face (typical coffee-bean) · **25. Coffeeae** (vol. 5,3)
- Calyx limb clearly present, usually persistent in fruit; seeds 1–several in each locule, not grooved · 12
12. Inflorescences lax or fasciculate; corolla tube often funnel-shaped, sometimes very short; ovules usually pendulous; seeds mostly with fibrous fingerprint-like striations; embryo with radicle superior · **23. Octotropideae** (vol. 5,3)
- Inflorescences always fasciculate; corolla tube cylindrical or funnel-shaped, never very short; ovules and seeds not as above; if radicle superior then corolla tube cylindrical with style arms below the anthers · 13
13. Style arms usually divergent, mostly glabrous, exserted or included; ovules 1–many per locule, often impressed on the placenta; seeds not as below (exotestal cells without thickenings) · · · · · · · · · · · · · · · · · · · **21. Gardenieae** subtribe **Diplosporinae** (vol. 5,3)
- Style undivided save at tip, pubescent, exserted; ovules 1 per locule, descendent; seeds finely transversely wrinkled-striate (exotestal cells with thickenings) · **22. Cremasporeae** (vol. 5,3)
14. Inflorescences elongate, often spike-like (sometimes capitate but not in the Flora Zambesiaca area); corolla hairy outside, not markedly showy; fruit berry-like; seeds numerous, angular · · · · · · · · · · · · · · · · **21. Gardenieae** subtribe **Bertierinae** (vol. 5,3)
- Inflorescences not as above; corolla if hairy outside usually showy; fruit various; seeds various or occasionally angular · 15
15. Inflorescences terminal (sometimes on reduced leafless branches, or sometimes these branches are further reduced so that the inflorescence is pseudo-axillary); corollas mostly salver-shaped or with reflexed lobes, occasionally showy; ovules 1–few or numerous in each of the 2 locules; seeds free, either with a ± circular excavation on ventral face or with a ruminate endosperm · **24. Pavetteae** (vol. 5,3)
- Inflorescences various but if terminal then corolla very showy and mostly campanulate or funnel-shaped and seeds numerous in pulpy tissue · 16
16. Flowers on very short branches above a single leaf (TAB. **46**/3, 4); corolla tube not more than 3 cm long; placenta pendulous with 1–2(3) embedded ovules; fruit fleshy, not more than 1.3 cm across; seeds lacking a testa; embryo with radicle superior · **20. Aulacocalyceae** (vol. 5,3)
- Inflorescence various, if as above then corolla and fruit usually larger; ovules and seeds mostly numerous · · · · · · · · · · · · · · · · · · **21. Gardenieae** subtribe **Gardeniinae** (vol. 5,3)

*Artificial key to genera**

1. Herbs, sometimes subshrubs (if climbing see Key 3); corolla lobes always valvate; rhaphides present in all save *Virectaria* and *Stipularia*, if raphides absent and inflorescences clearly axillary go to Key 2 · **Key 1**

---

* including Rubioideae updated from part 1. Choice is not critical for plants with ambiguous habit as these have been included in more than one key.

   – Plants woody, or sometimes with ± herbaceous shoots from a woody rootstock (pyrophytes); corolla lobes valvate, contorted or sometimes imbricate; rhaphides present or absent · · 2

2. Climbing and/or scandent plants · **Key 3** (p. 222)
   – Plants not climbing, but sometimes scrambling · 3

3. Plants subshrubby to shrubby, or occasionally with ± herbaceous shoots from a woody or rhizomatous rootstock, with the herbaceous shoots 0.05–2 m tall and mostly burnt off by annual fires (pyrophytes) · **Key 2** (p. 220)
   – Shrubs or trees (0.75)2–20 m tall · **Key 4** (p. 222)

Key 1
(Herbaceous genera)

1. Leaves and stipules similar, in whorls of 4–8; herbs and herbaceous climbers with ± rotate corollas (or tubular in introduced plants); ovules solitary in each locule; fruits bilobed, indehiscent; plants often adhesive due to prickles and harsh hairs (*Rubieae*) · · · · · · · · · 2
   – Leaves clearly different from the stipules, opposite, or less often in whorls of 3(5); corolla mostly clearly tubular · 4

2. Calyx with 4–6 distinct teeth; flowers in small terminal involucrate heads; corolla tube cylindrical · **Sherardia arvensis** L. (introduced)
   – Calyx limb reduced to a rim; inflorescence not involucrate; corolla rotate · · · · · · · · · · 3

3. Petioles very well developed, 1–8 cm long (in one species ± winged and difficult to tell from young leaf blades); corolla usually 5-merous · · · · · · · · · · · · **36. Rubia** (see 5,1: 193)
   – Petioles very short or leaves sessile; corolla usually 4-merous · **37. Galium** (see 5,1: 195)

4. Leaf blades 10–34 cm long, densely felted underneath with cream-coloured hairs; stipules red, 3–4 cm long; ovary (2)3–5-locular with numerous ovules/seeds · · · · · · · · · · · · · · ·
· **64. Stipularia africana** Beauv. (vol. 5,3)
   – Leaves and stipules much smaller; if ovules/seeds numerous then locules 2 · · · · · · · · · 5

5. Ovules (or seeds) 2–many in each locule · 6
   – Ovules (or seeds) solitary in each locule, or flowers male · · · · · · · · · · · · · · · · · · · 19

6. Corolla lobes very narrowly lanceolate and about as long as, or longer than, the corolla tube; stigma capitate; flowers not heterostylous, both style and stamens exserted; capsule often with 1 persistent valve and 1 falling valve; small shrubs and herbs without rhaphides
· **62. Virectaria major** (K. Schum.) Verdc. (vol. 5,3)
   – Corolla lobes usually not so narrow and often shorter than the tube; stigma usually bilobed; flowers often heterostylous, with either stamens or style included; capsule mostly opening at the beak; rhaphides present (usually easily visible in ovary or fruit wall) · · · · · · · · · · 7

7. Inflorescences consisting of very lax elongated cymes, axillary (lateral) but only one per node; flowers 5-merous and leaves uninerved; decumbent herbs of wet places with small white or blue flowers · **25. Pentodon** (see vol. 5,1: 116)
   – Inflorescences not as above; if flowers 5-merous, then other characters not so combined · 8

8. Flowers mostly 4-merous; leaf-blades frequently linear to narrowly elliptic and uninerved, or sometimes elliptic to ovate with lateral nerves ± discernible · · · · · · · · · · · · · · · · · 9
   – Flowers mostly 5-merous; leaf-blades mostly elliptic to ovate with obvious lateral venation, but if uninerved then a decumbent subshrub · 17

9. Flowers solitary, or fascicled at numerous nodes forming long interrupted spike-like inflorescences; leaves linear-subulate, rather sparse; corolla tube 1.5–2 mm long · · · · · · ·
· **24. Manostachya** (see vol. 5,1: 113)
   – Features not combined as above · 10

10. Beak of capsule as long as, or longer than, the rest of the capsule · · · · · · · · · · · · · · 11
   – Beak of capsule shorter than the rest of the capsule · 12

11. Robust subshrubby herbs with distinctly discolorous leaves; capsule rounded at the base
· **21. Hedythyrsus** (see vol. 5,1: 105)
   – Delicate herbs; leaves not distinctly discolorous; capsule cordate at base · · · · · · · · · · · ·
· **20. Mitrasacmopsis** (see vol. 5,1: 103)

12. Corolla tube cylindrical, at least 2 cm long; anthers included and style exserted; flowers not heterostylous · **19. Conostomium** (see vol. 5,1: 101)
   – Corolla tube cylindrical or funnel-shaped, less than 2 cm long (sometimes with included anthers and exserted style but then usually heterostylous), or if 2 cm long then with both anthers and stigmas included · 13

13. Anthers and stigmas included, the latter always overtopped by the former; corolla tube narrowly cylindrical · **18. Kohautia** (see vol. 5,1: 85)
 – Anthers and/or stigmas exserted or if both included then anthers overtopped by the stigma; corolla tube cylindrical or funnel-shaped · 14
14. Stems strict, erect, usually unbranched; leaves linear or filiform; stipule sheath tubular, truncate or with 2 minute teeth; seeds dorsiventrally flattened ·
 · **23. Amphiasma** (see vol. 5,1: 111)
 – Stems lax or erect, mostly branched; leaves sometimes linear; stipular sheath scarcely tubular; seeds angular, subglobose or ovoid-trigonous · 15
15. Capsule with a thick bony wall and a solid beak, tardily dehiscent; prostrate perennial herbs often forming small mats · **26. Lelya** (see vol. 5,1: 118)
 – Capsule with a thinner horny wall, with or without a beak but never solid, early dehiscent; plants upright or prostrate · 16
16. Capsule opening both septicidally and loculicidally; erect perennial herbs or subshrubs; inflorescences many-flowered, corymbose or subglobose heads; seeds ovoid-trigonous
 · **22. Agathisanthemum** (see vol. 5,1: 107)
 – Capsule opening loculicidally; herbs annual or perennial, erect or prostrate, rarely subshrubby; inflorescences various but seldom many-flowered; seeds angular or subglobose
 · **27. Oldenlandia** (see vol. 5,1: 120)
17. Flowers solitary, axillary or pseudo-terminal; subshrub of rock crevices; leaves small revolute · **16. Batopedina** (see vol. 5,1: 82)
 – Flowers in more extensive inflorescences; habit of plants not as above; leaves larger, not revolute · 18
18. Inflorescences capitate, or lax much branched compound cymes, not elongating into simple "spikes" in fruit, although individual branches sometimes become spicate; corolla usually without a showy eye, the bases of the lobes not connate; fruit globose or obpyramidal, less often ovoid-oblong · · · · · · · · · · · · · · · · · **14. Pentas** (see vol. 5,1: 66)
 – Inflorescences capitate, later elongating into a long simple "spike", rarely with axillary spikes from the upper axils and frequently with solitary flowers at the lower nodes; corolla with a crimson hairy eye formed by the connate bases of corolla lobes; fruits oblong · **15. Otomeria** (see vol. 5,1: 78)
19. Style topped by a pollen presenter which is globose, cylindrical or mitriform and often ridged, 2–several-lobed at the apex when mature; flowers never heterostylous; inflorescences always axillary; fruit with 2–several woody pyrenes · · · · · · · · · · · · · · · · ·
 · **Vanguerieae** - *go to Vanguerieae key, couplet 14, p. 229*
 – Style lacking a pollen presenter; flowers frequently heterostylous; fruit occasionally with pyrenes (but then inflorescence terminal) or sometimes with one many-seeded stone (putamen) · 20
20. Plants usually ericaceous (heath-like) in appearance, often dioecious; flowers unisexual, wind-pollinated; stigmas long and feathery; corolla markedly inconspicuous; disk absent; fruit dehiscing into 2 mericarps · 21
 – Plants generally not ericaceous in appearance; flowers hermaphrodite, insect-pollinated; stigmas not as above; corollas inconspicuous or moderately showy; disk present; fruit various · 22
21. Inflorescence made up of axillary, ± sessile flower clusters; fruit supported by a carpophore
 · **29. Anthospermum** (see vol. 5,1: 150)
 – Inflorescence terminal, paniculate to thyrso-paniculate, fruit never supported by a carpophore · **30. Galopina** (see vol. 5,1: 160)
22. Corolla tube filiform or narrowly funnel-shaped; calyx lobes unequal, 1 or more enlarged; flowers not heterostylous, both stamens and style exserted; ovary bicarpellate, but often with one carpel reduced in fruit · · · · · · · · · · · · · · · · **28. Otiophora** (see vol. 5,1: 142)
 – Corolla tube not markedly narrow; calyx lobes usually equal; flowers frequently heterostylous; ovary 2–5-locular; carpel reduction unusual in fruit · 23
23. Ovules pendulous from near apex of locule; fruit dry or somewhat fleshy, indehiscent or tardily splitting into cocci · 24
 – Ovules erect from the locule base, or attached near middle of the septum; fruit drupe-like, or dry and variously dehiscent · 25
24. Slender annual herb; flowers 4-merous, white; corolla tube 3 mm long; fruit breaking off to leave a small cup-like structure which is covered in masses of rhaphides (TAB. 13) · · · · ·
 · **11. Paraknoxia** (see vol. 5,1: 57)

- Plants often perennial and more robust; flowers 5-merous, usually bright blue; corolla tube exceeding 3 mm long; cup-like structure not developed under the fruit · **12. Pentanisia** (including *Chlorochorion*) (see vol. 5,1: 57)

25. Ovule erect from the locule base; flowers always terminal, often pedicellate; fruit drupe-like; creeping or sometimes shrubby herbs, often on forest floors · · · · · · · · · · · · · · 26
- Ovule attached near middle of the septum; flowers sessile in axils, or in terminal heads; fruit dry (often splitting) or less often succulent; herbs or subshrubs, usually in open situations · 27

26. Leaf blades seldom cordate at the base; pyrenes not dehiscent; seeds with a red-brown or purplish testa; endosperm often ruminate; subshrubby herbs · **1. Psychotria** (partly) (see vol. 5,1: 9)
- Leaf blades mostly cordate at the base; pyrenes with a ± well-marked line of dehiscence; seed with a pale testa; endosperm not ruminate; usually small creeping herbs · **2. Geophila** (see vol. 5,1: 32)

27. Ovary 3-locular; stigma 3-lobed; fruit with 3 cocci · · · · · · **35. Richardia** (see vol. 5,1: 190)
- Ovary 2-locular; stigmas 2-lobed or capitate; fruits indehiscent, capsular with 2 valves, with 2 cocci or circumscissile · 28

28. Fruit circumscissile about its middle, the top coming off like a lid; flowers in dense globose nodal clusters, minute (corolla tube up to 2 mm long); seeds with a ventral impressed X-like pattern · **34. Mitracarpus** (see vol. 5,1: 188)
- Fruit indehiscent or opening by longitudinal slits, or 2-coccous; flowers in axillary or terminal clusters, usually larger; seeds lacking an X-like pattern · · · · · · · · · · · · · · · · · 29

29. Succulent creeping plant of the seashore with imbricate leaves; stipules quite broad, sheathing, with very short processes; stems rooting at the nodes; fruits indehiscent; seeds not lobed · **31. Phylohydrax** (see vol. 5,1: 162)
- Plant not a littoral succulent, or if somewhat so (*Diodia* subgen. *Pleiaulax*) then leaves not imbricate; stipules with longer processes; stems not rooting at the nodes; fruits dividing into cocci; seeds lobed · 30

30. Fruit with 2 indehiscent or ± indehiscent cocci · · · · · · · · · · **32. Diodia** (see vol. 5,1: 162)
- Fruit opening from apex to base or base to apex, or splitting into 2 cocci, one or both dehiscent · **33. Spermacoce** (see vol. 5,1: 165)

Key 2
(Pyrophytic genera)

1. Calyx tubes joined together; fruits fleshy, fused into syncarps; corolla lobes valvate; rhaphides present (especially in ovary and fruit walls) · **7. Morinda angolensis** (R. Good) F. White (see vol. 5,1: 47)
- Calyx tubes always free; fruit not compound; corolla lobes various; rhaphides present or absent · 2

2. Ovules (or seeds) few–numerous in each locule · 3
- Ovules (or seeds) solitary in each locule · 7

3. Corolla lobes valvate; flowers heterostylous; rhaphides present; capsules dehiscent; stipules fimbriate · **14. Pentas** (partly) (see vol. 5,1: 66)
- Corolla lobes contorted; flowers not heterostylous; rhaphides absent; fruit indehiscent; stipules not fimbriate · 4

4. Flowers clearly axillary in both axils; inflorescence fasciculate with cupular bracteoles; corolla tube less than 12 mm long, ± equal to the lobes in length; fruit 5–15 mm in diameter · 5
- Flowers terminal; inflorescence not as above; corolla tube mostly longer and exceeding the lobes; fruit mostly larger · 6

5. Calyx limb 4–5 mm long, entire in bud, splitting into 2–3 lobes in mature flowers; anthers basifixed; fruit 10–15 mm across; linear bacterial nodules present on midrib and petioles · **83. Sericanthe suffruticosa** (Hutch.) Robbr. (vol. 5,3)
- Calyx limb not as above; anthers dorsifixed; fruit 6–8 mm across; bacterial nodules absent · **82. Tricalysia cacondensis** Hiern (vol. 5,3)

6. Flowers 1–3; corollas pubescent outside, lobes (4)5(6); seeds free, with a small excavation on the ventral face; stipules 6–16 mm long · · · · · · · · · · **92. Leptactina** (partly) (vol. 5,3)
- Flowers solitary; corollas glabrous outside, lobes 5 or 6–8; seeds in a pulpy placenta, not excavated; stipules not more than 7 mm long · · · · · · · · **74. Gardenia** (partly) (vol. 5,3)

7. Corolla lobes contorted; corolla tube distinctly longer than the lobes; flowers not heterostylous, anthers and style always exserted; rhaphides absent; seeds with excavation on ventral face; inflorescence terminal on leafless lateral branches (sometimes reduced or even entirely suppressed); bacterial nodules present in leaf blades · · · · · · · · · · · · · · ·
· **94. Pavetta** (partly) (vol. 5,3)
− Corolla lobes valvate; corolla tube subequal or a little longer than the lobes, if longer then usually bright blue; rhaphides absent or present; flowers often heterostylous; seeds not excavated; inflorescences terminal or axillary; bacterial nodules absent in leaf blades except for some species of *Psychotria* · 8
8. Flowers heterostylous; stigma bifid; pyrene-walls cartilaginous to crustaceous; inflorescences terminal · 9
− Flowers not heterostylous; style topped with a pollen presenter which is globose, cylindrical or mitriform, often ridged, 2–several-lobed at the apex when mature; pyrene-walls crustaceous to woody; inflorescences axillary · 11
9. Corolla blue, white, lilac or purple; corolla tube longer than the lobes; ovules pendulous; fruit ± dry · · · · · · · · · · · · · · · **12. Pentanisia schweinfurthii** Hiern (see vol. 5,1: 59)
− Corolla white; corolla tube ± equalling the lobes or a little longer; ovules erect; fruit fleshy · 10
10. Inflorescence axes not white; pyrenes without well-marked dehiscence; testa red-brown or purplish; endosperm often ruminate; stipules entire or bifid, often caducous, never becoming corky · **1. Psychotria** (partly) (see vol. 5,1: 9)
− Inflorescence axes white, tinged purple; pyrenes with ± well-marked dehiscence; testa pale; endosperm not ruminate; stipules ± truncate, often becoming corky · · · · · · · · · · · · ·
· · · · · · · · · · · · · · **4. Chassalia umbraticola** Vatke (see vol. 5,1: 38)
11. Ovary 2-locular · 12
− Ovary 3–8-locular, a few flowers sometimes with 2-locular ovaries · · · · · · · · · · · · · · 14
12. Corolla coriaceous, drying wrinkled, glabrous; buds obtuse; leaves glabrous, drying the characteristic yellow-green of an aluminium-accumulator; pyrenes very thickly woody, irregularly ridged with lines of dehiscence apparent ·
· **49. Multidentia concrescens** (Bullock) Bridson & Verdc.
− Corolla not coriaceous nor drying wrinkled; buds acuminate; leaves hairy or velvety, the blades not drying yellow-green; pyrenes not so thickly woody · · · · · · · · · · · · · · · · 13
13. Small subshrub 15–30 cm tall; calyx lobes narrowly oblong, elliptic or lanceolate, 0.8–5 mm long; leaves hairy beneath but not so densely as to hide the surface · · · · · · · · · · · · · · · ·
· **48. Pygmaeothamnus zeyheri** (Sond.) Robyns
− Shrub or subshrub 0.6–5 m tall; calyx lobes linear, 1.5–4 mm long; leaves velvety beneath, surface obscured · · · · · · · · · · · · · · · · · **45. Rytigynia decussata** (K. Schum.) Robyns
14. Corolla tube 1.7–2.5 cm long, sometimes curved, rusty tomentose outside; calyx lobes 5–13 mm long, lanceolate-triangular; fruit c. 1.4 cm across, with a rusty coloured velvety pubescence; leaf blades drying dark, with a fine network of pale yellowish nerves on upper surface and pale velvety beneath · · · · · · · · · · · · · · **38. Ancylanthos rubiginosus** Desf.
− Corolla tube much shorter; other characters not as combined above · · · · · · · · · · · · · 15
15. Plants glabrous to pubescent; leaf blades sometimes with fine white pubescence, finely velvety beneath or occasionally felted beneath (if in doubt try both leads) · · · · · · · · · 16
− Plants with a conspicuous indumentum; leaf blades densely velvety or felted on lower surface · 18
16. Stems branched; leaf blades often acuminate at apex; calyx lobes reduced or dentate; corolla never large; lobes blunt, apiculate or distinctly tailed · · · · · **45. Rytigynia** (partly)
− Stems mostly unbranched; leaf blades infrequently acuminate; calyx and corolla various
· 17
17. Leaves mostly in whorls of 3–5; calyx limb truncate, dentate or lobed; corolla large or small, lobes blunt, apiculate or infrequently subulate (*F. chlorantha*) · · · · · · · · · · · · **43. Fadogia**
− Leaves in pairs, rarely in whorls of 3; calyx limb with linear or linear-oblong lobes; corolla small, lobes distinctly apiculate · **40. Pachystigma** (partly)
18. Calyx lobes 0.5–1 mm long; calyx tube and corolla similarly covered by dense, pale, curled hairs; fruit c. 10 mm across, drying black, ± glabrous or with scattered cottony hairs; indumentum on young stems and lower surface of leaves especially densely felted; small shrubs 0.3–4 m tall, with branched stems · **44. Fadogiella**
− Calyx lobes longer; calyx tube indumentum often differing in density and/or type from that on the corolla; fruit often larger, mostly yellow, orange or reddish becoming brownish

when dry, often with golden velvety hairs; indumentum on stems and leaves mostly velvety but sometimes with curled hairs; branched shrubs or single-stemmed subshrubs · **42. Tapiphyllum** (partly)

Key 3
(Climbing and/or scandent genera)

1. Plants covered with curved prickles; leaves and stipules similar to each other, in whorls of (2)4(8); corolla rotate to subcampanulate; fruit fleshy, 2-seeded · **36. Rubia** (see vol. 5,1: 193)
 – Plants not prickly but spines sometimes present; leaves usually paired, very dissimilar to stipules; corolla distinctly tubular ·2
2. Flowers showy, 1–several, appearing axillary on alternate sides of successive nodes (TAB. 46/6); calyx lobes reddish, oblong-lanceolate, contorted in bud; corolla silky pubescent outside; fruit large, up to 10 × 5.5 cm · · · · · **78. Sherbournia bignoniiflora** (Welw.) Hua (vol. 5,3)
 – Flowers and inflorescences not as above, or if flowers showy then corolla not silky; fruit much smaller ·3
3. Plant foetid; leaves paired or in whorls of 3, long-petiolate and cordate at the base; fruits flattened, the outer pericarp falling off to expose 2, compressed, winged pyrenes attached by long filiform stalks ·**13. Paederia** (see vol. 5,1: 63)
 – Plant not evil-smelling; leaves not as above; fruit not of this characteristic structure · · · · 4
4. Inflorescences terminal, sometimes with additional axillary inflorescences near apex of main stem ·5
 – Inflorescences axillary ·8
5. Corolla lobes valvate; flowers heterostylous; stipules 2-lobed · · · · · · · · · · · · · · · · · · 6
 – Corolla lobes contorted; flowers never heterostylous, always both stigma and anthers exserted; stipules entire, fimbriate, or awned · 7
6. Corolla lobes reduplicate-valvate, showy often with coloured eye near throat; ovules/seeds numerous; rhaphides absent · · · · · · · · · · · · · · · · · · · **65. Mussaenda** (partly) (vol. 5,3)
 – Corolla lobes simply valvate, not showy; ovules/seeds 1 per locule; raphides present · **1. Psychotria ealaensis** De Wild. (see vol. 5,1: 17)
7. Ovules 1–several, impressed on a placenta attached to the septum; corolla with 5 reflexed lobes; seeds several per fruit, with a small cavity and entire endosperm; stipules often triangular-ovate, dark (dark green when fresh, turning black when dry) · **93. Tarenna** (partly) (vol. 5,3)
 – Ovule single, erect from the base of each locule; corolla with 4 or 5 spreading lobes; seed 1 per fruit, spherical with strongly ruminate endosperm; stipules fimbriate, or shortly sheathing and aristate, not darkened · · · · · · · · · · · · · · · · · · **96. Rutidea** (vol. 5,3)
8. Stigma divided into 4–5 lobes or arms; ovary 4–5-locular; ovules/seeds numerous · **63. Sabicea** (vol. 5,3)
 – Pollen presenter cylindrical, 2-lobed at the tip; ovary 2-locular; ovules/seeds 1 per locule · · 9
9. Leaves coriaceous, drying bright green, glabrous; stipules with a strongly keeled lobe; inflorescences not pedunculate; calyx limb reduced to a rim, shorter than the disk; anthers reflexed; fruit didymous; seeds with endosperm not streaked by granules · **53. Psydrax** subgen. **Phallaria**
 – Leaves chartaceous to coriaceous, not drying bright green; stipules lacking a keeled lobe; inflorescence pedunculate; calyx limb at least equalling the disk, often dentate, occasionally lobed; anthers exserted but not reflexed; fruit ± obcordate in outline; seed with endosperm streaked with granules · **54. Keetia**

Key 4
(Woody genera)

1. Leaf blades 10–34 cm long, with white felted tomentum beneath; stipules 3–4 cm long, reddish; inflorescences axillary, each enclosed in an involucre · **64. Stipularia africana** Beauv. (vol. 5,3)
 – Leaf blades, stipules and inflorescences not as above ·2
2. Flowers in globose or capitate heads ·3
 – Flowers solitary or in lax to compact or spike-like inflorescences · · · · · · · · · · · · · · 11

3. Calyces ± joined together; fruits a fleshy syncarp ···························· 4
   – Calyces not joined together; fruits not forming a syncarp ······················ 5
4. Inflorescences 4–9 mm in diameter (excluding corollas), 5–20-flowered; flowers heterostylous; corolla lobes valvate, later spreading; ovules/seeds 1 per locule; rhaphides present ·································· **7. Morinda** (see vol. 5,1: 45)
   – Inflorescences 30–50 mm in diameter (excluding corollas), distinctly more than 35-flowered; flowers not heterostylous; corolla lobes imbricate, remaining erect; ovules/seeds numerous; rhaphides absent ·········· **55. Sarcocephalus pobeguinii** Pobég. (vol. 5,3)
5. Inflorescence heads perfectly spherical; bracts caducous or insignificant; interfloral bracteoles absent or present ·· 6
   – Inflorescence heads not perfectly spherical, often subtended by bracts ············ 9
6. Ovules solitary and pendulous in each locule; fruit fleshy ·····················
   ····························· **58. Cephalanthus natalensis** Oliv. (vol. 5,3)
   – Ovules numerous; fruit dry ·· 7
7. Corolla lobes valvate; stipules conspicuous, (3)3.5–8 × 2.5–5 cm, free or joined near base, often reddish; flowering heads 3–many; seeds with small wing ····· **59. Hallea** (vol. 5,3)
   – Corolla lobes imbricate; stipules not so markedly conspicuous, never red; flowering heads solitary; seeds not winged ··· 8
8. Leaves paired, broadly elliptic; inflorescence terminal, at least 2–4 cm across (including corollas); calyx lobes spathulate, persistent; wood bright orange; stipules 3–5.6 × 1.5–3.5 cm, oblanceolate to broadly oblanceolate ······ **56. Burttdavya nyasica** Hoyle (vol. 5,3)
   – Leaves in whorls of 3–4, lanceolate; inflorescence lateral, up to 2.5 cm across (including corollas); calyx lobes oblong, eventually deciduous; wood not bright orange; stipules smaller, ± triangular, bifid ····· **57. Breonadia salicina** (Vahl) Hepper & Wood (vol. 5,3)
9. Inflorescence heads borne in axils, few-flowered (mostly not more than 5); ovules/seeds numerous ····················· **63. Sabicea dinklagei** K. Schum. (vol. 5,3)
   – Inflorescence heads terminal, several- to many-flowered; ovules/seeds 1 per locule ··· 10
10. Corolla lobes valvate; flowers heterostylous; inflorescence with bracts surrounding base of inflorescence; bacterial nodules absent from leaves; rhaphides present ···············
   ································· **1. Psychotria** sect. **Involucratae** (see vol. 5,1: 11)
    – Corolla lobes contorted; flowers with style and anthers always exserted; inflorescence with bracts small or sometimes larger and ± membranous; bacterial nodules present in leaves; rhaphides absent ····························· **94. Pavetta** (partly) (vol. 5,3)
11. Inflorescences spike-like, racemose or narrowly thyrsoid ····················· 12
    – Inflorescences not as above ··· 13
12. Inflorescences subtended by large red leaf-like paired stipitate venose bracts; corolla cylindrical below, funnel-shaped or narrowly campanulate above with erect lobes, glabrous outside; fruit a capsule; seeds conspicuously winged ·························
   ········· **60. Hymenodictyon floribundum** (Hochst. ex Steud.) B.L. Robinson (vol. 5,3)
    – Inflorescences not subtended by conspicuous bracts; corolla ± salver-shaped, pubescent outside; fruit a berry; seeds not winged ····· **70. Bertiera angusiana** N. Hallé (vol. 5,3)
13. Calyx limb often white, eccentric, entire or ± shallowly lobed, venose and accrescent, up to 2.8 cm wide in fruit; fruit dry and often tardily dehiscent; corolla tube narrowly tubular, 3–5 cm long ····································· **17. Carphalea** (see vol. 5,1: 82)
    – Calyx limb not as above, usually with 4–5 distinct lobes or teeth, or reduced to a rim ·· 14
14. At least a few calyx lobes large and leaf-like in each inflorescence, mostly exceeding 2 cm long and wide, white, cream or red ·································· 15
    – Calyx lobes equal or slightly unequal and green, or if 1–2 enlarged into a coloured lamina then not attaining 2 cm in length or width ····························· 16
15. Fruit ± succulent or at least indehiscent; corolla lobes reduplicate-valvate, apiculate ·····
   ····································· **65. Mussaenda** (partly) (vol. 5,3)
    – Fruit dry, dehiscing at the apex; corolla lobes induplicate-valvate, bearing subapical filiform appendages to 2 mm long ···················· **66. Pseudomussaenda** (vol. 5,3)
16. Small tree or shrub growing in the littoral zone just above high tide level, the stems covered with large leaf scars; leaves large, crowded at the ends of the branchlets; calyx limb truncate; corolla white, salver-shaped, velvety outside, the tube ± 2.5 cm long with 4–9 lobes; ovary 4–9-locular; fruit globose, fibrously woody, up to 3.5 cm in diameter, containing one multi-seeded stone ···················· **98. Guettarda speciosa** L. (vol. 5,3)
    – Not growing in the littoral zone, or if so then calyx limb, corolla, fruit and habit not as above ··· 17

17. Corollas valvate, mostly white and not markedly showy; flowers heterostylous; rhaphides present · 18
 – Corollas valvate, contorted or sometimes imbricate; flowers mostly with secondary pollen presentation, occasionally heterostylous or homostylous (if both valvate and heterostylous then corolla usually coloured or pubescent); rhaphides absent · · · · · · · · · · · · · · · · · 25
18. Inflorescences axillary or supra-axillary (TAB. **46**/8, 9) · 19
 – Inflorescences terminal (TAB. **46**/1) · 21
19. Inflorescences corymbose to subumbellate, at least some flowers pedicellate; ovules numerous in each locule; fruit orange or red when ripe · · **9. Pauridiantha** (see vol. 5,1: 51)
 – Inflorescences congested, pedicels usually absent; ovule solitary in each locule; fruit not orange or red · 20
20. Leaves with numerous arching lateral nerves and closely parallel scalariform tertiary venation; locules 4–12; ovules erect; fruit blue when ripe, with 4–12 pyrenes; inflorescences axillary, sessile or pedunculate · **6. Lasianthus** (see vol. 5,1: 43)
 – Leaves with venation mostly closely reticulate; locules 2; ovules pendulous; fruit probably green to black when ripe, 1-seeded; inflorescences supra-axillary or less often axillary, always pedunculate · **10. Craterispermum** (see vol. 5,1: 53)
21. Stipules entirely sheathing, c. 10 mm long, several-fimbriate; ovary superior; fruit apex devoid of calyx limb remains · **5. Gaertnera** (see vol. 5,1: 42)
 – Stipules sheathing only at the base, not fimbriate; ovary inferior; fruit crowned by calyx limb or with apical scar · 22
22. Styles divided into 6–8 filiform lobes; ovary with 4–10 locules, each containing 2(3) collateral erect ovules, but only 1 develops into a seed in each locule; corolla-tube c. 10 mm long, woolly tomentose outside or at least with pubescent lines; drupes red, with woody putamen; plants mainly littoral or in coastal bushland · · **8. Triainolepis** (see vol. 5,1: 51)
 – Styles 2-fid; ovary/fruit 2-locular; other characters not as combined above · · · · · · · · · 23
23. Pyrenes without a well-marked dehiscence; testa red-brown or purplish; endosperm often ruminate; bacterial nodules sometimes present in the leaves; stipules entire or bifid · **1. Psychotria** (see vol. 5,1: 9)
 – Pyrenes with ± well-marked dehiscence; testa pale; endosperm not ruminate; bacterial nodules never present in the leaves; stipules never lobed, frequently becoming corky · · · 24
24. Inflorescence axes not white; pyrenes opening by 2 marginal slits; flower-buds never winged; shrub with leaves mostly developing after the flowers have matured; stems corky · **3. Chazaliella** (see vol. 5,1: 35)
 – Inflorescence axes often white tinged purple; pyrenes opening by 1 dorsal slit; flower-buds and corolla lobes often with longitudinal narrow wing-like keels (but not present in some species); flowers contemporary with mature leaves; stems not markedly corky · **4. Chassalia** (see vol. 5,1: 38)
25. Corolla lobes valvate; style topped with a globose, cylindrical or mitriform pollen presenter, often ridged; pollen presenter 2- to several-lobed at apex when mature, hollow with style attached internally or occasionally with style attached at the base; flowers never heterostylous; locules 2–10; pyrene-walls woody or sometimes crustaceous; inflorescences always axillary · · · · **Vanguerieae** (see dichotomous key p. 227, or multiaccess key p. 232)
 – Corolla lobes contorted or imbricate, or if valvate then pollen presenter not as above or flowers heterostylous; locules 2, rarely several to many; if the fruit is a pyrene then walls scarious to crustaceous; inflorescences, terminal, lateral or axillary · · · · · · · · · · · · · · 26
26. Spines present; corolla showy; fruit large · 27
 – Spines absent; corolla and fruit as above or not · 29
27. Lateral spurs (brachyblasts) always absent; flowers supra-axillary, 2 per node (TAB. **46**/8); corolla trumpet-shaped, longitudinally ribbed; seeds 8–16 per fruit, c. 10 mm in diameter · · · · · · · · · · · · · · · · · · · **79. Didymosalpinx norae** (Swynnerton) Keay (vol. 5,3)
 – Lateral spurs absent or present; flowers not supra-axillary; corolla not trumpet-shaped; seeds numerous, up to 5 mm in diameter · 28
28. Corolla tube much shorter than the lobes; calyx lobes oblong, ovate or spathulate, persistent and ± accrescent in fruit; leaves very often pubescent · · · · · **72. Catunaregam** (vol. 5,3)
 – Corolla tube distinctly longer than the lobes; calyx limb with short triangular teeth, persistent in fruit; leaves always glabrous · · · · · · · · · · · · · · **75. Hyperacanthus** (vol. 5,3)
29. Plants deciduous; flowers and fruit clustered on mature stems with the leaves restricted to apical and lateral shoots (TAB. **46**/10); bracts/bracteoles chaffy; calyx with limb-tube completely absent and lobes lanceolate; corolla funnel-shaped; fruit 1.2–2 cm in

diameter with large scar; (if a deciduous plant with chaffy bracts then calyx limb shortly tubular and persistent in fruit, see couplet 49) ·· **91. Feretia aeruginescens** Stapf (vol. 5,3)

– Plants evergreen, or if deciduous then other characters not combined as above ····· 30

30. Inflorescences axillary, in both axils of a node (sometimes at naked nodes) or sometimes supra-axillary, pedunculate or sessile (TAB. **46**/8–10), if on brachyblasts (contracted lateral spurs) see next lead of this key; corolla lobes always contorted ················ 31

– Inflorescences terminal on main and lateral branches (occasionally appearing axillary by extreme reduction of lateral branches in which case bacterial nodules are present in the leaves) or lateral (i.e. alternate, with inflorescence in only one axil at each node); sometimes with a few axillary inflorescences in addition to the terminal one (TAB. **46**/1–7), or borne on brachyblasts (contracted lateral spurs) (TAB. **46**/11–13); corolla lobes contorted, valvate or sometimes imbricate ···································· 41

31. Flowers solitary, supra-axillary (TAB. **46**/8); corolla trumpet-shaped, longitudinally ribbed; fruit 2–3.3 cm long, thin walled; seeds with fibrous coat ·······················
··················· **79. Didymosalpinx norae** (Swynnerton) Keay (vol. 5,3)

– Flowers 1–many, axillary (TAB. **46**/9); corolla, fruit and seeds not as above ········ 32

32. Inflorescences pedunculate, distinctly and rather laxly branched; flowers always pedicellate ·· 33

– Inflorescences 1- to few-flowered, or congested glomerules, sometimes shortly pedunculate; flowers often pedicellate ·· 35

33. Stipules often blackening when dry; corolla tube cylindrical ± equal to lobes in length; seeds with a small ± circular excavation, or a partly ruminate endosperm ············
································· **93. Tarenna** (partly) (vol. 5,3)

– Stipules not tending to blacken when dry; corolla tube and seeds not as above ······ 34

34. Corolla subrotate; tube c. 1 mm long; endosperm ruminate, visible as longitudinal striations on seed surface ········· **89. Galiniera saxifraga** (Hochst.) Bridson (vol. 5,3)

– Corolla funnel-shaped to campanulate; endosperm entire ·······················
································· **90. Kraussia floribunda** Harv. (vol. 5,3)

35. Stipules 1–1.2 cm long, attenuate from a triangular base, eventually caducous; flowers not pedicellate, mostly borne at naked nodes (TAB. **46**/10); both anthers and style included well below the corolla throat; seeds 2–4 per fruit ·······························
················· **86. Belonophora hypoglauca** (Welw. ex Hiern) A. Chev. (vol. 5,3)

– Stipules smaller, awned or more broadly triangular; inflorescences mostly borne at leafing nodes (TAB. **46**/9); flowers often pedicellate; anthers and style both exserted, or less often either or both included at or near the throat; seeds 1, or few to several ··········· 36

36. Calyx limb reduced to a rim, usually shorter than disk, but occasionally truncate and ± equalling the disk; corolla tube ± equal to lobes; anthers and style exserted; fruit 2-seeded; seeds with a well-defined groove on ventral face (typical coffee-beans); inflorescence 1–several-flowered ························· **97. Coffea** (vol. 5,3)

– Calyx limb mostly at least equalling the disk, truncate, dentate or lobed; fruit 2–many-seeded; seeds lacking a ventral groove ·································· 37

37. Stipules triangular-acuminate, sometimes shortly aristate, often caducous, sometimes obscured by inflorescences even at apical node; small round often cordate leaves often present at base of lateral branches; style undivided, shortly bifid at apex or sometimes distinctly bifid, usually hairy; flowers several in sessile or pedunculate glomerules, very rarely pedicellate ·· 38

– Stipules shortly sheathing at base, distinctly aristate; modified leaves not present at base of lateral branches; style clearly 2-armed, usually glabrous; flowers mostly fewer, stalked or subsessile ·· 39

38. Flowers 5(6)-merous; anthers exserted; bracts and bracteoles triangular-acuminate not forming cupules; seeds with endosperm entire and testa wrinkled-striate ············
·············· **87. Cremaspora triflora** (Thonn.) K. Schum. (vol. 5,3)

– Flowers 4-merous; anthers included, or only tips showing; bracts and bracteoles cupular, often thinly chaffy; seeds with ruminate endosperm and testa with fingerprint-like striations ·· **88. Polysphaeria** (vol. 5,3)

39. Leaves with acumen very distinctive, 0.8–4 cm long; calyx limb-tube truncate, 2.5–7 mm long, characteristically split down one side; anthers locellate; ovule/seeds one per locule; testa apparently absent ···· **84. Calycosiphonia spathicalyx** (K. Schum.) Robbr. (vol. 5,3)

– Leaves with acumen much shorter; calyx limb seldom as above; anthers not locellate; ovules/seeds mostly more than one per locule; testa present ·················· 40

40. Anthers ± medifixed, the connective not enlarged; thecae ± contiguous and subparallel; bacterial nodules absent · **82. Tricalysia** (vol. 5,3)
– Anthers basifixed, the connective enlarged so that the thecae diverge; linear bacterial nodules often present along the midrib and petiole · · · · · · · · · · **83. Sericanthe** (vol. 5,3)
41. Subshrubs; corolla lobes valvate, very narrowly lanceolate, subequal to tube in length, suberect; both capitate stigma and stamens exserted; calyx lobes not persistent in fruit; capsule often with 1 persistent and 1 falling lobe; seeds numerous, unwinged · **62. Virectaria major** (K. Schum.) Verdc. (vol. 5,3)
– Mostly shrubs or trees; corolla lobes and fruit not as above · · · · · · · · · · · · · · · · · · ·42
42. Corollas tubular below, funnel-shaped to narrowly campanulate above with short erect lobes; anthers included; style long exserted; capsules oblong or subcylindric, lenticellate; seeds with wing bifid at one end · · · · · · · **60. Hymenodictyon parvifolium** Oliv. (vol. 5,3)
– Corollas not as above; fruit if capsular then ± globose and not lenticellate; seeds unwinged, or with a fimbriate wing · 43
43. Corollas with a coloured star-shaped eye at throat; corolla lobes valvate or imbricate; flowers heterostylous or isostylous; ovules numerous in each locule; fruit berry-like, slightly fleshy to fleshy · 44
– Corollas lacking eye as above; corolla lobes contorted; flowers mostly with secondary pollen presentation; ovules 1–numerous in each locule; fruit berry-like (often large and ± woody), drupe-like or sometimes capsular · 45
44. Each stipule almost completely divided into 2 separate lobes, rather small and caducous; corolla lobes imbricate (quincuncial) (TAB. 1/2); calyx lobes ± leaf-like, often somewhat spathulate, persistent in fruit; fruit globose, slightly fleshy but soon dry · **67. Heinsia** (vol. 5,3)
– Each stipule entire or if divided not completely 2-lobed; corolla lobes reduplicate-valvate (TAB. 1/1); calyx lobes triangular, at the most slightly spathulate; fruit fleshy, indehiscent, often lenticellate · **65. Mussaenda** (partly) (vol. 5,3)
45. Inflorescences consisting of several- to many-flowered cymes, one borne at every other node, the flowering nodes strongly anisophyllous (TAB. 46/7); flowers small, corolla tube c. 6 mm long; fruit less than 10 mm in diameter · **71. Aidia micrantha** (K. Schum.) White (vol. 5,3)
– Inflorescences not as above and the flowering nodes not anisophyllous; flowers and fruit small to large · 46
46. Inflorescences strictly lateral (TAB. 46/6); stipules green, moderately conspicuous, triangular or ovate, scarcely sheathing; seeds not held together by pulpy tissue, seed surface with a fingerprint-like pattern of striations · 47
– Inflorescences and stipules not as above; seeds 1 or few–numerous, often held together by pulpy tissue; seed surface lacking a fingerprint-like pattern · · · · · · · · · · · · · · · · · · 48
47. Shrubs or small trees; seldom less than 1 m tall; corolla tube narrowly cylindrical · **80. Oxyanthus** (vol. 5,3)
– Small shrubs, or less often a tree, 0.3–1.8(8) m tall; corolla tube cylindrical at base and funnel-shaped above · · · · · · · · · · · · · · · · · · **81. Mitriostigma axillare** Hochst. (vol. 5,3)
48. Plants with brachyblasts (contracted lateral spurs) or cushion shoots (TAB. 46/11–13) · · 49
– Plants without brachyblasts or cushion shoots, lateral branches often short but with leaves clearly spaced · 52
49. Flowers precocious, subtended by a series of chaffy bracts/bracteoles; corolla tube narrowly cylindrical, 1–4.2 cm long; calyx limb c. 1.5 mm long, shortly tubular, repand to shortly dentate, persistent in fruit; fruit 5 mm in diameter, 2-seeded · **85. Argocoffeopsis eketensis** (Wernham) Robbr. (vol. 5,3)
– Flowers contemporary with the leaves; bracts/bracteoles not chaffy; calyx limb-tube 1–2.5 mm long, distinctly lobed; fruit few–many-seeded ·50
50. Corolla tube usually at least 15 mm long; calyx lobes narrowly triangular to lanceolate · **74. Gardenia brachythamnus** (K. Schum.) Launert (vol. 5,3)
– Corolla tube up to 7 mm long; calyx lobes ovate, often spathulate, separated by broad sinuses · 51
51. Corollas rotate, tube scarcely exceeding calyx limb-tube; calyx lobes 4–7(13) mm long; fruit yellow-green or brown when ripe, 1.8–3.3 × 1.7–2.5 cm · · · · **72. Catunaregam** (vol. 5,3)
– Corollas funnel-shaped, tube longer than the lobes; calyx lobes 0.5–1.5(2.5) mm long; fruit black, 5–7 mm in diameter · · · · · · · **73. Coddia rudis** (E. Mey. ex Harv.) Verdc. (vol. 5,3)
52. Inflorescence usually borne on a very short shoot above a single leaf (TAB. 46/3, 4); corolla tube variously shaped, hairy outside · 53

- Inflorescence essentially terminal, sometimes overtopped by the growing apex, or pseudoaxillary (due to suppression of lateral branches) (TAB. **46**/1, 2); corolla tube mostly cylindrical to narrowly funnel-shaped and glabrous or hairy outside · · · · · · · · · · · · 55
53. Flowers markedly showy; corolla tube 2.5–23 cm long, narrowly cylindrical at base, campanulate or funnel-shaped above; corolla lobes contorted to the left or right; fruit ± ovoid, 1.8–8.5 cm long; ovules numerous; seeds numerous in pulpy tissue; testa present · **76. Rothmannia** (vol. 5,3)
- Flowers moderately showy; corolla up to 3 cm long; corolla lobes contorted to the left; fruit globose, c. 13 mm in diameter; ovules 2–10; seeds few, free; testa absent · · · · · · · · · · · 54
54. Corolla campanulate · · · · · · · · · · · · · · **68. Heinsenia diervilleoides** K. Schum. (vol. 5,3)
- Corolla salver-shaped, often split along one side · · · · · · · · · · · **69. Aulacocalyx** (vol. 5,3)
55. Flowers 1 or up to 3, markedly showy, sometimes overtopped by the stem apex; fruit large; seeds numerous in pulpy tissue · 56
- Flowers if showy then more numerous, never overtopped by stem apex; fruit mostly small to moderately large; seeds 1–few, or if numerous then seeds free · · · · · · · · · · · · · · · 58
56. Deciduous shrub; corolla yellow-green, purple-spotted; corolla tube much shorter than lobes; calyx limb with corky ridges protecting bud in dry season, not persistent in fruit but leaving a bull's-eye-like scar · · · · · · · · · · · · **77. Phellocalyx vollesenii** Bridson (vol. 5,3)
- Evergreen shrub; corolla white or cream, unmarked; corolla tube clearly longer than lobes; calyx limb not as above; fruit usually bearing persistent calyx limb · · · · · · · · · · · · · · 57
57. Branches and leaves often ternate; stipules sheathing, often truncate but sometimes triangular, with sticky exudate when young; flowers solitary; calyx limb-tube truncate or often with decurrent lobes; corolla lobes contorted to the left; fruit thick-walled · **74. Gardenia** (partly) (vol. 5,3)
- Branches and leaves not ternate; stipules ovate, tightly inrolled at apex and readily falling, lacking a sticky exudate; flowers 1–3; calyx lobes not decurrent; corolla lobes contorted to the right; fruit fairly thin-walled · · · · · · · · · · · · · · · · · **75. Hyperacanthus** (vol. 5,3)
58. Fruit a capsule, ± globose; seeds with a fimbriate wing; corolla distinctly pubescent outside; lobes 4–6; anthers ± two thirds exserted · **61. Crossopteryx febrifuga** (Afzel. ex G. Don) Benth. (vol. 5,3)
- Fruit not dehiscent; seeds never winged; corolla pubescent or more often glabrous outside; lobes 4 or 5(6); anthers usually fully exserted · 59
59. Calyx lobes subfoliaceous, 0.8–3 cm long, persistent in fruit; corolla large, tube 2–11 cm long, pubescent outside; fruit large, 0.8–2.2 cm in diameter; seeds numerous; stipules conspicuous, either triangular and erect or ± rounded and reflexed · · · · · · · · · **92. Leptactina** (vol. 5,3)
- Calyx lobes not nearly as conspicuous as above; corolla generally smaller; seeds 1–several; stipules less conspicuous, never rounded and reflexed · 60
60. Flowers 5(6)-merous; fruit 1–several-seeded; seeds variously shaped, entire or ruminate; stipules often with central area darkened (dark green when fresh, or black when dry) but difficult to observe in taxa with reduced branches · · · · · · · · · · · **93. Tarenna** (vol. 5,3)
- Flowers 4-merous; fruit 2-seeded; seeds ± hemispherical with an excavation on the ventral face; stipules never with central area darkened · 61
61. Bacterial nodules usually present in the leaf blades, either scattered or arranged along the midrib; inflorescence sessile, laxly corymbose to subcapitate, never with rhachis articulated, sometimes inflorescence pseudo-axillary due to reduction of lateral branches; bracteoles stipule-like; pollen presenter entire; seeds rugulose, moderately shiny · **94. Pavetta** (vol. 5,3)
- Bacterial nodules absent; inflorescence sessile or pedunculate; rhachis sometimes white or tinted, often articulated; bracteoles linear, short; pollen presenter with 2 divaricate arms; seeds dull, often rusty coloured, not rugulose · · · · · · · · · · · · · · · · · · **95. Ixora** (vol. 5,3)

Tribe 13. **VANGUERIEAE**

Key to the genera of the Vanguerieae *

1. Spines present · 2
- Spines absent (rarely some side branches spine-like in *Vangueria randii* and *Cuviera semseii*) · 8

* see also multi-access key on page 232

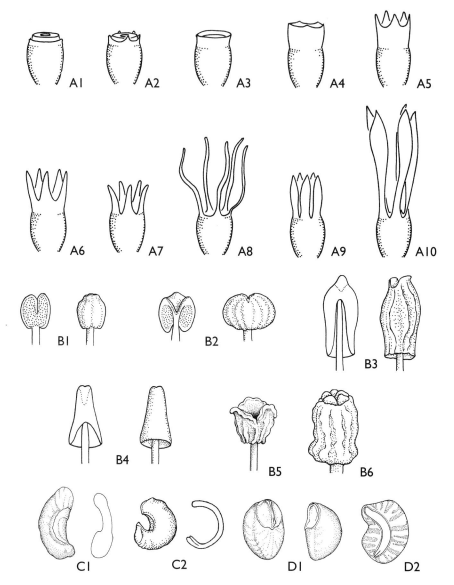

Tab. 47. VANGUERIEAE. A. —**calyx limbs, diagrammatic**. A1, reduced rim; A2, denticulate rim; A3, limb-tube ± equal to disk; A4, limb-tube exceeding disk; A5, limb-tube exceeding disk but longer than lobes; A6, limb-tube shorter than lobes; A7, limb-tube ± reduced, lobes linear-oblong; A8, limb-tube ± reduced, lobes linear separated by broad sinuses; A9, limb-tube ± reduced, lobes oblong-elliptic; A10, limb-tube ± reduced, lobes spathulate. B. —**pollen presenters**. B1, PYROSTRIA LOBULATA (× 10), from *Carmichael* 535; B2, CANTHIUM OLIGOCARPUM (× 10), from *Troupin* 2751; B3, KEETIA GUEINZII (× 10), from *Parnell* 2160; B4, PSYDRAX KRAUSSIOIDES (× 10), from *Torre & Correia* 18674; B5, FADOGIA STENOPHYLLA subsp. ODORATA (× 6½), from *Watermeyer* 138; B6, VANGUERIA MADAGASCARIENSIS (× 12), from *Haarer* 1752. C. —**seed with embryo**. C1, MULTIDENTIA CRASSA (× 1½), from *Angus* 185; C2, PSYDRAX LIVIDA (× 3), from *Orpen* 50/50. D. —KEETIA ZANZIBARICA. D1, pyrene, 2 views (× 3), from *Shabani* 267; D2, longitudinal section through seed, (× 3), from *Kibuwa* 2499. From F.T.E.A. and Kew Bull. A & B drawn by Sally Dawson, C & D drawn by Diane Bridson.

2. Calyx lobes conspicuous, leaf-like, 5–9 mm long, c. 7 times as long as the calyx tube, exceeding the corolla tube, persistent in fruit; spines supra-axillary on the branches, or present on trunks of saplings, often quite robust; pyrenes 3–5 · **39. Lagynias lasiantha** (Sond.) Bullock
  – Calyx lobes not so conspicuous; other characters not combined as above · · · · · · · · · · · 3
3. Spines present on trunks of young trees, or restricted to young or coppice shoots and ternately arranged, or less often paired from leaf axils; ovary 2- or 3-locular; corolla tube lacking ring of deflexed hairs inside; corolla lobes not apiculate; fruit 1.2–2.5 cm long · **50. Canthium** subgen. **Lycioserissa**
  – Spines positioned above lateral branches (sometimes reduced cushion shoots), or from leaf axils of more mature stems; other characters not combined as above · · · · · · · · · · · · · · 4
4. Calyx lobes linear, 1–3.5 mm long, persistent in fruit; ovary always 2-locular · **45. Rytigynia bugoyensis** (K. Krause) Verdc.
  – Calyx lobes triangular, up to 1 mm long, acute; ovary 2- or 3–5-locular · · · · · · · · · · · · · 5
5. Inflorescences many-flowered, corymbose; peduncles 8–35 mm long · **51. Plectroniella armata** (K. Schum.) Robyns
  – Inflorescences few-flowered, or flowers solitary; peduncles seldom reaching 8 mm in length · 6
6. Stipules with a few hairs inside; pedicels 8–15 mm long; corolla tube 2.5–4 mm long · **50. Canthium kuntzeanum** Bridson
  – Stipules with tuft of hairs inside; pedicels shorter; corolla tube not exceeding 2 mm in length · 7
7. Spines only occasionally arising above reduced cushion shoots; calyx limb 0.25–0.6 mm long, truncate or shortly toothed; locules 2–3(5); fruit subglobose · · · · · **45. Rytigynia** (partly)
  – Spines always arising above reduced cushion shoots (TAB. **46**/12); calyx limb 0.5–1 mm long, divided almost to base into triangular or narrowly triangular lobes; locules always 2; fruit somewhat dorsiventrally flattened · · · · · · · · · · · · **50. Canthium** subgen. **Canthium**
8. Inflorescences entirely enclosed in bud by paired connate persistent bracts, umbellate or sometimes 1-flowered; pollen presenter solid with style attached to the base (TAB. **47**/B1); leaves usually with tertiary venation obscure; corolla throat densely congested with hairs · **52. Pyrostria**
  – Inflorescences never enclosed by paired bracts; pollen presenter hollowed at the base with the style attached internally (TAB. **47**/B2–B4); other characters not combined as above · · · · 9
9. Corolla lobes 2–4 cm long, linear-lanceolate, greatly exceeding the tube, erect, tomentose or pubescent outside; fruit 2–4 cm long with calyx remains triangular to linear; pyrenes 2 · **47. Vangueriopsis lanciflora** (Hiern) Robyns
  – Corolla lobes not as above; fruit smaller, or if large then with more than 2 pyrenes, or with calyx remains cupular · 10
10. Climbing and/or scandent plants · 11
  – Trees, shrubs (sometimes subscandent), or subshrubby herbs · · · · · · · · · · · · · · · · 12
11. Leaves coriaceous, drying bright green, glabrous; stipules with a strongly keeled lobe; inflorescences not pedunculate; calyx limb reduced to a rim, shorter than the disk; anthers reflexed; fruit didymous; seeds with endosperm not streaked with granules · **53. Psydrax** subgen. **Phallaria**
  – Leaves chartaceous to coriaceous, not drying bright green, glabrous or pubescent; stipules lacking keeled lobe; inflorescences pedunculate; calyx limb at least equalling the disk, often dentate, occasionally lobed; anthers exserted but not reflexed; fruit ± obcordate in outline; seeds with endosperm streaked with granules (TAB. **47**/D2) · · · · · · · · · · · · · **54. Keetia**
12. Ovary/fruit 2-locular (rarely with occasional 3-locular ones as well); pollen presenter 2-lobed at tip, (TAB. **47**/B1–B4) · 13
  – Ovary/fruit 3–8-locular (rarely with occasional 2-locular ones as well); pollen presenter 3–8-lobed at tip, (TAB. **47**/B5, B6) · 26
13. Subshrubby herbs or single-stemmed shrubs from a woody rootstock, up to 2 m tall · · · 14
  – Shrubs or trees (1)2–20 m tall · 16
14. Leaves coriaceous; corolla coriaceous, drying wrinkled; pyrenes very thickly woody; inflorescences frequently supra-nodal · **49. Multidentia concrescens** (Bullock) Bridson & Verdc.
  – Leaves not coriaceous; corolla not coriaceous; pyrenes not thickly woody; inflorescences axillary · 15

15. Plants small, 15–30 cm tall; leaves glabrous or with sparse bristly hairs on both surfaces; calyx lobes somewhat leaf-like, narrowly oblong, elliptic or lanceolate; fruit 1.5–2 cm across (TAB. **60**) · **48. Pygmaeothamnus zeyheri** (Sond.) Robyns
  – Plants larger, at least 50 cm tall; leaves velvety beneath, discolorous; calyx lobes linear; fruit up to 1 cm across · · · · · · · · · · · · · · · · · · · **45. Rytigynia decussata** (K. Schum.) Robyns
16. Styles usually at least twice as long as the corolla tube; pollen presenter ± cylindrical, (TAB. **47**/B3, B4); stipules glabrous within; disk sometimes pubescent; inflorescences usually subtended by leaves, (TAB. **46**/9); seed with cotyledons perpendicular to ventral face, (TAB. **47**/C2, D2) · 17
  – Styles usually only slightly longer than the corolla tube (save in *Multidentia exserta* which has stipules hairy within); pollen presenter mostly as broad as long, (TAB. **47**/B2); stipules hairy or glabrous within; disk glabrous; inflorescences often subtended by leaves; seeds with cotyledons parallel to ventral face, (TAB. **47**/C1) · 18
17. Trees or shrubs, sometimes scandent; leaves typically subcoriaceous to coriaceous, drying light green, or occasionally chartaceous in deciduous species; if plants scandent, then stipules triangular to truncate at the base with a strongly keeled lobe; calyx limb consisting of a dentate to repand rim, usually much shorter than the disk, only occasionally equalling the disk, (TAB. **47**/A1, A2); anthers usually reflexed; fruit not or scarcely indented at the apex except when ± didymous; pyrene cartilaginous to woody with shallow apical crest · **53. Psydrax**
  – Scandent bushes with lateral branches set at right-angles to the main ones, often subtended by modified leaves; leaves chartaceous to subcoriaceous, rarely coriaceous; stipules lanceolate to ovate or triangular, acuminate; calyx limb repand to dentate, equalling or sometimes exceeding the disk, (TAB. **47**/A3–A5); anthers usually erect; fruit strongly or slightly indented at the apex, typically obcordate in outline; pyrene woody with lid-like area surrounding the crest (either positioned on ventral face, (TAB. **47**/D1) or across the apex, (TAB. **72**/9)) · **54. Keetia**
18. Calyx lobes linear-oblong or oblong-spathulate, 2.3–4 mm long, persistent on fruit; inflorescence clearly branched, bearing linear or subfoliaceous bracts and bracteoles; corolla lobes acuminate or shortly appendaged · · · · · · · · · · · · · · · **46. Cuviera** (partly)
  – Calyx lobes if present, not as above; inflorescence branched or unbranched, but bracts and bracteoles usually inconspicuous; corolla lobes blunt, or if acuminate or appendaged then inflorescence ± unbranched · 19
19. Calyx limb-tube cupular, repand or lobed but lobes rarely exceeding it, (TAB. **47**/A4, A5); fruit large; pyrenes very thickly woody, strongly irregularly ridged with lines of dehiscence apparent, (TAB. **61**/13); leaves discolorous, mostly with a conspicuous network of tertiary nerves; corolla tube with a ring of deflexed hairs inside, lobes never apiculate; pollen presenter spherical to elongate-ellipsoid, ribbed · · · · · · · · · · · · · · · · · **49. Multidentia**
  – Calyx limb-tube obsolete or short, lobes absent or, if present, usually greatly exceeding the tube; pyrenes not so thickly woody, nor ridged as above; other characters not combined as above · 20
20. Flowers mostly in many-flowered, pedunculate dichasial or complicated branched cymes, occasionally subumbellate by reduction; leaves often restricted to new growth but not strictly so ·21
  – Flowers solitary, or in few- to several-flowered fascicles, subumbellate or less often with rudimentary branches, peduncles mostly, but not always, suppressed; leaves well spaced along branches or restricted to cushion shoots · 22
21. Stipules seldom sheathing at the base when mature, often becoming corky outside, if lobed then lobe not decurrent and often caducous; leaves strictly restricted to new growth, (TAB. **46**/10); inflorescence with flowers usually arranged to one side of the ultimate inflorescence branch; calyx limb ± obsolete, (TAB. **47**/A1, A2); fruit obcordate in outline and strongly indented at the apex, or obovoid · · · · · · · · **50. Canthium** subgen. **Afrocanthium** (partly)
  – Stipules sheathing at base, bearing a linear to subulate often decurrent lobe; leaves occasionally restricted to new growth; inflorescence not as above; calyx limb lobed to the base or almost so, (TAB. **47**/A7), lobes often somewhat unequal; fruit slightly indented at the apex · **50. Canthium** subgen. **Lycioserissa**
22. Leaves restricted to very short reduced branchlets or cushion shoots, hence pseudo-verticillate in appearance, (TAB. **46**/11); inflorescences always at leafless nodes · **50. Canthium** subgen. **Afrocanthium** (partly)
  – Leaves not restricted to short branchlets; inflorescences in axils of normal leaves, (TAB. **46**/9) · 23

23. Flowers functionally unisexual, the female inflorescence 1-flowered, the male few- to many-
flowered; stipules not pubescent inside; corolla tube very short, 1–2 mm long; lobes erect
· **50. Canthium** subgen. **Bullockia**
 – Flowers not functionally unisexual; stipules pubescent to villous inside; corolla tube usually,
but not always, more than 2 mm long; corolla lobes mostly spreading · · · · · · · · · · · 24
24. Corolla tubes glabrous or with few hairs within; calyx limb lobed to the base, (TAB. 47/A7);
young stems scarcely lenticellate; fruit 13–25 mm across; corolla lobes neither acuminate
nor appendaged · **50. Canthium** subgen. **Lycioserissa**
 – Corolla tubes with a ring of deflexed hairs within, or occasionally glabrous; calyx limb shortly
tubular, sometimes dentate, (TAB. 47/A3–A5); young stems often conspicuously lenticellate,;
fruit up to 14 mm across; corolla lobes blunt, acuminate or appendaged · · · · · · · · · · · 25
25. Leaf blades drying blackish, up to 4 cm long, acute to obtuse at apex; stipules with few hairs
inside; corolla lobes acute · · · · · · · · · · · · · · · · · · **50. Canthium kuntzeanum** Bridson
 – Leaf blades not drying blackish, up to 6.5 cm long, narrowly acuminate or attenuate at
apex; stipules pubescent within; corolla lobes appendaged · · · · · · · · **45. Rytigynia** (partly)
26. Calyx lobes conspicuous, subfoliaceous up to 6 mm wide; cymes branched with scattered ±
leaf-like bracts and bracteoles; leaves drying blackish, restricted to non-woody apical stems;
pyrenes with pronounced keels, (TAB. 58/11) · · · · · · · · · · · **46. Cuviera semseii** Verdc.
 – Calyx lobes ± absent or various, but if as broad then other characters not combined as above
· 27
27. Calyx lobes oblanceolate, spathulate, oblong-elliptic to oblong or occasionally narrowly
oblong, (1)1.5–5.5 mm wide, always densely velvety, (TAB. 47/A9, A10) · · · · · · · · · · · 28
 – Calyx lobes ± absent, or if present short, narrowly triangular or linear, occasionally narrowly
oblong, often velvety, (TAB. 47/A1–A8) · 30
28. Corolla bright orange or orange-yellow, rusty tomentose outside; tube 17.5–25 mm long,
greatly exceeding calyx lobes; fruit c. 14 mm across with rusty coloured velvety pubescence;
leaf blades drying dark with a fine network of pale yellowish nerves above and pale velvety
beneath · **38. Ancylanthos**
 – Corolla cream or yellow-green, indumentum not rusty outside; tube much shorter, often
exceeded by calyx lobes, or if as long, then curved; fruit 15–45 mm across, ± glabrous; leaf
blades lacking a reticulum of nerves · 29
29. Inflorescence compact, few–several-flowered; pedicels ± obsolete to 7 mm long; leaf blades
scarcely discolorous · · · · · · · · · · · · · · **40. Pachystigma macrocalyx** (Sond.) Robyns
 – Flowers solitary or inflorescence lax, few or several-flowered; pedicels always clearly present;
leaf blades distinctly discolorous · **39. Lagynias**
30. Calyx lobes short, 0.5–1 mm long; calyx tube and corolla both entirely covered by dense
pale curled hairs; fruit c. 10 mm across, drying black, ± glabrous or with scattered cottony
hairs; indumentum on young stems and lower surface of leaves especially densely felted;
shrubs 0.3–4 m tall with branched stems · **44. Fadogiella**
 – Calyx lobes seldom as short; if indumentum present on calyx and flower buds then differing
in density and/or type, and mostly with straight hairs; fruit large or small, lacking cottony
hairs; small trees, shrubs, or pyrophytes · 31
31. Suffrutescent plants with herbaceous or subshrubby mostly (but not always) unbranched
stems from a woody rootstock, up to 1.2 m tall (pyrophytes) · · · · · · · · · · · · · · · · · · 32
 – Subshrubs, shrubs or small trees, if under 1.2 m tall, then stems branched · · · · · · · · · 33
32. Leaves mostly (but not always) in whorls of 3–5; calyx limb truncate, dentate or lobed; corolla
large or small; lobes blunt, apiculate or infrequently subulate (*F. chlorantha*) · · · **43. Fadogia**
 – Leaves in pairs or rarely in whorls of 3; calyx limb with linear or linear-oblong lobes; corolla
never large; lobes apiculate · **40. Pachystigma** (partly)
33. Flowers solitary or 2–10(15) umbellate; peduncle usually clearly present; fruit never large;
calyx lobes reduced or dentate; corolla lobes blunt, apiculate or distinctly tailed · · · · · · ·
· **45. Rytigynia** (partly)
 – Flowers 5–many, in pedunculate branched inflorescences or sessile contracted
inflorescences; fruit large, 0.8–5 cm across, often with velvety hairs; calyx lobes linear,
narrowly-oblong or shortly so, in which case they can be somewhat triangular; corolla lobes
clearly apiculate · 34
34. Plants glabrous, pubescent or densely pubescent; inflorescence a dichasial cyme with
flowers scattered along the arms, usually lax and many-flowered but a few species with few-
flowered or compact inflorescences; corolla glabrous or hairy outside · · · · **41. Vangueria**
 – Plants densely pubescent, often velvety or woolly; inflorescences never branched; corolla
densely hairy outside · **42. Tapiphyllum** (partly)

Non-Dichotomous Guide to Genera of the *Vanguerieae*

Since the delimitation of genera in the *Vanguerieae* is problematical and the distinguishing characters are for the most part weak, variable or need to be used in combination, it is thought useful to provide a non-dichotomous guide to genera. This guide will be especially helpful where material is incomplete; it will indicate the genus, or at least greatly restrict the choice between genera.

Numbers in **bold** indicate genus number, and numbers in parentheses (non-bold) indicate the species number.

38	(1)	*Ancylanthos*
39	(1–3)	*Lagynias*
40	(1–5)	*Pachystigma*
41	(1–9)	*Vangueria*
42	(1–7)	*Tapiphyllum*
43	(1–16)	*Fadogia*
44	(1–2)	*Fadogiella*
45		*Rytigynia*
45.1	(1–16, 18–21)	*Rytigynia* subgen. *Rytigynia*
45.2	(17)	*Rytigynia* subgen. *Fadogiopsis*
46	(1–3)	*Cuviera*
47	(1)	*Vangueriopsis*
48	(1)	*Pygmaeothamnus*
49	(1–5)	*Multidentia*
50		*Canthium*
50.1	(1)	*Canthium* subgen. *Canthium*
50.2	(2–11)	*Canthium* subgen. *Afrocanthium*
50.3	(12–13)	*Canthium* subgen. *Lycioserissa*
50.4	(14–15)	*Canthium* subgen. *Bullockia*
50	(16)	*Canthium kuntzeanum*
51	(1)	*Plectroniella*
52	(1–6)	*Pyrostria*
53		*Psydrax*
53.1	(1–15)	*Psydrax* subgen. *Psydrax*
53.2	(16)	*Psydrax* subgen. *Phallaria*
54	(1–4)	*Keetia*

This guide is divided into two parts: the first (**Guide 1**) lists only *restricted characters*, i.e. characters which do not occur in more than three genera, while the second (**Guide 2**) lists *widely distributed characters*.

Guide 1 only includes a selection of genera, and is intended to aid quick recognition of the more characteristic ones. Check the list of characters against the specimen and note the generic number/s listed against any character/s it possesses. If more than one number is noted either proceed to Guide 2 (using the generic numbers noted as your starting point) or refer to the dichotomous key and search for couplets that may help the choice.

In **Guide 2** each character is considered in two or more states. Each character state is listed against the corresponding generic numbers. If the character is variable or intermediate it will be listed for both states.

In order to use Guide 2 proceed as follows:

1. Select from the list *any* character clearly present on the specimen and note all the generic numbers corresponding to it.

2. Next, working in *any order*, proceed to a second character. Refer to the list of generic numbers you previously noted, then delete any that are not indicated for the new character state. Ignore any additional numbers.

3. Continue until no more can be eliminated. If more than one number remains refer to the dichotomous key and search for couplets that may help the choice between the short-listed genera.

Parentheses have been used in the number-lists to indicate the following:

a) the character state is present only in the species indicated, e.g. **46** (1) — occurs in *Cuviera semseii* but not in any other species of *Cuviera*.

b) the character state occurs only infrequently or partially in the genus or species, e.g. (**52**) or (**39** (1)).

Guide 1. Restricted characters

− leaves subcoriaceous to coriaceous, shiny above, drying yellowish-green ·· **48**; **49** (4,5); **53**

− inflorescence supra-axillary ································· **43**(14); **49**(5)

− bracts paired, connate with silky hairs inside, completely enclosing the
  young umbellate inflorescence, persistent, (TAB. **68**/3) ············ **52**

− bracts linear-lanceolate, conspicuous; inflorescence cymose, (TAB. **58**/1) · **46**

− flowers unisexual or functionally so ······················· **50.4**; **52**(2–5)

− corolla subcoriaceous, wrinkled when dry ····················· **49**(4,5)

− corolla tube deep red ································· **43**(15)

− corolla lobes linear-lanceolate, distinctly longer than tube, (TAB. **59**/6) ·· **47**

− pollen presenter attached at base, (TAB. **47**/B1) ················· **52**

− pollen presenter distinctly widened at base, (TABS. **47**/B4 & **71**/11) ···· **46**(1); **47**; **53.2**

− pyrenes with lid-like area, (Tabs. **47**/D1 & **72**/9); endosperm streaked
  with granules, (TABS. **47**/D2 & **72**/11) ······················ **54**

− cotyledons positioned parallel to ventral face of seed, (TAB. **47**/C2) ···· **53**; **54**

− cotyledons 2–4 times longer than the radicle, (TAB. **48**/14) ·········· **38**

Guide 2. Widely distributed characters

General habit characters

a) climbers ······························· (**50**(16)), (**53.1** (13,14)), **53.2, 54**

b) subshrubby herbs or shrubs (mostly single-stemmed)
   from a woody rootstock, up to 1.5(2) m tall ·········· **38**, (**39**), **40** (1–3), **42, 43, 44, 45,
   48, 49** (5)

c) shrubs to small trees, branched, sometimes scandent ··· **38, 39, 40** (4), **41, 42, 44, 45, 46,
   47, 49** (1–4), **50, 51, 52, 53.1**, (**54**)

d) trees over 10 m tall ······················· **41, 46, 49** (2), **50.2** (4), **53** (1–4)

Spines

a) spines present ······························· (**39**(1)), **45**(2,10,16), **46**(1), **50.1,
   50.3, 50**(16), **51**

b) spines absent ····························· **38, 39, 40, 41, 42, 43, 44, 45, 46,
   47, 48, 49, 50.2**, (**50.3**), **50.4**
   (**50**(16)), **52, 53, 54**

Inflorescence position

a) inflorescences subtended by mature leaves (occasionally
   also a few at nodes where leaves have fallen, especially
   in fruiting stage), (TAB. **46**/9) ·················· **40**(3), **43, 44, 45**, (**46**), **49**(5),
   50.3, 50.4, 50(16), **52, 53, 54**

b) inflorescences always borne at nodes from which the leaves
   have fallen; immature or mature leaves borne at apex of
   stem, or absent at time of flowering, (TAB. **46**/10) ···· **38, 39, 40**(1,2,4), **41, 42, 43, 44,
   45, 46, 47, 48, 49, 50.1, 50.2,
   (50.3), 50**(16), (**52**), **53**(5–7)

c) inflorescences borne on brachyblasts below immature or
   mature leaves or leaves absent at time of flowering,
   (TAB. **46**/11, 12) ····················· **45**(2,16), **50.1, 50.2**(2,3), **51**

Young stems

a) conspicuously lenticellate · **41, 45, (46), 49, (50.1), (50.2), 51**

b) lenticels inconspicuous · **38, 39, 40, 41, 42, 43, 44, 45, 46,
47, 48, (50.1), 50.2, 50.3, 50.4,
50**(16)**, 51, 52, 53, 54**

Leaves

a) always paired · **38, 39, 40, 41, 42, (43), 44, 45,
46, 47, 48, 49, 50, 51, 52, 53, 54**

b) some at least in whorls of 3–6 · · · · · · · · · · · · · · · · · · **38, (42), 43, (44), 49**(5)**, 50.3,
53**(15)

Leaf indumentum

a) always glabrous, save for domatia · · · · · · · · · · · · · · · · **39**(1)**, 40, 41, 43, 45, 46, 49, 50,
51, 52, 53, 54**

b) sparsely to densely pubescent, but leaf-surface visible
beneath · **(38), 40, 41, 43, 45, 48, 49**(4)**,
50.1, 50.2, 50.4, 53**(5–7)**, 54**

c) densely velvety or felted beneath, leaf-surface entirely
obscured · **38, 39**(2,3)**, 40**(3,4)**, 41, 42, 43,
44, 45.1**(8,21)**, 45.2, 46**(3)**, 47**

Stipules

a) conspicuously hairy within · **38, 40, 41, (43), 44, 45, 46, 47,
49, 50.1, 50.2, 50.3, 51**

b) hairs absent or inconspicuous within · · · · · · · · · · · · · · **39, 40, 41, 42, 43, 44, 45, 46, 47,
48, (49), 50.4, 50**(16)**, 52, 53, 54**

Inflorescence

a) flowers always solitary · **39**(3)**, 45, 50.4**(female)**, 50**(16)**,
52**(4,5)

b) peduncle absent (occasionally very short); inflorescence
unbranched, (1)few–many-flowered, usually pedicellate · **41**(5)**, 42, 45, 50.4**(male)**,
53**(7,11–16)

c) peduncle short or well-developed; inflorescence
unbranched or slightly branched, c. 2–10-flowered · · · · · **38, 40, 41, 42, 43, 44, 45, 48, 50.1,
(50.2), 50.3, (50**(16)**), 52**

d) peduncle short or well-developed; inflorescence
distinctly branched, many-flowered · · · · · · · · · · · · · · · **39**(1,2)**, 41, 46, 47, 48, 49, 50.2,
(50.3), 51, 53**(1–6,8–10)**, 54**

Calyx limb

a) reduced to a rim shorter than the disk, sometimes
dentate, (TAB. **47**/A1, A2) · **45, 50.1, 50.2, (51), 52, 53**

b) limb-tube present, small (at least equalling disk) or well-
developed, not lobed or with lobes at most equalling
tube, (TAB. **47**/A3–A5) · **43, 44, 45, 47, 49, 50.1, (50.2),
50.4, 50**(16)**, 51, 53, 54**

c) limb-tube present (at least 1.5 mm long); lobes always
exceeding tube, triangular, triangular-ovate, (TAB. **47**/A6) **40**(1–3)**, 43, 45, 48, 49**(5)

d) limb-tube ± undeveloped, up to 1 mm; lobes short or
long, linear, linear-oblong or, less often, triangular or
sometimes unequal, touching or apart at base,
(TAB. **47**/A7, A8) · **41, 42, 43**(11)**, 45**(16,17)**,
46**(2,3)**, 50.3, 52**

e) limb-tube ± undeveloped or small; lobes long or very
long, narrowly oblong, spathulate, oblong ovate or
subfoliaceous, (TAB. 47/A9, A10) · · · · · · · · · · · · · · · **38, 39, 40**(4), **46**(1)

Corolla tube size

a) large, 10–30 mm long · **38, 39**(3), **42**(6,7), **43**(13–15),
44(2), **49**(2)

b) medium, 5–10 mm long · **39**(2), **40, 43, 45, 47, 49**(2)

c) small, 2–5 mm long · **39**(1), **40, 41, 42, 43, 44**(1), **45,
46, 47, 48, 49, 50.2**(6–11), **50.3,
50**(16), **51, 52**(1–4), **53, 54**

d) very small, 1–1.75 mm long · · · · · · · · · · · · · · · · · **45**(10), **50.1, 50.2**(2–5), **50.4,
52**(5), **53**(5,6,8,10)

Corolla tube indumentum outside

a) glabrous outside (at most with a few hairs near apex) · · · **39**(1), **40, 41, 43, 45, 46, 48, 49,
50, 51, 52, 53, 54**

b) with fine or coarse hairs outside but surface visible · · · · **40, 41, 43, 45**(8), **46, 48, (50.1),
(50.2**(9))

c) densely velvety, tomentose or felted outside · · · · · · · · **38, 39**(2,3), **40**(4), **42, 44, 47**

Corolla tube hairs within

a) ring of deflexed hairs present inside · · · · · · · · · · · · · · **38, 39, 40, 41, 42, 43, 44, 45, 46,
47, 48, 49, 50.1, 50.4, 50**(16), **51,
53, 54**

b) ring of deflexed hairs absent inside · · · · · · · · · · · · · · **45, 50.2, 50.3, 52, 53**(5–8)

Corolla lobes

a) blunt to acute · **41, 43, 44, 45, 49, 50, 51, 52, 53.1,
54**

b) acuminate to apiculate · **39, 40, 41, 42, 43, (44), 45, 46,
47, 53.2**

c) distinctly tailed · **38, 41, 42, 43, 45, 46, 48**

Disk

a) glabrous · **38, 39, 40, 41, 42, 43, 44, 45, 46,
47, 48, 49, 50, 51, 52, 53**(5–16)

b) hairy · **53**(1–4), **54**

Style

a) equalling or shortly exceeding corolla tube · · · · · · · · · **38, 39**(3), **40, 41, 42, 43, 44, 45,
48, 49, 50, 52**

b) exceeding corolla tube by at least twice · · · · · · · · · · · **39**(1,2), **46, 47, 49, 53, 54**

Pollen presenter

a) ± as wide as long or somewhat longer than wide,
(TAB. 47/B1, B2 & B5) · **40, 41, 42, 43, 44, 45, 49, 50, 51, 52**

b) distinctly longer than wide, (TAB. 47/B3, B4 & B6) · · · **38, 39, 41, 42, 46, 47, 48, 49, 53, 54**

Stigmatic lobes

a) 2, only exposed when mature, (TAB. 47/B1–B4)   **45, 46**(2,3), **47, 48, 49, 50, 52, 53, 54**

b) (2)3–5, either always exposed,
(TAB. 47/B5), or exposed only when mature,
(TAB. 47/B6) · **38, 39, 40, 41, 42, 43, 44, 45, 46**(1), **51**

Locules

a) 2 · **45, 46**(2,3), **47, 48, 49, 50, 52, 53, 54**

b) (2)3–5 · **38, 39, 40, 41, 42, 43, 44, 45, 46**(1), **51**

Fruit size

a) small to medium, up to 1 cm across · · · · · · · · · · · · · · **42, 43, 44, 45, 50, 52, 53, 54**

b) large, 1–5 cm across · **38, 39, 40, 41, 42, 43, 44, 46**(1), **47, 48, 49, 50**(16), **51**

Fruit shape

a) ± globose, often 3–5-lobed when dry   **38, 39, 40, 41, 42, 43, 44, 45, 46**(1), **48, 49, 51**

b) somewhat dorsiventrally flattened, clearly 2-lobed · · · · · **45, 46**(2,3), **47, 49, 50, 52, 53, 54**

38. ANCYLANTHOS Desf.

Ancylanthos Desf. in Mém. Mus. Paris **4**: 5, t. 2 (1818). —Robyns in Bull. Jard. Bot. État **11**: 324 (1928) (as "*Ancylanthus*"). —Bridson in Kew Bull. **51**: 343 (1996).

Shrubs, slightly to much branched from a woody rootstock. Leaves usually opposite or sometimes in whorls of 3, shortly petiolate; stipules connate into a sheath below, densely hairy within, distinctly subulate-caudate at the apex, at least the base persistent. Flowers conspicuous, in few- to several-flowered simple or branched peduncled pubescent cymes; bracts small. Buds elongate, ferruginous pubescent outside. Calyx tube campanulate; limb-tube very reduced; lobes 5, well-developed and distinctly leaf-like. Corolla tube cylindrical, elongate, slightly to distinctly curved, open and glabrous at the throat but with a ring of deflexed hairs inside just above the base, and some scattered long hairs elsewhere; lobes triangular-ovate, reflexed, much shorter than the corolla tube, appendaged. Stamens inserted at the throat, the anther tips exserted. Ovary globose or subglobose, 4–5-locular, each locule with a pendulous ovule; style slender, glabrous, shortly exserted; pollen presenter large, cylindrical, 4–5-lobed at the apex. Fruit globose or lobed when dry, fleshy, crowned with calyx lobes, containing up to 5 pyrenes. Pyrenes thinly woody, smooth, scarcely crested around apex, truncate at point of attachment. Embryo with exceptionally long cotyledons.

A small, central African genus comprised only of *Ancylanthos rubiginosus* and possibly also a poorly known taxon from Angola. Fruits edible.

Ancylanthos rubiginosus Desf. in Mém. Mus. Paris **4**: 5, t. 2 (1818) (as "*rubiginosa*"). —De Candolle, Prodr. **4**: 468 (1830). —Hiern in F.T.A. **3**: 158 (1877); Cat. Afr. Pl. Welw. **1**: 484 (1898). —Robyns in Bull. Jard. Bot. État **11**: 326 (1928). —F. White, F.F.N.R.: 400, fig. 67, I (1962). —Verdcourt in Kew Bull. **42**: 131, fig. 2E (1987). —Bridson in Kew Bull. **51**: 350 (1996). TAB. **48**. Type from Angola.
   Ancylanthus ferrugineus Welw. in Andr. Murray J. Trav. Nat. Hist. **1**: 29 (1868). Type from Angola.
   Ancylanthus bainesii Hiern in F.T.A. **3**: 160 (1877). —Robyns in Bull. Jard. Bot. État **11**: 329 (1928). —Miller in J. S. African Bot. **18**: 82 (1952). —F. White, F.F.N.R.: 400 (1962). Type: Botswana, without locality, *Baines* (K, lectotype, chosen by Robyns).

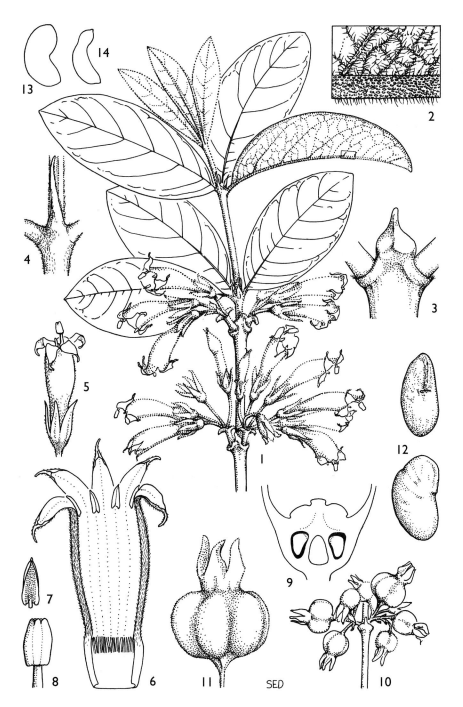

Tab. 48. ANCYLANTHOS RUBIGINOSUS. 1, apical portion of stem (× ²/₃); 2, detail from lower surface of leaf blade (× 6); 3, mature stipule (× 3), 1–3 from *Reed* ?97; 4, juvenile stipule (× 3), from *Richards* 16933; 5, flower (× 1); 6, opened corolla (× 2); 7, dorsal view of anther (× 4); 8, pollen presenter (× 4); 9, longitudinal section through ovary (× 8), 5–9 from *Angus* 539; 10, fruiting node (× ²/₃), from *Richards* 16933; 11, fruit (× 2); 12, pyrene, 2 views (× 2); 13, seed (× 2); 14, embryo (× 2), 11–14 from *Lawton* 1044. Drawn by Sally Dawson.

Ancylanthus fulgidus Welw. ex Hiern in F.T.A. **3**: 159 (1877); Cat. Afr. Pl. Welw. **1**: 484 (1898). —K. Schumann in Warburg, Kunene-Samb.-Exped. Baum: 390 (1903). —Durand, Syll. Fl. Congol.: 271 (1909). —De Wildeman, Comp. Kasai: 424 (1910); Études Fl. Katanga [Ann. Mus. Congo, Sér. IV, Bot.] **2**: 153 (1913); Notes Fl. Katanga **2**: 74 (1913); Contrib. Fl. Katanga: 214 (1921); Pl. Bequaert. **3**: 204 (1925). —Robyns in Bull. Jard. Bot. État **11**: 326, figs. 33 & 34 (1928). —F. White, F.F.N.R.: 400 (1962). Type from Angola.

Suffrutex 0.3–1.5 m tall. Stems erect, single or several shoots from a woody rootstock, mostly burnt off every year; young parts densely covered with rather rough short spreading pale yellowish hairs; older stems ridged with reddish-brown glabrous bark, eventually scaling in small pieces. Leaves opposite or in whorls of 3, 3.5–15.5 × 1.2–7.5 cm, ovate, broadly to narrowly elliptic or elliptic-lanceolate, rounded, acute or shortly acuminate at the apex, rounded to cuneate at the base, very discolorous, rather roughly pubescent above, roughly velvety beneath and obscuring the surface, or only densely pubescent on all the venation and not obscuring the surface; venation drying pale and yellowish above contrasting with brownish-green, and raised closely reticulate and brownish beneath contrasting with green; petiole 0.6–1.3 cm long; stipules c. 2 mm long, triangular, with subulate appendage c. 5 mm long. Inflorescences often borne at leafless nodes, (1)3–7(10 or more)-flowered; peduncle up to 7 mm long; secondary inflorescence branches sometimes present; pedicels 2–12 mm long, bracts c. 6 × 3.5 mm, oblong. Calyx tube 3 mm long, campanulate, densely yellowish-brown pubescent; limb-tube c. 1 mm long; lobes 5–13 × 2–5.5 mm, triangular to ovate- or oblong-lanceolate, pubescent. Corolla apiculate in bud; orange-yellow or bright orange, densely ± velvety pubescent; tube 1.7–2.5 cm long; lobes 5–6 × 2.5–4 mm, triangular-ovate, often distinctly veined in dry state, distinctly appendaged. Style exserted 5–6 mm, the pollen presenter greenish-yellow, 2–3 mm long, cylindrical, lobed at the apex. Fruit yellowish, 1.8 × 1.4 cm, ellipsoid, or 1.3 cm in diameter and subglobose, shortly yellowish-brown velvety pubescent, ± lobed when dry, crowned with persistent calyx.

Botswana. N: Chobe National Park, between Serondela (Serondella) and Negwezumba R., fl. 17.x.1972, *Pope et al.* 811 (K; SRGH). **Zambia**. B: Zambezi (Balovale), fl. 1.xi.1952, *Gilges* 233 (K; PRE). N: Lake Chishi, bud 13.ix.1958, *Fanshawe* 4819 (K; NDO). W: Mwinilunga, Luakera Bridge, fl. 2.x.1937, *Milne-Redhead* 2525 (K). S: Livingstone, fl. 3.xii.1955, *Gilges* 504 (K; PRE; SRGH). **Zimbabwe**. W: Victoria Falls, fl. xi.–xii.1904, *C.E.F. Allen* 97 (K). S: Hwange Game Reserve, west of Dom Pan, x.1960, *Pringle* 11/60 (K; SRGH).

Also in Angola and Namibia. Mostly on Kalahari Sand, in *Afzelia–Terminalia–Parinari*, *Diplorhynchus–Terminalia–Uapaca–Hymenocardia*, *Baikiaea–Brachystegia*, *Burkea–Baikiaea–Pterocarpus* woodlands, also in *Acacia erioloba–Combretum* thicket and *Terminalia sericea* scrub, occasionally on steep rocky places on Karroo Sandstone and pan and dambo edges; 900–1400 m.

39. LAGYNIAS E. Mey. ex Robyns

Lagynias E. Mey. ex Robyns in Bull. Jard. Bot. État **11**: 312 (1928).

Shrubs or understorey trees, or sometimes ± scandent, glabrous or velvety hairy, with or without spines. Leaves opposite, petiolate, mostly discolorous when dry; stipules subulate from a very short triangular base, soon falling. Flowers small, or large in one species, pale coloured, solitary or in opposite, few- to many-flowered lateral cymes; peduncles short and pedicels mostly elongate. Calyx tube subglobose; limb-tube ± absent; lobes 5, much longer than the tube and usually almost as long as or longer than the corolla tube (except in long-flowered species), elongate-spathulate, narrowly oblong-spathulate or oblong, obtuse. Corolla in bud elongate with a cylindrical tube and clavate limb, 5-apiculate; tube narrowly cylindrical, curved in long-flowered species, inside glabrous or sparsely hairy at the throat but with a ring of deflexed hairs below the middle; lobes 5, triangular-lanceolate, tailed at the apex, usually reflexed. Ovary 5-locular; ovules solitary in each cell. Style filiform, mostly distinctly exserted; pollen presenter cylindrical, 5-ribbed, the apex very shortly 4–5-lipped. Stamens 5, inserted at the throat; anthers short, usually completely exserted. Fruit varying in shape according to number of pyrenes developed, asymmetrically ellipsoid, pyriform or subglobose; pyrenes 1–5, with woody walls.

A small genus of 5 well defined species restricted to eastern and south-eastern Africa as far as South Africa (North Prov. and KwaZulu-Natal); close to *Pachystigma*, *Cuviera* and *Vangueria* but with a characteristic facies. One specimen of *L. lasiantha* (Sond.) Bullock from Mozambique, Inhaca Island is definitely spiny.

1. Flowers solitary; corolla tube 15–25 mm long, somewhat curved; petiole 0–2 mm long · · ·
  · 3. *monteiroi*
 – Flowers few–many; corolla tube 3–6 mm long; petioles at least 2 mm long and often more · 2
2. Leaves almost glabrous, 1.8–10 cm long; calyx lobes about 1.5 mm wide · · · · 1. *lasiantha*
 – Leaves densely velvety beneath, 1–5.5 cm long; calyx lobes about 2.5–4.5 mm wide · · · · ·
  · 2. *dryadum*

1. **Lagynias lasiantha** (Sond.) Bullock in Bull. Misc. Inform., Kew **1931**: 274 (1931). —
   Verdcourt in Kew Bull. **11**: 450 (1957). —J.G. Garcia in Mem. Junta Invest. Ultramar **6**
   (sér.2): 35 (1959) [Contrib. Conhec. Fl. Moçamb. IV (1959)]. —K. Coates Palgrave, Trees
   Southern Africa, ed. 3, rev.: 877 (1988). —Pooley, Trees of Natal, Zululand & Transkei:
   472, figs. (1993). Type from South Africa (KwaZulu-Natal).
   Pachystigma lasianthum Sond. in Linnaea **23**: 55 (1850).
   Vangueria lasiantha (Sond.) Sond. in F.C. **3**: 14 (1865). —Bews, Fl. Natal & Zululand: 198
   (1921). —De Wildeman in Bull. Jard. Bot. État **8**: 57 (1922).
   Cuviera australis K. Schum. in Bot. Jahrb. Syst. **28**: 78 (1899). Type: Mozambique,
   Delagoa Bay, *Schlechter* 11958 (B†, holotype; COI; K; LE; PRE; WAG).
   Lagynias discolor E. Mey. in Drège, Zwei Pflanzengeogr. Dokum.: 159 (1843) (nomen). —
   Robyns in Bull. Jard. Bot. État **11**: 313, figs. 29 & 30 (1928). Syntypes from South Africa.

Shrub, small tree or woody scrambler 2–6 m tall, bark grey, smooth. Stems rarely(?) with stout paired spines; youngest shoots slightly to densely appressed pubescent, later glabrous and with grey or reddish-brown peeling bark, longitudinally wrinkled. Leaves 1.8–10 × 0.8–3.8 cm, elliptic, oblong or elliptic-lanceolate, narrowly rounded at the apex, cuneate to rounded at the base, thinly coriaceous, very markedly discolorous, glaucous beneath, glabrous or with sparse pubescence on nerves beneath; petiole 5–10 mm long; stipules forming a short sheath 1–3 mm long, with subulate appendages 1.5–3.5 mm long, densely hairy inside. Flowers in 5–many-flowered dichasially branched cymes with pubescent axes; peduncle 3–10 mm long; pedicels 2–8 mm long; bracts 1.5–2 mm long, ovate, acuminate; bracteoles c. 0.5 mm long. Calyx tube c. 1.5 mm long, cup-shaped, pubescent; lobes 5–9 × 1.2–1.5 mm, linear-spathulate. Corolla with 5 short tails in bud, green or ± yellow, pubescent; tube 3–3.5 mm long, ± cylindrical, puberulous outside, sparsely pilose at throat and with a ring of deflexed hairs within; lobes 6 × 1.5 mm, oblong-lanceolate, acuminate to shortly tailed at apex. Style exserted 4–5 mm; pollen presenter cylindrical-mitriform, scarcely 1 mm long, minutely 5-lobed at apex. Ovary 5-locular. Fruit yellowish-brown, 1.5–4.5 cm in diameter, subglobose, with 3–5 pyrenes.

Mozambique. N: Nampula, km 8 toward Namatil (Nametil), fl. 7.xii.1967, *Torre & Correia* 16389 (LISC; LMA; PRE; WAG). GI: Gaza Prov., 12 km from Chibuto on road to Alto Changane, young fr. 12.ii.1959, *Barbosa & Lemos* 8392 (K; LISC; LMA; PRE). M: between Costa do Sol and Marracuene, Muntanhane, fl. 10.xi.1960, *Balsinhas* 224 (K; BM; COI; LISC; LMA; PRE). Z: Gurué, Mt. Currarre, along R. Loussi, fr. 11.ii.1964, *Torre & Paiva* 10536 (FHO; LISC; LMA; MO).
  Also in South Africa (KwaZulu-Natal). Coastal ridge woodland and thicket and dune vegetation, with *Dialium*, *Afzelia*, *Garcinia*, *Strychnos* etc., sandy flats, rocky places, often on red sand; 0–750 m.

2. **Lagynias dryadum** (S. Moore) Robyns in Bull. Jard. Bot. État **11**: 315 (1928). —Hutchinson,
   Bot. S. Africa: 318 (1946). —J.G. Garcia in Mem. Junta Invest. Ultramar **6** (sér.2): 36 (1959)
   [Contrib. Conhec. Fl. Moçamb. IV (1959)]. —Drummond in Kirkia **10**: 276 (1975). —K.
   Coates Palgrave, Trees Southern Africa, ed. 3, rev.: 877 (1988). —A.E. Gonçalves in Garcia
   de Orta, Sér. Bot. **5**: 196 (1982). TAB. 49. Type: Mozambique, Gazaland, Madanda Forest,
   Swynnerton 1030 (BM, holotype; K).
   Vangueria dryadum S. Moore in J. Linn. Soc., Bot. **40**: 93 (1911). —De Wildeman in Bull.
   Jard. Bot. État **8**: 51 (1922).

Straggling shrub or small tree 2.5–5 m tall. Bark pale grey to grey-brown, smooth or very finely fissured; branches long and arching, densely velvety with yellow-brown pubescence when young, later glabrous, blackish or very dark reddish-brown,

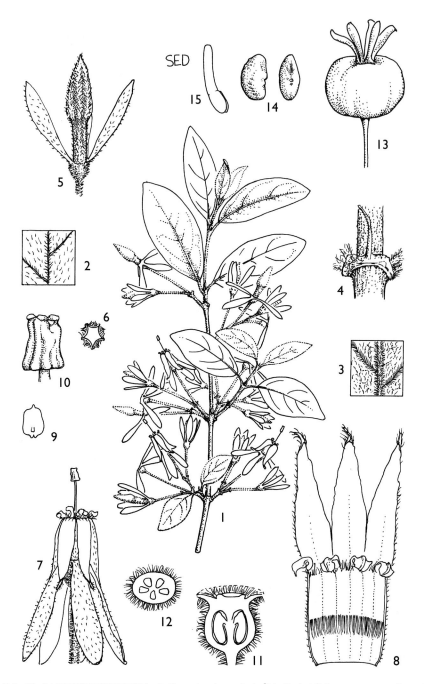

SED

Tab. 49. LAGYNIAS DRYADUM. 1, flowering branch (× ⅔); 2, detail from upper surface of leaf blade (× 2); 3, detail from lower surface of leaf blade (× 2); 4, stipule (× 3); 5, flower bud (× 2); 6, section through corolla bud (× 2); 7, flower (× 2); 8, part of opened corolla (× 4); 9, dorsal view of anther (× 4); 10, pollen presenter (× 8); 11, longitudinal section through ovary (× 6); 12, cross section through ovary (× 6); 1–12 from *Müller & Pope* 1864; 13, fruit (× 1); 14, pyrene, 2 views (× 1); 15, embryo (× 2), 13–15 from *Chase* 6827. Drawn by Sally Dawson.

minutely longitudinally fissured. Leaves 1–5.5 × 0.5–2 cm, oblong-elliptic, ± rounded at both ends, densely velvety pubescent on both surfaces, discolorous, rather thick; petiole 2–3 mm long, velvety; stipules broad, c. 1 mm long with a subulate appendage 2–4 mm long, hairy within. Flowers sweetly scented, in 2–few-flowered dichasial cymes, or sometimes solitary; peduncle 5–8(12) mm long; pedicels 1.5–2 cm long; bracts 2–3 mm long, oblong; all parts densely yellowish-grey velvety pubescent. Calyx tube c. 2 mm in diameter, subglobose, densely pubescent; lobes 7–14 × 2.5–4.5 mm, oblanceolate, spathulate or narrowly oblong-elliptic, densely pubescent. Corolla shortly tailed in bud, greenish-yellow; tube 5–6 mm long, cylindrical, densely pubescent outside, sparsely hairy at the throat and with a ring of deflexed hairs inside towards the base; lobes 7 × c. 2 mm, oblong-lanceolate, shortly apiculate, hairy outside. Style exserted 5–6 mm; pollen presenter 1.3 mm long, cylindrical, 5-lobed. Ovary 5-locular. Fruit brown, 2–2.5 cm in diameter, with 3–5 pyrenes, mostly crowned with persistent calyx lobes, edible; pyrenes 1.5 cm long, rugulose, with a dark red dendritic sculpture and scattered microscopic yellow glands.

Zambia. E: Chadiza Hill, fl. 1.xii.1958, *Robson* 795 (BM; K; LISC). **Zimbabwe**. N: Mutoko (Mtoko), Nyangadzi R. crossing to Chikore Community Land, fl. i.1972, *Davies* 3349 (K; LISC; SRGH). E: Mutare Distr., Zimunya's Reserve, road to Bazeley Bridge, fl. 16.xii.1956, *Chase* 6283 (K; LISC; SRGH). S: Masvingo Distr., Mutirikwi (Mtilikwe) R., Bangara Falls, fl. 13.xii.1953, *Wild* 4373 (K; LISC; SRGH). **Mozambique**. T: Cahora Bassa, Songo, fl. & fr. 28.i.1973, *Torre et al.* 18904 (EA; FHO; J; LMA; P). MS: between Muanza and Inhaminga, fl. 4.xii.1971, *Müller & Pope* 1864 (K; LISC; SRGH).

Also in South Africa (North Prov.). *Brachystegia tamarindoides*, *Androstachys* and other mixed woodlands, riverine thicket, dry river courses, usually on sand and on hillsides with granite boulders; 200–990 m.

3. **Lagynias monteiroi** (Oliv.) Bridson in Kew Bull. **51**: 351 (1996). Type: Mozambique, Delagoa Bay, *Monteiro* 50 (K, holotype).

    Ancylanthos monteiroi Oliv. in Hooker's Icon. Pl. **8**: 7, t. 1208 (1877). —Robyns in Bull. Jard. Bot. État **11**: 325 (1928). —J.G. Garcia in Mem. Junta Invest. Ultramar **6** (sér.2): 37 (1959) [Contrib. Conhec. Fl. Moçamb. IV (1959)]. —K. Coates Palgrave, Trees Southern Africa, ed. 3, rev.: 890 (1988). —Pooley, Trees of Natal, Zululand & Transkei: 478, figs. (1993).

Small shrub 1.2–1.5 m tall. Stems at first densely ± ferruginous velvety pubescent, later glabrous; bark grey-brown, minutely cracking. Leaves 1–3.5 × 0.5–2.3 cm, round to ovate, elliptic or obovate, rounded to subacute at apex, rounded to cuneate at the base, discolorous, densely velvety grey pubescent on both surfaces but surface not entirely obscured save in young leaves; petiole 0–2 mm long; stipules 1–1.5 mm long, short and broad, with a subulate decurrent appendage 1–3 mm long, pubescent. Flowers solitary; pedicels 6 mm long, densely grey or ferruginous pubescent. Calyx pubescent; tube 2 mm long, globose to campanulate; lobes 4(6) × 1.2–2.5 mm, linear-oblong to oblong, rounded at the apex. Corolla yellow, densely pubescent outside; tube about 1.5–2.5 cm long, 8–9 mm wide at apex, 5 mm wide at the base, usually slightly or distinctly curved, with a ring of deflexed hairs just above the base inside; lobes 6–8 × 2–3 mm, narrowly triangular to oblong-triangular, acuminate with short appendages. Style exserted 3–5 mm, the pollen presenter 2 mm long, cylindrical, shortly 4-lobed at the apex. Fruits 1.5–2 cm in diameter, globose, sparsely pubescent; pyrenes c. 11 mm long, ± smooth.

Mozambique. M: Matutuíne (Bela Vista), Tinonganine, Santaca, fr. 7.xii.1961, *Lemos & Balsinhas* 243 (BM; COI; K; LISC; LMA; PRE). GI: Mandlakaze (Muchopes) between Chiducuane and Chicomo, fl. 7.xii.1944, *Mendonça* 3317 (LISC; LMU; MO).

Also in South Africa (KwaZulu-Natal). Ecology unknown, probably forest on coastal sand dunes, etc.; 30 m.

40. PACHYSTIGMA Hochst.

Pachystigma Hochst. in Flora **25**: 234 (1842). —Robyns in Bull. Jard. Bot. État **11**: 117 (1928). —Verdcourt in Kew Bull. **36**: 541–547 (1981).

Subshrubby herbs or small shrubs, glabrous or less often pubescent. Leaves opposite, rarely in whorls of 3, shortly petiolate; stipules connate, hairy within, often at length deciduous. Flowers fairly small, in axillary usually opposite cymes or fascicles, often few-flowered; bracts and bracteoles present. Calyx lobes 5, linear to ovate, erect, obtuse, ± leaf-like, sometimes exceeding the corolla tube. Corolla apiculate to distinctly appendiculate in bud; green or greenish-yellow to greenish-white, glabrous or pubescent; tube cylindrical with a ring of deflexed hairs inside, and the throat usually slightly hairy; lobes reflexed, often apiculate. Stamens inserted at the throat, the anthers exserted. Ovary 3–5-locular with one pendulous ovule per loculus; style slender, exserted, the stigma cylindrical, smooth or ± sulcate, obscurely 5-lobed. Fruit globose or obovoid with 2–5 pyrenes crowned by the persistent calyx limb.

A poorly defined and probably polytypic genus of about 10 species in southern and central Africa, but also including some species from eastern Africa. At present 5 species in the Flora Zambesiaca area are maintained in *Pachystigma* although one may be confused with *Tapiphyllum* (species 2), one with *Fadogia* (species 3) and two with *Lagynias* (species 4 & 5).

1. Herbs or small subshrubs 5–60 cm tall; calyx lobes not, or scarcely as long as, the corolla tube · 2
 – Shrub or tree 0.6–3 m tall; calyx lobes longer than the corolla tube (where known) · · · 4
2. Leaves small, 1–3 × 0.5–1.4 cm; cymes 1–2-flowered, in lower axils (a poorly known species from Barotseland, Zambia) · 3. *albosetulosum*
 – Leaves larger, 3–16 × 1–4.7 cm; cymes 3–20-flowered · 3
3. Perennial subshrubby herb 5–15(25) cm tall; inflorescences usually few and borne near the ground; pyrenes up to 1.5 × 0.8 cm · 1. *pygmaeum*
 – Taller subshrubs 15–60 cm tall; inflorescences often at several nodes, sometimes as many as 6, and borne well above ground level; pyrenes much smaller · · · · · · · · · · · · 2. *micropyren*
4. Leaves obtuse to rounded at the apex and base, densely velvety · · · · · · · · · · 4. *macrocalyx*
 – Leaves acute at the apex and cuneate at the base, sparsely shortly pubescent · · · · 5. *sp. A*

1. **Pachystigma pygmaeum** (Schltr.) Robyns in Bull. Jard. Bot. État **11**: 122, figs. 17 & 18 (1928). —Codd & Voorendyk in Bothalia **8** (Suppl. 1): 49, figs. 51, 52, 53 & map 4 (1965). — Fanshawe & Hough, Poisonous Pl. Zambia: 10 (1967). —Verdcourt in Kew Bull. **36**: 541 (1981); in F.T.E.A., Rubiaceae: 765, fig. 135 (1991). TAB. **50**. Type from South Africa.
 Vangueria pygmaea Schltr. in J. Bot. **35**: 342 (1897). —Bews, Fl. Natal & Zululand: 198 (1921). —De Wildeman in Bull. Jard. Bot. État **8**: 61 (1922). —Burtt Davy in Bull. Misc. Inform., Kew **1925**: 37 (1925). —Marloth, Fl. S. Africa **3**, 2: 193, figs. 83, 84 (1932).
 Vangueria setosa Conrath in Bull. Misc. Inform., Kew **1908**: 224 (1908). —De Wildeman in Bull. Jard. Bot. État **8**: 64 (1922). Type from South Africa.
 Vangueria rhodesiana S. Moore in J. Bot. **47**: 130 (1909). —De Wildeman in Bull. Jard. Bot. État **8**: 61 (1922). Type: Central Zimbabwe, *Rand* 1349 (BM, holotype).
 Pachystigma rhodesianum (S. Moore) Robyns in Bull. Jard. Bot. État **11**: 119 (1928).

Perennial, erect or prostrate, subshrubby herb, 5–15(25) cm tall, usually with a long rhizome. Stems at first sparsely to densely setose-hairy but becoming glabrous; bark brown, finely fissured. Leaves paired, very rarely in whorls of 3; blades 4–9.5(16) × 1.6–4.7 cm, oblong to oblong-elliptic, elliptic or oblanceolate, obtuse, acute or shortly acuminate at the apex, cuneate at the base, glabrous to covered on both surfaces with dense seta-like hairs which do not in any way hide the surface, often quite scabrid; petiole 0–3(7) mm long; stipules scarious to thicker with a subulate appendage 2–6 mm long, from a base 1–2 mm long, densely setose or glabrescent. Inflorescences axillary, simple, 3–7(many)-flowered, often at ground level, densely setose; peduncles 5–6(12) mm long; pedicels 5–8 mm long, bracts scarious, usually several at base of inflorescence and conspicuous, 2–3 mm long and wide, obovate, obtuse. Calyx tube 1.5 mm long; limb-tube 0–2 mm long; lobes erect, 2–8.5 mm long, narrowly oblong or linear, usually setose. Corolla filiform-apiculate in bud, appendages up to 3 mm long; tube white or pale greenish-cream, sparsely setose, 4–8 mm long, subcampanulate to quite narrowly cylindrical; lobes green outside, 4–6 mm long including the filiform appendage, 2 mm wide, narrowly triangular. Anthers brown, exserted, 1–1.7 mm long; filaments short, about 1 mm long. Ovary 5-locular. Style slender, 5–9 mm long; pollen presenter pale green, cylindrical or coroniform, 0.75–1 mm long, slightly 5-lobulate. Fruit black or yellow-brown, 1.5–2 cm in diameter, subglobose, pear-shaped or oblique, with (1)2–5 pyrenes; pyrenes 1.1–1.5 cm × 6–8 mm, ellipsoid, obtusely subtrigonous, with a sharp point above point of attachment, rugulose.

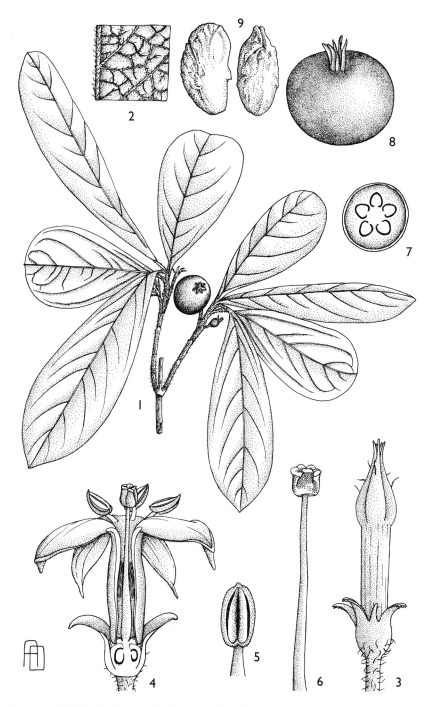

Tab. 50. PACHYSTIGMA PYGMAEUM. 1, habit (× ²/₃); 2, detail from lower surface of leaf blade (× 2), 1 & 2 from *Fanshawe* 10077; 3, flower bud (× 4); 4, flower with section removed (× 4); 5, stamen (× 8); 6, style and pollen presenter (× 6); 7, transverse section through ovary (× 10), 3–7 from *Milne-Redhead & Taylor* 8693; 8, fruit (× 1⅓); 9, pyrene, 2 views (× 2), 8 & 9 from *Fanshawe* 10077. Drawn by Ann Farrer, with 9 by Sally Dawson. From F.T.E.A.

Zambia. N: Mbala Distr., escarpment above Lufubu R. Plain, fl. 19.ii.1957, *Richards* 8623 (K). W: Luanshya, fl. 13.i.1956, *Fanshawe* 2704 (K; NDO). S: Mazabuka Distr., 1.6 km below confluence of Siamambo and Bunchele streams, fl. 6.ii.1960, *White* 6882 (FHO; K). **Zimbabwe**. C: Makoni Distr., near Maidstone, fl. 5.i.1931, *Fries, Norlindh & Weimarck* 4102 (K; LD; PRE; SRGH). E: Chimanimani (Melsetter), fl. 24.xii.1948, *Chase* 1238 (BM; K; SRGH). **Malawi**. N: Chitipa Distr., Mafinga Hills (Mts.), 2.iii.1982, *Brummitt et al.* 16247 (K). C: Dedza Distr., Chongoni Forest Reserve, fr. 6.ii.1968, *Jeke* 14 (K; SRGH) (see note).

Also in southern Tanzania, Swaziland and South Africa (North Prov., North-West Prov. and Mpumalanga). Open ground, short grassland, dambos, often on sandy ground, also laterite pans and on termite mounds; 1170–2280 m.

Although Robyns made a new subgenus *Pseudopachystigma* for *P. rhodesianum* the characters do not hold and it is clearly a synonym of *P. pygmaeum*. *Jeke* 14 cited above has leaves up to 16 × 7 cm, but is only a form of this species. Glabrous-leaved and hairy-leaved specimens occur in the same populations in many areas, but the similar South African species *P. thamnus* Robyns which is always glabrous has a different facies and is a distinct species.

Well known as a plant poisonous to sheep and other stock in South Africa but not toxic in Zimbabwe (see Shone & Drummond, Poisonous Pl. Rhod.: 1 (1965)).

2. **Pachystigma micropyren** Verdc. in Kew Bull. **42**: 139 (1987). Type: Zambia, Kitwe, *Mutimushi* 2394 (K, holotype; NDO).

Subshrub 15–60 cm tall, with several erect or ± procumbent slender woody shoots from an underground rhizome; shoots at first scabrid yellowish-pubescent, later ± glabrous, dark purplish-red. Leaves paired, not in whorls, 3–10 × 0.9–3.2 cm, elliptic, oblong, obovate-oblong or oblanceolate, rounded to subacute or obscurely acuminate at the apex, cuneate at the base, setulose-pubescent to scabrid on main nerves, particularly beneath, and also on the margins; petiole c. 5 mm long; stipules 2 mm long, triangular, with appendages 2 mm long. Inflorescences borne at (1)3–6 nodes, 10–20-flowered, bristly pubescent; the inflorescence pair at each node forming a dense cluster up to 3.5 cm wide; peduncles and pedicels 3–5 mm long; bracts 3 × 2 mm, ± oblong. Calyx tube 1.2 mm long, globose, densely pubescent; lobes 4–6.5 × 0.7–1 mm, linear, more sparsely pubescent. Corolla 5-tailed at apex in bud; tube 5.5 mm long, cylindrical to funnel-shaped, glabrous at base, densely bristly pubescent above; lobes 2–3 × 1.5–1.8 mm, triangular, with a terminal appendage 1.5–2 mm long. Pollen presenter shortly exserted, cylindrical, 10-grooved, 5-lobed at apex. Fruit yellow to orange, 1.2 cm in diameter, globose, said to be glossy, very sparsely pubescent; pyrenes small 6 × 5 × 3 mm, the keel at dehiscent end about two thirds the length of the pyrene, ± smooth.

Zambia. W: Luanshya, 12.iii.1955, *Fanshawe* 2132 (BR; K; NDO).
Also in Angola. *Brachystegia* woodland and derived associations, sometimes on laterite outcrops; ?1200 m.

Pachystigma micropyren appears close to *Tapiphyllum cistifolium* var. *latifolium*. The latter differs in stipule shape and by the leaf blades which are densely tomentose beneath and have the venation impressed above. Although these two taxa do appear to be different species they could perhaps be congeneric.

3. **Pachystigma albosetulosum** Verdc. in Kew Bull. **42**: 139 (1987). Type: Zambia, Kalabo, Liuwa Plain, *Drummond & Cookson* 6461 (K, holotype; SRGH).

Subshrub with several stems 10–13 cm tall from a slender woody rhizome. Stems densely covered with white seta-like hairs. Leaves small, 1–3 × 0.5–1.4 cm, oblong-elliptic, obtuse to subacute at the apex, more or less rounded at the base, discolorous, drying dark above with a dense pubescence of short white hairs similar to the stem; venation beneath slightly raised, reticulate and very densely covered with, but not totally obscured by, short white setae; the areas between the tertiary veins are whitish thus giving a characteristic appearance to the surface; petioles very short, scarcely 1 mm long; stipules 1 mm long, triangular, with a subulate appendage 2 mm long, densely setulose-pubescent. Cymes 1–2-flowered, in lower axils; peduncle about 5 mm long; pedicels 1–2 mm long; bracts and bracteoles 0.5–1.5 mm long, narrowly triangular. Calyx densely spreading setulose-pubescent; tube c. 1.5 mm in diameter, subglobose; lobes 1.5–2 mm long, oblong-triangular. Corolla yellow, pubescent outside; tube 5 mm long, ± funnel-shaped, inside with a median ring of deflexed bristly hairs and some tangled hairs in upper part above this; lobes 5–6 × 2 mm, oblong-lanceolate, with appendages scarcely 1 mm long. Ovary presumed 5-locular. Style 7–8 mm long; pollen presenter coroniform, 1 mm long, 5-lobed at apex. Fruits not known.

Zambia. B: Kalabo, Liuwa Plain, Paramount Chief's Game Reserve, 48 km north of Kalabo, fl. 14.xi.1959, *Drummond & Cookson* 6461 (K; SRGH).
Apparently endemic to this area. Grassy plain near margin of patch of woodland.

4. **Pachystigma macrocalyx** (Sond.) Robyns in Bull. Jard. Bot. État **11**: 130 (1928). —Palmer & Pitman, Trees Southern Africa **3**: 2111 (1973), pro parte excl. *P. bowkeri* in syn. —K. Coates Palgrave, Trees Southern Africa, ed. 3, rev.: 889 (1988), pro parte excl. *P. bowkeri* in syn. — Pooley, Trees of Natal, Zululand & Transkei: 478, figs. (1993). Type from South Africa (KwaZulu-Natal).
 Vangueria macrocalyx Sond. in Linnaea **23**: 59 (1850); in F.C. **3**: 14 (1865). —Sim, For. Fl. Col. Cape Good Hope: 244 (1907). —Bews, Fl. Natal & Zululand: 198 (1921). —De Wildeman in Bull. Jard. Bot. État **8**: 58 (1922).
 Canthium macrocalyx (Sond.) Baill. in Adansonia **12**: 191 (1878).

Shrub or small tree 0.6–3 m tall, sometimes bearing spine-like remains of axillary branches on the trunk; branchlets mostly densely yellow- or golden-ferruginous pubescent, becoming glabrous or ± entirely glabrous; bark greyish. Leaves paired, 1.5–4.5 × 1–2.5(3) cm, ovate, broadly elliptic or oblong, rounded to obtuse at the apex, rounded to cuneate at the base, densely to very densely yellow-velvety; petiole up to 4 mm long; stipules 2–3 mm long, triangular, with a filiform appendage up to 6 mm long. Inflorescences few- to many-flowered, velvety; peduncle up to 3 mm long; pedicels obsolete, up to 2(3) mm long; bracts 2 × 1 mm, oblong. Calyx tube 1.5 mm long, glabrous to densely pubescent; lobes c. 10 × 1–3 mm, much longer than the corolla tube, linear to narrowly oblong or oblanceolate, glabrous to densely pubescent inside and out. Corolla tailed or only acuminate in bud, cream-coloured; tube c. 4 mm long, broadly funnel-shaped; lobes 4.5 × 2 mm, triangular, acuminate with appendages c. 1 mm long. Fruit orange-yellow, about 20 mm long, globose or ovoid, velvety or sparsely so when fully mature, crowned by the calyx lobes; pyrenes 10–20 mm long.

Mozambique. M: Maputo (Lourenço Marques), sterile viii.1918, *Rogers* 21360 (K).
Also in South Africa (Mpumalanga, KwaZulu-Natal and Eastern Cape Prov.) and Swaziland; 30 m.
Pachystigma bowkeri Robyns, a ± glabrous species of forests in South Africa, was much confused with *P. macrocalyx*. The distinction between these two taxa was recently clarified [Kok, Boshoff & van Wyk in S. Afr. J. Bot. **55**: 560–563 (1989)]. Only one sterile fragment of *P. macrocalyx* is known from the Flora Zambesiaca area. In many ways *P. macrocalyx* resembles *Lagynias*, especially *L. monteiroi*.

5. **Pachystigma sp. A**

Shrub to c. 2 m tall. Stems slender, brown, ± densely pubescent with curled hairs. Leaves 2–4.5 × 0.8–2 cm, elliptic or narrowly rhombic-elliptic, acute at the apex and cuneate at the base, sparsely shortly pubescent. Fruits c. 1.4 cm in diameter, sparsely pubescent with short hairs; calyx lobes persistent, 5 × 1.5 mm, lanceolate; pyrenes 3–4.

Mozambique. GI: Bilene Distr., 8 km from Praia de S. Martinho towards Macia, fr. 5.xi.1969, *Correia & Marques* 1388 (WAG).
Not known elsewhere. From dense *Syzygium, Albizia, Ozoroa, Apodytes* etc. forest.
The above material is inadequate, but this specimen will probably prove to be an undescribed *Pachystigma* or *Lagynias*.

41. VANGUERIA Comm. ex Juss.

Vangueria Comm. ex Juss., Gen. Pl.: 206 (1789). —Robyns in Bull. Jard. Bot. État **11**: 273 (1928). —Verdcourt in Kew Bull. **36**: 547–556 (1981).

Deciduous shrubs or small trees, unarmed or rarely spiny, with glabrous to densely velvety foliage. Leaves paired, never whorled, often quite large; stipules broad and connate into a short sheath at the base, produced into a linear or subulate appendage at the apex, hairy inside, deciduous or more or less persistent. Inflorescences usually many-flowered, in axillary, divaricately branched cymes at opposite sides of nodes, often at leafless nodes and sometimes flowering before leaves are fully developed; cyme branches ± elongate with regularly spaced shortly petiolate flowers. Calyx tube hemispherical or depressed campanulate; calyx limb divided into 5(6) triangular, oblong, linear or ligulate lobes. Corolla small, often distinctly apiculate in bud (due to lobe appendages), usually yellow or greenish-yellow; corolla tube shortly cylindrical or campanulate, glabrous or hairy outside, with a ring of deflexed hairs inside and densely hairy at the throat; lobes 5, narrowly triangular, mostly reflexed, sometimes each with a distinct narrow appendage at the apex. Anthers ovate or oblong, apiculate, shortly exserted. Ovary 5(6)-locular; ovule solitary in each locule, pendulous. Style filiform-cylindrical, glabrous, shortly exserted; pollen presenter cylindrical, very shortly 5-lobed at the apex. Disk annular, slightly raised. Fruit large, indehiscent and fleshy, ± globose, often lobed or angular in the dry state, (4)5(6)-locular, containing (4)5(6) pyrenes, sometimes bearing the remains of calyx lobes, glabrous, often edible.

Species delimitation in this genus is difficult, and it is clear that too many species have been recognised in the past. Robyns recognised 27 species from tropical and South Africa and Madagascar; his circumscriptions depending on characters (such as leaf and corolla indumentum, length of calyx lobes and degree of development of corolla lobe appendages) used in various combinations. Many will probably consider that not enough reduction has been made in this present treatment and might reduce the genus to two species based on calyx length.

The genus is rare in West Africa. The inflorescences are very attractive to many kinds of insects and the fruits of several species are edible. The monotypic subgenus *Itigi* Verdc. accommodates the Tanzanian species *V. praecox*.

1. Leaves small, 1–2(4) × 0.4–1.3(3) cm, ovate or almost round, densely velvety; calyx lobes short, 1.5–2 mm long; corolla tube short, 0.8–2.5 mm long (SE Botswana) ·· 5. *parvifolia*
 – Leaves larger, glabrous to velvety; calyx lobes 0.5–8 mm long; corolla tube usually longer, (2–3)3–5 mm long ·2
2. Calyx lobes* triangular-oblong to narrowly-oblong, 0.5–1.5(2–3) mm long** · · · · · · · · ·3
 – Calyx lobes more elongate, oblong to linear (1.2)3–7(8) mm long · · · · · · · · · · · · · ·7
3. Leaf blades almost glabrous, or ± densely pubescent above and on venation below; petioles slender, 5–9(15) cm long, spreading pubescent; stems slender; inflorescences c. 7-flowered, borne on a peduncle 9–15 mm long with a conspicuous bracteate ring at ± mid-point; corolla glabrous or with very few hairs, with a tube separated from limb by a constriction in bud, at least when dry · 6. *esculenta*
 – Leaves entirely glabrous, pubescent to densely velvety or seldom sparsely pubescent; petioles shorter, or if as long and both slender and pubescent then tending to dry blackish; stems robust or slender; inflorescences few- to many-flowered, subsessile to pedunculate; bracteate ring less evident; corolla glabrous or pubescent ·4
4. Leaves glabrous, or essentially so ·5
 – Leaves pubescent to velvety ·6
5. Leaves with (7)8–10 pairs of main lateral nerves, blades stiffly papery, never drying dark or blue-black; tertiary nerves finely reticulate, usually drying dark (but not blue-black) below and pale above; petioles cream-coloured; stems mostly robust, yellow-brown to mid-brown in colour; inflorescences 15–30-flowered, borne on peduncles up to 10 mm long · 1. *madagascariensis*
 – Leaves with 5–6 pairs of main lateral nerves, blades thinly papery, often drying blue-black; tertiary nerves not particularly finely reticulate, frequently drying blue-black; petioles often drying blue-black; stems slender, red or purplish, rather powdery; inflorescences 5–9-flowered, subsessile or on peduncles up to 2.5 mm long · · · · · · · · · · · · · 3. *cyanescens*
6. Indumentum velvety and shorter, often ferruginous or yellowish; leaves not drying bluish; 4–30 × 2.5–18 cm; petioles 3–10 mm long; stipule appendage 3–20 mm long; inflorescences up to 30-flowered, borne on peduncles 6–8 mm long · · · · · · · · · · · · · · · · · 2. *infausta*

* The calyx lobes are often shed in mature fruit; if in doubt try both couplets.
** the undescribed species treated as number 10. *Vangueria sp.* below, based on 3 rather varied sheets from Lengwe Game Reserve in Malawi, will not key out further.

– Indumentum, particularly on buds, longer, mostly greyish; leaves often drying dark bluish or with blue-black nerves, 1–8(12.5) × 0.5–3.8(6) cm; petioles 1–6 mm long; stipule appendage 2–7 mm long; inflorescences up to 13-flowered, subsessile or borne on peduncles up to 5(10) mm long · 4. *proschii*
7. Foliage hairy; buds hairy · 8. *volkensii*
– Foliage glabrous, or almost so in the Flora Zambesiaca area; buds glabrous · · · · · · · · · 8
8. Leaves with a coarse reticulum of tertiary nerves; stems dull, greyish- to mid red-brown, sparsely lenticellate; corolla c. 6 mm long usually not clavate in bud; tube 2–3 mm long not markedly hairy at throat; lobes distinctly tailed · 9. *randii*
– Leaves with a fine reticulum of tertiary nerves; stems shiny, plum-coloured and conspicuously lenticellate; corolla c. 8–9 mm long, clavate in bud; tube 4–5 mm long, the throat tufted with projecting hairs; lobes acuminate but not tailed · · · · · · · · 7. *apiculata*

1. **Vangueria madagascariensis** J.F. Gmel., Syst. Nat., ed. 13, **2**: 367 (1791). —Webster, Food Plants of the Philippines: 196, t. 47b (1924). —J.G. Garcia in Mem. Junta Invest. Ultramar **6** (sér. 2): 32 (1959) [Contrib. Conhec. Fl. Moçamb. IV (1959)]. —Verdcourt in Kew Bull. **36**: 548 (1981); in Kew Bull. **42**: 830, fig. 3A (1987); in Fl. des Mascareignes, fam. 108, Rubiacées: 108, t. 32 (1989); in F.T.E.A., Rubiaceae: 849, figs. 131/22, 132/15 & 150 (1991). —Beentje, Kenya Trees, Shrubs Lianas: 551 (1994). TABS. 47/B6 & **51**. Type from Madagascar.
Vavanga chinensis Rohr in Skr. Naturhist.-Selsk. **2**: 207 (1792). Type from West Indies.
Vavanga edulis Vahl in Skr. Naturhist.-Selsk. **2**: 208, t. 7 (1792). Type as for *V. chinensis*, nom. illegit.
Vangueria edulis (Vahl) Vahl in Symb. Bot. **3**: 36 (1794).
Vangueria edulis Lam., Tabl. Encycl. **5**, 2: 235 (1819); **1**, 2: t. 159 (1792) without specific epithet. Type from Mauritius.
Vangueria acutiloba Robyns in Bull. Jard. Bot. État **11**: 286 (1928). —Brenan, Check-list For. Trees Shrubs Tang. Terr.: 537 (1949). —Dale & Greenway, Kenya Trees & Shrubs: 478 (1961). Type from Tanzania.
Vangueria floribunda Robyns in Bull. Jard. Bot. État **11**: 285 (1928). Type from South Africa (Mpumalanga Prov.).
Vangueria venosa Robyns in Bull. Jard. Bot. État **11**: 290 (1928). —F.W. Andrews, Fl. Pl. Anglo-Egypt. Sudan **2**: 466 (1952) non Sond. Type from Ethiopia.
Vangueria robynsii Tennant in Kew Bull. **22**: 443 (1968). Type as for *V. venosa*.
Vangueria cyanescens sensu Pooley, Trees of Natal, Zululand & Transkei: 470, fig. (1993), non Robyns.

Shrub or small tree, 1.5–15 m tall, often multi-stemmed and sometimes with a spreading crown; stems mostly robust, glabrous, longitudinally ridged, with pale to mid-brown bark, mostly smooth and unpeeling (but peeling or powdery in one southern Tanzanian variant). Leaf blades 8–28 × 3.2–15 cm, narrowly to broadly elliptic or elliptic-lanceolate, acute to shortly acuminate at the apex, cuneate to rounded or less often ± subcordate at the base, entirely glabrous or sometimes very young leaves pilose beneath and adult ones sparsely pubescent; lateral nerves in (7)8–10 main pairs; tertiary nerves finely reticulate, often drying distinctly dark below and pale above; petiole 0.3–1.8 cm long, often drying cream-coloured; stipules with a broad base 3–5 mm long, and a narrow apex 0.4–1.8 cm long; glabrous or pubescent. Inflorescence c. 30-flowered, pubescent; branches 1–4.5 cm long, 7–10-flowered; main peduncle c. 10 mm long; pedicels about 2 mm long, save those of the central flower of the inflorescence which measures 4 mm long. Calyx tube 1.2–3 mm long; lobes 0.5–1.5 mm long, triangular-oblong to narrowly oblong, ± pubescent. Corolla slightly to usually markedly acuminate in bud, or apiculate due to the corolla lobe appendages, greenish-yellow, yellow or cream, glabrous or rarely with a few hairs; tube 3–4.5 mm long; lobes 3.5–4.5 mm long with appendages up to 0.5 mm long. Style 7–8 mm long; pollen presenter yellow, 1.2–1.5 mm long, cylindrical. Fruits green to brownish, 2.5–5 cm diameter, subglobose; pyrenes 4–5, each c. 20 × 12 × 8 mm with thick woody walls 1–2 mm thick. Seeds c. 16 × 6 × 4.5 mm, narrowed at one end.

Malawi. C: Salima, Lake Nyasa Hotel, fr. 15.ii.1959, *Robson* 1611 (BM; K; SRGH).
Mozambique. Z: Maganja da Costa, Floresta de Gobene, 45 km from Olinga (Maganja), fr. 14.ii.1966, *Torre & Correia* 14616 (FHO; LISC; LMU; MO; SRGH). T: Cahora Bassa (Cabora Bassa), 2 km from dam, R. Mucangádzi, fr. 8.ii.1973, *Torre et al.* 19035 (C; COI; LISC; LMA; MO; WAG). MS: Buzi, Mucheve Forest Reserve, fl. 28.x.1963, *Carvalho* 698 (LMU; PRE). M: between Goba and Catuane, fl. & fr. 24.x.1940, *Torre* 1862 (LD; LISC; LMU).

Tab. 51. VANGUERIA MADAGASCARIENSIS. 1, apical portion of flowering branch (× ²/₃), from *Haarer* 1752; 2, stipules (× 2), from *Greenway & Kanuri* 13468; 3, flower bud (× 6); 4, flower (× 6); 5, part of corolla opened out, with style and pollen presenter (× 6); 6, pollen presenter (× 12); 7, calyx (× 8); 8, longitudinal section through ovary (× 12); 9, transverse section through ovary (× 12), 3–9 from *Haarer* 1752; 10, fruit (× ²/₃), from *Lindsay* s.n.; 11, pyrene (× 1), from *Richards* 20808. Drawn by M.E. Church. From F.T.E.A.

Also in Nigeria, Cameroon, Zaire (Dem. Rep. Congo), Central African Republic, Sudan, Ethiopia, Uganda, Kenya, Tanzania, Swaziland and South Africa (North Prov., Gauteng, Mpumalanga and KwaZulu-Natal) and Madagascar. Also widely cultivated throughout the tropics for its edible fruits. In grassveld on sandy rocky soil, or sand banks at edge of lakes and beaches; 90–470 m.

The material cited above is directly comparable with the South African material previously called *Vangueria floribunda*, which consistently has pale robust stems, short cream-coloured petioles and large stiffly chartaceous leaves. It is close to material from Madagascar, but that mostly has slightly longer petioles. Material from further north in mainland Africa is more variable. In southern Tanzania (Ufipa and Iringa Districts) *V. madagascariensis* has proved difficult to separate from *V. apiculata*; see note after that species for further remarks concerning specimens from Zambia N. Specimens of *V. madagascariensis* with some pubescence on either leaves and/or corolla are not uncommon and *V. infausta* could well be considered a pubescent form of this species. However, when one compares typical *V. infausta* with typical *V. madagascariensis* and considers that all but a very few specimens are easily named and that the rest are abundantly distinct, combining the two would be impractical. In Tanzania hairy and glabrous variants have been found growing together, e.g. *Brummitt & Polhill* 13617 and 13617A in Tanzania, Iringa District. Robyns treats *V. venosa* as published by Richard but the latter only lists it in synonymy.

2. **Vangueria infausta** Burch., Trav. S. Africa **2**: 258, fig. on 259 (1824). —De Candolle, Prodr. **4**: 454 (1830). —Sonder in F.C. **3**: 13 (1865). —Hiern in F.T.A. **3**: 147 (1877). —Engler, Pflanzenw. Ost-Afrikas **C**: 384 (1895). —Sim, For. Fl. Port. E. Africa: 75 (1909). —Robyns in Bull. Jard. Bot. État **11**: 307 (1928). —Marloth, Fl. S. Africa **3**, 2: 193, pl. 49 fig. 82 (1932). —Codd, Trees & Shrubs Kruger Nat. Park: 177, fig. 164 (1951). —Miller in J. S. African Bot. **18**: 85 (1952). —J.G. Garcia in Mem. Junta Invest. Ultramar **6** (sér. 2): 33 (1959) [Contrib. Conhec. Fl. Moçamb. IV (1959)]. —von Breitenbach, Indig. Trees Southern Africa: 1130, fig. on 1129 (1965). —Palmer & Pitman, Trees Southern Africa **3**: 2079, figs. on 2080 (1973). —van Wyk, Trees Kruger Nat. Park **2**: 571, t. 702 (1974). —Drummond in Kirkia **10**: 276 (1975), pro parte. —Palmer, Field Guide Trees Southern Africa: 293 (1977). —Bridson & Troupin in Fl. Pl. Lign. Rwanda: 610, fig. 207.2 (1982); in Fl. Rwanda **3**: 230, fig. 73.2 (1985). —K. Coates Palgrave, Trees Southern Africa, ed. 3, rev.: 873, pl. 301 (1988), pro parte. —Verdcourt in Kew Bull. **36**: 549 (1981); in F.T.E.A., Rubiaceae: 851 (1991). — Pooley, Trees of Natal, Zululand & Transkei: 470, figs. (1993). Type from South Africa.

Shrub or small tree 1.5–8 m tall, very similar to previous species save for the indumentum; trunk smooth, grey or eventually rough and ridged; branches often robust, ridged, pale or dark, the young parts densely ferruginous, pubescent or velvety. Leaf blades 4–30 × 2.5–18 cm, elliptic to oblong-elliptic, ovate or sometimes almost round or lanceolate, rounded, subacute or ± acute to shortly acuminate at the apex, rounded to cuneate or rarely ± subcordate at the base, often discolorous, densely pubescent to usually densely softly velvety on both surfaces; the hairs often yellowish or ferruginous in dry state; lateral nerves in 6–8 main pairs; petiole 3–10 mm long, similarly hairy; stipules with base 2–4 mm long and apical part 3–20 mm long, hairy. Inflorescences densely hairy, sometimes extensive and much branched but typically rather short; branches 1.5–3.5 cm long, each 5–10-flowered; peduncles 6–8 mm long; pedicels 1–2.5 mm long, but those of the central flower of each dichasial element c. 3.5 mm long. Calyx tube 0.75–1.2 mm long; lobes obtusely triangular to narrowly oblong, 1–1.25(2–2.5) mm long. Corolla rounded or apiculate in bud, green or yellow-green, typically densely spreading hairy outside but glabrous in some variants; tube 3–4.5 mm long; lobes 3–4 mm long with appendages up to 0.5 mm long, or practically obsolete. Style 4.5–6 mm long; pollen presenter yellow, 1 mm long. Fruit green, usually ripening to dull orange-brown or purplish, 1.5–4.7 cm in diameter, depressed subglobose. Pyrenes up to 13–20 × 6–13 × 5–8 mm.

Leaves often immature at time of flowering; inflorescences not extensively branched; corolla
   densely spreading hairy outside; lobes not at all apiculate · · · · · · · · · · · subsp. *infausta*
Leaves mostly fully mature at time of flowering; inflorescences often extensively branched;
   corolla glabrous or spreading hairy outside; lobes apiculate · · · · · · · · · subsp. *rotundata*

Subsp. **infausta** —Verdcourt in Kew Bull. **36**: 549 (1981); in F.T.E.A., Rubiaceae: 852 (1991).
   Vangueria velutina Hook. in Bot. Mag. **57**: t. 3014 (1830). —Sim, For. Fl. Port. E. Africa: 75 (1909). —Robyns in Bull. Jard. Bot. État **11**: 306 (1928). Type grown at Liverpool from Madagascan material.

Vangueria tomentosa Hochst. in Flora **25**, 1: 238 adnot. (1842). —Robyns in Bull. Jard. Bot. État **11**: 308 (1928). —Brenan, Check-list For. Trees Shrubs Tang. Terr.: 538 (1949), pro parte. —F. White, F.F.N.R.: 424 (1962). —Gomes e Sousa, Dendrol. Moçamb.: 679, t. 227 (1967). —Verdcourt in Kew Bull. **36**: 549 (1981). —A.E. Gonçalves in Garcia de Orta, Sér. Bot. **5**: 211 (1982). Type from South Africa.

Vangueria munjiro S. Moore in J. Linn. Soc., Bot. **40**: 92 (1911). Type: Mozambique, Kurumadzi R., Jihu, *Swynnerton* 1293 (BM, holotype; K, fragment).

Leaves usually toward the lower end of the scale of size given above and often not fully developed at flowering time, always densely velvety. Inflorescences not extensively branched. Corolla rounded at apex and not at all apiculate in bud; densely spreading hairy outside. Fruit up to 3.5 cm in diameter.

Botswana. SE: about 30 km WNW of Lobatse on Kanye road, 18.i.1960, *Leach & Noel* 178 (K; SRGH); 2 km east of Kanye, fl. 2.ix.1978, *O.J. Hansen* 3449 (C; GAB; K; PRE; SRGH; WAG). **Zambia.** N: Kawambwa, fl. 10.xi.1957, *Fanshawe* 3891 (K; NDO). W: Bwana Mkubwa, fl. 10.xi.1953, *Fanshawe* 489 (K; NDO). C: Kasanka National Park, eastern side of Musande River, c. 1.5 km S of Musande Tent Camp, fl. 18.xi.1993, *Harder et al.* 1954 (K; MO). E: Chadiza, fl. 27.xi.1958, *Robson* 736 (BM; K; LISC; SRGH). **Zimbabwe.** W: Matopos Distr., Plumtree, fr. iii.1949, *Davies* 21 (K; SRGH). C: Harare, fl. 16.xi.1921, *Eyles* 3218 (K; SRGH). E: Chirinda, Chipeta Forest, fl. 22.x.1947, *Wild* 2138 (K; SRGH). S: Masvingo Distr., 3.2 km north of Masvingo (Fort Victoria), Kapota Hill, fl. ix.1956, *Miller* 3656 (K; SRGH). **Malawi.** N: Mzimba, 14.4 km north of Loudon towards Edingeni, fl. 27.xii.1975, *Pawek* 10582 (K; MAL; MO). S: Zomba, Balaka, Rivi-Rivi Farm, fl. 18.xii.1957, *G. Jackson* 2125 (K). **Mozambique.** N: Malema R., fl. xi.1931, *Gomes e Sousa* 789 (K). Z: Morrumbala, fl. 10.x.1949, *Andrada* 1956 (COI). T: Tete (Tette), xi.1858, *Kirk* (K). MS: Manica, Mavita, Moçambize Valley, fl. 25.x.1944, *Mendonça* 2611 (BR; LISC; LMA; MO). GI: Chipenhe, Régulo Chiconela, Chirindzeni Forest, fl. 13.x.1957, *Barbosa & Lemos* 8049 (COI; K; LISC; LMA). M: Maputo (Lourenço Marques), fl. 30.xi.1897, *Schlechter* 11550 (K).

Also in Rwanda, Tanzania, Swaziland and South Africa, ?Namibia and ?Angola. Grassland, thicket and open woodlands, mainly *Combretum, Brachystegia, Ostryocarpus–Pterocarpus–Adansonia, Acacia–Colophospermum–Kirkia–Terminalia, Uapaca, Syzygium–Strychnos–Mimusops* woodlands, often on termite mounds, in rocky places and even dunes; 0–1500 m.

Subsp. **rotundata** (Robyns) Verdc. in Kew Bull. **36**: 549 (1981); in F.T.E.A., Rubiaceae: 852 (1991). —Beentje, Kenya Trees, Shrubs Lianas: 550 (1994). Type from Tanzania.

Vangueria rotundata Robyns in Bull. Jard. Bot. État **11**: 300, figs. 27 & 28 (1928). —Brenan, Check-list For. Trees Shrubs Tang. Terr.: 538 (1949). —Dale & Greenway, Kenya Trees & Shrubs: 479 (1961).

Vangueria tomentosa sensu Dale & Eggeling, Indig. Trees Uganda, ed. 2: 358 (1952), pro parte. —F.W. Andrews, Fl. Pl. Anglo-Egypt. Sudan **2**: 466 (1952). —Dale & Greenway, Kenya Trees & Shrubs: 479 (1961) non Hochst.

Vangueria campanulata sensu Dale & Greenway, Kenya Trees & Shrubs: 478 (1961), pro parte non Robyns.

Leaves often attaining dimensions towards the upper end of the scale of size given above and often more developed at flowering time, densely pubescent to densely velvety. Inflorescences often extensively branched. Corolla acute to distinctly apiculate at the apex in bud, glabrous to densely spreading hairy outside. Fruit up to 4.7 cm in diameter.

Var. **rotundata** —Verdcourt in Kew Bull. **36**: 550 (1981); in F.T.E.A., Rubiaceae: 852 (1991).

Corolla pubescent to densely spreading hairy outside.

Zambia. N: Mbesuma, Chambeshi R., fl. 11.xii.1961, *Astle* 1104 (K). **Zimbabwe.** E: Chipinge Distr., north of Chirinda, Houtberg Farm, fl. xi.1967, *Goldsmith* 131/67 (K; SRGH). **Malawi.** N: Nkhata Bay Distr., Sanga, 16 km S of Mzuzu–N8 road junction, fl. 17.xii.1972, *Pawek* 6105 (K; MAL; MO). C: Dowa, Mwera Hill Station, fl. 31.x.1950, *G. Jackson* 220 (K). S: Zomba, Residency Grounds, *Whyte* (K). **Mozambique.** N: Eráti, 10 km from Namapa towards Nacaroa, fl. 11.xii.1963, *Torre & Paiva* 9509 (BR; LISC; LMA; SRGH). Z: Lugela-Mocuba Distr., Namagoa Estate, fl. xi, *Faulkner* in PRE 66 (BR; COI; K; LISC; PRE). MS: Mossurize, Mts. of Espungabera (Spungabera), fl. 13.xi.1943, *Torre* 6191 (LISC).

Also in Uganda, Kenya and Tanzania. Grassland with forest patches, and in *Combretum–Terminalia, Adansonia–Acacia, Cussonia–Erythrina* woodlands and thickets, also lake edge; 60–1700 m.

Forbes from Delagoa Bay, Mozambique, has the pointed buds of this subspecies but other specimens from the same area do not.

Var. **campanulata** (Robyns) Verdc. in Kew Bull. **36**: 549 (1981); in F.T.E.A., Rubiaceae: 852 (1991). Type from Kenya.

    Vangueria campanulata Robyns in Bull. Jard. Bot. État **11**: 293 (1928). —Dale & Greenway, Kenya Trees & Shrubs: 478 (1961).

    Vangueria sp. of Brenan, Check-list For. Trees Shrubs Tang. Terr.: 557 (1949).

Corolla glabrous or almost so save for slight pubescence above.

Zambia. E: Makutus, fl. 27.x.1972, *Fanshawe* 11560 (K; NDO). **Malawi**. N: Mafinga Hills (Mts.), Chisenga, fl. 8.xi.1958, *Robson & Fanshawe* 513 (K; LISC; NDO; SRGH).

Also in ?Cameroon, Kenya and Tanzania. *Brachystegia–Uapaca* woodland, *Syzygium–Ficalhoa* riverine woodland, and beside streams in montane forest; 1500–2000 m.

General Note. The division of *V. infausta* into subspecies and varieties seems reasonable when comparing South African populations to material in the Flora of Tropical East Africa area but in most parts of the Flora Zambesiaca area it is not worth trying to make a distinction. Determination as *V. infausta* sensu lato will be enough for most purposes. However, subspecies *rotundata* with its two varieties is included here as these taxa may have some validity in the northern and eastern parts of the Flora Zambesiaca area. They are difficult to determine in the absence of corollas and there is some overlap in the distribution ranges.

E. Phillips 4515 [Malawi: Nyika Plateau, Chintheche] has rather long calyx lobes but seems to be a variety of *V. infausta*; the lobes are not long enough for *V. volkensii* var. *kyimbilensis*, and the stems are rather different. Some Zambian Copperbelt specimens [e.g. *Harder et al.* 2627 (K; MO) from Mwekera Forest College, fr. 5.ii.1995; and *Fanshawe* 5976 (K) from Ndola, young fr. 14.xii.1953] also have longer calyx lobes but otherwise resemble *V. infausta*.

V. infausta could be considered a pubescent variety of *V. madagascariensis*. Indeed some robust specimens from Mozambique are very like *V. madagascariensis* except for the pubescence. There is also some confusion with *V. proschii*, see notes after that species.

3. **Vangueria cyanescens** Robyns in Bull. Jard. Bot. État **11**: 284 (1928). Type from Namibia.

Shrub or small tree, 0.8–5 m tall, often multi-stemmed and sometimes with a spreading crown; stems slender, glabrous, with wine-red bark, smooth when young but soon becoming powdery. Leaves 6.2–11(15.5) × 2.2–4.5 cm, narrowly to broadly elliptic, acute to shortly acuminate at the apex, cuneate or less often rounded at the base, entirely glabrous or sometimes with a few scattered hairs; lateral nerves in 5–6 main pairs; tertiary nerves moderately finely reticulate, often drying blue-black; petiole 2–5 mm long, often drying blackish; stipules on young stems with base c. 2 mm long and appendage c. 5 mm long; older stipules congested with dingy-coloured hairs inside. Inflorescences often appear before the leaves are fully expanded, sparsely pubescent to pubescent, 5–13-flowered; main peduncle up to 2.5 mm long; pedicels about 2 mm long, save those of central flower of inflorescence which may be longer. Calyx tube c. 1 mm long, pubescent or glabrous; lobes 1.2–1.8 mm long, narrowly oblong, usually glabrous. Corolla obtuse to slightly acuminate in bud, greenish-cream, glabrous; tube c. 4 mm long; lobes 2.5–3.5 mm long, acuminate. Style 5–6 mm long; pollen presenter yellow, 1.2 mm long, cylindrical. Fruits green to brownish, 1.8–2 cm diameter, subglobose; pyrenes 4–5, each c. 20 mm long.

Botswana. N: Ngamiland, 1930–31, *Curson* 739 (PRE). **Zambia**. B: 14.4 km ESE of Kaoma (Mankoya), fl. 21.xi.1959, *Drummond & Cookson* 6700 (K; SRGH); Sesheke, Masese, fr. 26.xii.1952, *Angus* 1034 (FHO; K). W: Ndola, fl. 14.xii.1953, *Fanshawe* 591 (K; NDO).

Also in Angola and Namibia. Kalahari Sand in *Baikiaea* mutemwa, in thicket on termite mounds, and in old gardens; 930–1110 m.

This taxon is closely related to the pubescent *V. proschii* and could equally well be considered a variety of it. Although the two taxa are more or less sympatric, *V. proschii* has a wider distribution. In the first specimen cited above there are a few scattered hairs on the venation beneath and a very few hairs on some corollas, but other "Barotseland" and Angolan material has glabrous leaves and corollas, like the type of *V. cyanescens*. *F. White* 2012, a scrappy sterile specimen from Senenga in Zambia, with large leaves c. 20 × 10 cm is perhaps an odd growth form of the species. Some specimens from the Ndola area in Zambia W: [fl. bud, 12.xii.1953, *Fanshawe* 576 (K; NDO); fl. bud 14.xii.1953, *Fanshawe* 596 (K; NDO); and fr.

7.iii.1954, *Fanshawe* 951 (K; NDO)] seem intermediate between *V. cyanescens* and *V. madagascariensis*. They have the powdery bark and densely tufted stipules of *V. cyanescens* but the inflorescence is more like that of *V. madagascariensis*. This material differs from *V. apiculata* as the stems are not shiny and conspicuously lenticellate, and both the petioles and calyx lobes are longer.

The name *V. cyanescens* has frequently been applied to specimens from South Africa (Mpumalanga, KwaZulu-Natal) and Swaziland, and has consequently been confused in literature with *V. madagascariensis* and/or *V. randii* subsp. *chartacea* (e.g. Pooley, Trees of Natal, Zululand & Transkei: 470, fig. (1993)).

4. **Vangueria proschii** Briq. in Annuaire Conserv. Jard. Bot. Genève **6**: 7 (1902). —De Wildeman in Bull. Jard. Bot. État **8**: 60 (1922). —Robyns in Bull. Jard. Bot. État **11**: 304 (1928). Type: Zambia, Barotseland, Sefula, *de Prosch* 22 (G, holotype, 2 sheets).

Vangueria lasioclados K. Schum. in Warburg, Kunene-Samb.-Exped. Baum: 387 (1903). —De Wildeman in Bull. Jard. Bot. État **8**: 57 (1922). —Miller in J. S. African Bot. **18**: 85 (1952) (as "*lasiocladus*"). —F. White, F.F.N.R.: 424 (1962). —Palmer & Pitman, Trees Southern Africa **3**: 2083 (1973). Type from Angola.

Vangueria rupicola Robyns in Bull. Jard. Bot. État **11**: 310 (1928). Type: Zimbabwe, Victoria Falls, *Allen* 184 (K, holotype; PRE).

Vangueria infausta sensu Drummond in Kirkia **10**: 276 (1975) quoad *V. lasioclados* and *V. rupicola* in syn.

Shrub 1.8–3 m tall; stems mostly slender, with grey or dark reddish-brown bark, at first with rather long spreading pubescence, later glabrescent or glabrous; branches often short. Leaves 1–8(12.5) × 0.5–3.8(6) cm, oblong, narrowly elliptic or elliptic-lanceolate, acute or acuminate at the apex, cuneate to rounded at the base, sometimes drying blue-black in colour, particularly the nervation, or stained blue-black in places, pubescent with rather long hairs, almost velvety beneath in young leaves but usually sparser, discolorous to some extent when young; lateral nerves in 5–7 main pairs; petiole 1–6 mm long; stipule bases 1.5 mm long, the appendage 2–7 mm long and filiform, the older stipules conspicuous with dingy hairs inside. Inflorescences often appearing before the leaves are fully expanded, short, 0.8–3 cm long, usually 7–13-flowered, spreading setulose pubescent; peduncle short or suppressed, up to 5(10) mm long; pedicels 0.5–2 mm long; bracts 1 × 1 mm, ovate. Calyx tube 1.5 mm long, broadly obconic, densely spreading pubescent; lobes (0.5)1.2–3.0 × 0.5–1.3 mm, triangular or narrowly oblong. Corolla obtuse in bud, green or yellow; tube 3–3.5 mm long, broadly cylindrical, with rather long spreading seta-like hairs save at the base of tube; lobes 2.5–5 × 1.5–2 mm, triangular. Style exserted (1)2–3 mm; pollen presenter 1.2 mm long, 5-lobed. Fruits 17(22) mm in diameter.

Botswana. N: Khwebe (Kwebe) Hills, fl. 18.xii.1897, *Mrs E.J. Lugard* 59 (K). **Zambia**. B: Zambezi (Balovale), Chavuma, fl. 12.x.1952, *Holmes* 946 (FHO; K). C: Luangwa Valley, Mfuwe, fl. 19.xii.1965, *Astle* 4195 (K). S: Namwala, fl. 17.x.1959, *Fanshawe* 5251 (K; NDO). E: Luangwa Valley, near Lusangazi R., fr. 18.iv.1966, *Astle* 4788 (K). **Zimbabwe**. N: Gokwe Distr., Sengwa Research Station, fl. 10.xii.1968, *N. Jacobsen* 321 (K; SRGH). W: Hwange National Park, Main Camp, young fr. 7.xii.1968, *Rushworth* 1326 (K; SRGH).

Also in Angola and Namibia. Riverine woodland and scrub, often in rocky places, and on river sandbanks, also in *Brachystegia* and *Baikiaea–Guibourtia* woodland on Kalahari Sand; 990–1380 m.

As previously noted, *V. cyanescens* could be considered a glabrous variety of *V. proschii*. Drummond sinks *V. lasioclados* and *V. rupicola* into *Vangueria infausta*. F. White however, kept the taxon distinct (as *V. lasioclados*) and this has been followed here. As White points out, the tendency to dry blue-black and the longer indumentum are distinctive. Nonetheless, some specimens have proved difficult to place and field observations along the Zambezi Valley, and especially in the Plumtree area of Zimbabwe W are needed to confirm this or otherwise.

Material from Zimbabwe W: Matopos District [Besna Kobila Farm, fl. bud ix.1953, *Miller* 1912 (K; SRGH); MOTH Shrine Kopje, young fr. 30.xi.1951, *Plowes* 1364 (K; SRGH); Matopos Research Station, fr. 15.i.1952, *Plowes* 1399 (K; SRGH); and Lucydale, young fr. 14.xii.1947, *West* 2496 (K; SRGH)] is not typical of *V. proschii*. In this material leaf blades tend to be broadly elliptic to almost round with a more or less rounded apex. These specimens have tentatively been determined as *V. cf. proschii*. One specimen from this area [Matobo, Farm Quaringa, bud x.1957, *Miller* 4611 (COI; SRGH)] has longer stipule lobes (c. 10 mm long) and calyx lobes (2 × 1 mm). More material and careful field studies are needed.

5. **Vangueria parvifolia** Sond. in Linnaea **23**: 58 (1850); in F.C. **3**: 14 (1865). —Burtt Davy in Bull. Misc. Inform., Kew **1908**: 173 (1908). —De Wildeman in Bull. Jard. Bot. État **8**: 60 (1922). Type from South Africa.

    Tapiphyllum parvifolium (Sond.) Robyns in Bull. Jard. Bot. État **11**: 110 (1928). —Miller in J. S. African Bot. **18**: 85 (1952). —Pooley, Trees of Natal, Zululand & Transkei: 470, figs. (1993). —K. Coates Palgrave, Trees Southern Africa, ed. 3, rev.: 874 (1988).

Shrub or small tree up to 6 m tall, much branched and spreading; bark grey, ± smooth or slightly flaking; young shoots densely velvety, sometimes ferruginous pubescent, later ± glabrous; lateral branches often nodulose with old stipule bases. Leaves sometimes borne on abbreviated lateral shoots, 1–4 × 0.3–3 cm, small, ovate to elliptic or almost round, rounded or very shortly obtusely acuminate at the apex, truncate to rounded at the base, scarcely discolorous, densely velvety on both surfaces; petiole up to 1(3) mm long; stipules connate, soon deciduous, 1–3 mm long, caudate, the appendage 1.5–2 mm long. Flowers in dense 3–7-flowered cymes; peduncle ± obsolete; bracts ± obscured by stipule-bases; pedicels 0–2 mm long. Calyx tube c. 1.2 mm long, subglobose; lobes 0.8–2 mm long, triangular to oblong. Corolla apiculate in bud, greenish-yellow, densely pubescent; tube 1.5–2.5 mm long, with a ring of deflexed hairs within; lobes 2.5 × 1.2, minutely apiculate. Style exserted about 1.2 mm; pollen presenter c. 1 mm long, ± cylindrical-coroniform, sulcate, 4–5-lobed at the apex. Fruit brown or reddish, 10–15 mm in diameter, subglobose, pubescent, with mostly 3–4 pyrenes.

Botswana. SE: foot of Otse (Ootze) Mt., fl. 22.x.1978, *O.J. Hansen* 3501 (C; GAB; K; PRE; SRGH; WAG).

Also in South Africa (North Prov., North-West Prov. and Gauteng). Thicket, on shallow soil of flat-topped ridges and rocky hills; 1050–1300 m.

This species was previously put in *Tapiphyllum*. However, Bridson (in Kew Bull. **51**: 349 (1996)) noted that it was atypical in many respects. *V. soutpansbergensis* N. Hahn [in Bothalia **27**: 45 (1997)], a glabrous species endemic to the Soutspansberg Mts., is very close to and certainly congeneric with *V. parvifolia*. Since this species pair falls outside the main distribution area for *Tapiphyllum* and glabrous species are unknown in this genus, *T. parvifolium* is here placed in *Vangueria*. Apart from the small leaves and the few-flowered dense inflorescence it fits this genus very well. Palmer and Pitman and Coates-Palgrave include Angola and Namibia in the distribution of this species. However, specimens at Kew from these areas seem better redetermined as *Vangueria proschii*. The reference to *T. parvifolium* in Pooley, Trees of Natal, Zululand & Transkei: 472, fig. (1993) presumably refers to some other taxon.

6. **Vangueria esculenta** S. Moore in J. Linn. Soc., Bot. **40**: 91 (1911). —De Wildeman in Bull. Jard. Bot. État **8**: 52 (1922). —Robyns in Bull. Jard. Bot. État **11**: 292 (1928). —Drummond in Kirkia **10**: 276 (1975). —K. Coates Palgrave, Trees Southern Africa, ed. 3, rev.: 872 (1988), pro parte. Type: Zimbabwe, Chirinda Forest, *Swynnerton* 65 (holotype BM; K).

    Vangueria esculenta var. *glabra* S. Moore in J. Linn. Soc., Bot. **40**: 91 (1911). —Robyns in Bull. Jard. Bot. État **11**: 293 (1928). Type: Zimbabwe, Chirinda Forest, *Swynnerton* 1307 (BM, holotype; K).

Small tree or shrub 1.8–6(12) m tall, with smooth bark; branchlets slender, at first covered with spreading yellowish-brown hairs, later ± glabrous, dull purplish, with finely ridged bark which flakes off in small pieces. Leaves membranous to thinly papery, 2–11 × 0.6–4.8 cm, elliptic or ovate-elliptic to elliptic-oblong, acuminate at the apex, rounded at the base, almost glabrous to densely pubescent above and on the nervation beneath, never velvety and the surface never obscured; lateral nerves in 6(7) main pairs; tertiary nerves moderately finely reticulate; petiole 5–9(15) mm long, spreading pubescent; stipules joined at base to form a cylindrical sheath 2 mm long with decurrent pubescent subulate appendage 5–7 mm long. Inflorescences pubescent, sweet-scented, often at leafless nodes, 1–2 cm long, usually of 7-flowered short dichasial cymes; peduncle 0.9–1.5 cm long with involucre of small brownish bracts at apex and another about one third the length below it or midway, presumably a true peduncle with a secondary axis but lateral branchlets suppressed; true secondary branches c. 2 mm long; pedicels 0–2 mm long. Calyx pubescent; tube c. 1 mm long, campanulate; lobes c. 1.5 mm long, triangular. Corolla rounded or ± acute in bud, distinctly contracted around the limb/tube junction in the dry state, white, yellow or pale green, glabrous or with a very few scattered hairs or with limb part pubescent outside, tube 4–5 mm long; lobes 3–3.5 × 1.5–1.8 mm, oblong-triangular, acute but not apiculate, strongly reflexed. Style exserted 1.5–2 mm; pollen presenter 0.7 mm long, ± cylindrical. Fruits deep yellow, 18–22 mm wide; pyrenes 3–5, each 18 × 8 mm with a conspicuous almost central notch.

Zimbabwe. E: Chirinda Forest, fl. 28.x.1947, *Wild* 2247 (K; SRGH). **Malawi**. S: Blantyre Distr., Bangwe Hill, 4 km east of Limbe, 23.xi.1977, *Brummitt et al.* 15161 (K). **Mozambique**. MS: 25 km SW of Lacerdónia, fl. 6.xii.1971, *Müller & Pope* 1908 (K; LISC; SRGH). Z: Serra Morrumbala, young fr. 13.xii.1971, *Müller & Pope* 2021 (K; LISC; SRGH).

Not known outside the Flora Zambesiaca area. Evergreen forest subcanopy; (200) 1110–1800 m.

The specimens cited from Malawi S and Mozambique Z are the only specimens known from outside Chirinda and adjacent areas. While they seem to be representative of *V. esculenta* more material is needed for critical study. In South Africa the name *V. esculenta* has been confused with *V. madagascariensis* and/or *V. randii* subsp. *chartacea* (e.g. Palmer & Pitman, Trees Southern Africa **3**: 2083 (1973) and Moll, Trees of Natal: 250 & 251 (1981)). The 'Chirinda Medlar'.

7. **Vangueria apiculata** K. Schum. in Engler, Pflanzenw. Ost-Afrikas **C**: 385 (1895). —Mildbraed, Wiss. Ergebn. Deutsch. Zentr.-Afrika Exped., Bot., part 4: 326 (1911). —Robyns in Bull. Jard. Bot. État **11**: 283 (1928). —Brenan, Check-list For. Trees Shrubs Tang. Terr.: 537 (1949). —Cufodontis in Phyton (Horn) **1**: 145 (1949). —Dale & Eggeling, Indig. Trees Uganda, ed. 2: 358 (1952). —F.W. Andrews, Fl. Pl. Anglo-Egypt. Sudan **2**: 466 (1952). —J.G. Garcia in Mem. Junta Invest. Ultramar **6** (sér. 2): 32 (1959) [Contrib. Conhec. Fl. Moçamb. IV (1959)]. —Dale & Greenway, Kenya Trees & Shrubs: 478 (1961). —Drummond in Kirkia **10**: 276 (1975). —Bridson & Troupin in Fl. Pl. Lign. Rwanda: 610, fig. 207.1 (1982); in Fl. Rwanda **3**: 230, fig. 73.1 (1985). —K. Coates Palgrave, Trees Southern Africa, ed. 3, rev.: 872 (1988). —Verdcourt in Kew Bull. **36**: 550 (1981); in F.T.E.A., Rubiaceae: 853 (1991). — Beentje, Kenya Trees, Shrubs Lianas: 550 (1994). Type from Kenya.

Vangueria longicalyx Robyns in Bull. Jard. Bot. État **11**: 278 (1928). —Pardy in Rhodesia Agric. J. **52**: 237 (1955). —J.G. Garcia in Mem. Junta Invest. Ultramar **6** (sér. 2): 32 (1959) [Contrib. Conhec. Fl. Moçamb. IV (1959)]. Type: Zimbabwe, Chirinda outskirts, *Swynnerton* 64 (K, holotype and isotype; BM).

Vangueria sp. near *randii* —Brenan, Check-list For. Trees Shrubs Tang. Terr.: 537 (1949).

Shrub, subscandent shrub or small spreading tree 1.8–15 m tall. Stems sometimes several, virgate with branching often horizontal; bark grey, brown or red-brown, more or less smooth or finely ridged; young shoots mostly dark plum-coloured and lenticellate, quite glabrous. Leaves 3–15(17) × 1.5–6(8) cm, elliptic, oblong, ovate or ovate-lanceolate to lanceolate, distinctly acuminate at the apex, rounded to cuneate or occasionally subcordate at the base, often discolorous and venose beneath when dry, quite glabrous or rarely slightly pubescent beneath; lateral nerves in (6)7–8 main pairs; tertiary nerves finely reticulate; petiole 7–10(14) mm long; stipules with filiform part 3–9.5 mm long from a short broad base 1.5–2 mm long. Inflorescences typically short, 1–3.5 cm long including peduncle, fairly lax to very condensed, the rhachis pedicels and calyx tube almost glabrous to usually pubescent or densely shortly hairy; peduncle 5–10 mm long, similarly hairy, with paired bracts near the apex 3 × 2 mm long; pedicels 2–4 mm long, pubescent. Calyx tube 1.5 mm in diameter, subglobose, glabrescent to densely shortly hairy; lobes (1.2)3–7 mm long, oblong to linear, sometimes slightly spathulate, glabrous or ciliolate. Corolla distinctly apiculate at the apex but not tailed in bud, glabrous outside, greenish-white, or green to yellow; tube 4–5 mm long with a conspicuous tuft of hairs projecting from the throat; lobes 4–4.5 × 2 mm. Style 6–8 mm long; pollen presenter 1.2 mm long. Fruit turning brown, 1.7–2.2 × 1.4–2.2 cm, subglobose, or sometimes irregularly ellipsoid where only one pyrene has developed, glabrous; pyrenes 3–5, 9–17 × 4–6 mm.

Zambia. N: Mbala Distr., Lake Chila, fl. 6.xi.1952, *R.G. Robertson* 221 (K; PRE) (see note). W: just S of Matonchi Farm, young fr. 2.i.1938, *Milne-Redhead* 3920 (K). **Zimbabwe**. N: Mazowe Distr., Mvurwi (Umvukwes), fl. 16.xii.1952, *Wild* 3920 (K; SRGH). C: Makoni, fl. 20.xii.1940, *Hopkins* in GHS 7851 (K; SRGH). E: Chimanimani (Melsetter), fl. 30.xi.1962, *West* 4346 (K; LISC; SRGH). S: Bikita Distr., Old Bikita, Danga, fl. 16.xii.1953, *Wild* 4397 (K; SRGH). **Malawi**. C: Dedza Mt., fr. 24.iv.1970, *Brummitt* 10103 (K). S: Ntcheu Distr., Chirobwe Forest, fl. 18.xii.1962, *Banda* 478 (K; SRGH). **Mozambique**. Z: Milange, Serra do Chiperone (Chiparene) subida pelo Chefe Marrega vertente Voltada a NW, fr. 1.ii.1972, *Correia & Marques* 2472 (BR; LMU). MS: Manica, Mt. Vumba, fr. 29.iv.1963, *Gomes e Sousa* 4799 (COI; K; PRE).

Also in Zaire (Dem. Rep. Congo), Rwanda, Sudan, Ethiopia, Uganda, Kenya and Tanzania. Riverine thicket and evergreen forest, and in mixed woodland, sometimes in rocky places, exposed hillsides and termite mounds; (850)960–2050 m.

Specimens from northern Zambia: *R.G.Robertson* 221 (Mbala, Lake Chila); *Dowsett-Lemaire* 73 (edge of Chowo Forest, Nyika Plateau); and *Nkhoma et al.* 137 (Nsumbu National Park) have shorter calyx lobes, but otherwise are a good match for *V. apiculata*. Similar specimens are known from the Ufipa and Iringa Districts of Tanzania. In many respects this material resembles the more northerly forms of *V. madagascariensis* and could equally well be classified with that species. The 'small wild medlar'.

8. **Vangueria volkensii** K. Schum. in Engler, Pflanzenw. Ost-Afrikas **C**: 384 (1895). —Robyns in Bull. Jard. Bot. État **11**: 298 (1928). —Brenan, Check-list For. Trees Shrubs Tang. Terr.: 538 (1949). —Verdcourt in Kew Bull. **36**: 551 (1981). —Bridson & Troupin in Fl. Pl. Lign. Rwanda: 610, fig. 207.3 (1982); in Fl. Rwanda **3**: 230, fig. 73.3 (1985). —Verdcourt in F.T.E.A., Rubiaceae: 854 (1991). —Beentje, Kenya Trees, Shrubs Lianas: 551 (1994). Type from Tanzania.

  Vangueria linearisepala K. Schum. in Bot. Jahrb. Syst. **33**: 351 (1903). —Robyns in Bull. Jard. Bot. État **11**: 296 (1928). —Brenan, Check-list For. Trees Shrubs Tang. Terr.: 537 (1949). —Dale & Greenway, Kenya Trees & Shrubs: 479 (1961). Type from Tanzania.

Spreading shrub or small tree 2.4–9(15) m tall, with dark grey, smooth or slightly fissured bark; shoots densely ferruginous-velvety when young, later dark and lenticellate. Leaves 3–17(26) × 1.5–10(14) cm, ovate-oblong or elliptic to ovate, acuminate at the apex, cuneate to rounded or subcordate at the base, densely pubescent to velvety all over with yellowish hairs; lateral nerves in 5–8 main pairs; tertiary nerves not markedly apparent; stipules with base 2 mm long and filiform part 5–12 mm long; petiole 5–13 mm long. Inflorescences short, 1–4 cm long, very densely shortly hairy; peduncle 6–20 mm long with bracts as in last species; pedicels 0.75–3 mm long, densely shortly hairy. Calyx tube 1–1.5 mm in diameter, subglobose, densely shortly hairy; calyx lobes 1.5–6(8) mm long, linear or oblong, densely hairy. Corolla distinctly apiculate in bud, often the 5 appendages very marked, bright green or yellow-green, greenish-cream inside, sparsely to densely hairy outside, sometimes only on the lobes, the tube more or less glabrous; tube 3.5–5.5 mm long; lobes 4–4.5 mm long, narrowly triangular. Fruit eventually turning brown, 2–2.5 cm long, subglobose, or sometimes asymmetric and ellipsoid if reduced to one pyrene by abortion; pyrenes 4–5 normally, up to 16 × 7–8 × 6 mm.

Var. **volkensii** —Verdcourt in Kew Bull. **36**: 551 (1981); in F.T.E.A., Rubiaceae: 854 (1991).

Buds distinctly pubescent. Rhachis, peduncle, pedicels, calyx tube and lobes all densely shortly pubescent. Inflorescences 1–3.5 cm long.

  Zimbabwe. E: Nyanga (Inyanga) road, *Gilliland* K991 (BM; K); west of Stapleford, Simla, *Gilliland* K1059 (BM; K; PRE). **Mozambique**. MS: Barûê, Serra de Choa, 15 km from Catandica (Vila Gouveia), fr. 28.iii.1966, *Torre & Correia* 15473 (COI; LISC; LMU; WAG).
  Also in Zaire (Dem. Rep. Congo), Rwanda, Ethiopia, Sudan, Uganda, Kenya and Tanzania. Forest and grassland; 1400 m.
  Vangueria volkensii could be treated as a hairy variant of *V. apiculata*, but since the two seem to be easily distinguished over a wide area of E Africa They have been maintained at specific rank. Although the three specimens cited above are disjunct from the main area of distribution, they do appear similar and are best treated as *V. volkensii* until the whole complex is better understood.

Var. **kyimbilensis** (Robyns) Verdc. in Kew Bull. **36**: 552 (1981); in F.T.E.A., Rubiaceae: 854 (1991). Type from Tanzania.

Buds entirely glabrous. Rhachis, peduncle, pedicels and particularly calyx tube and lobes sparsely hairy or glabrescent. Inflorescences 3–4 cm long.

  Zambia. E: Nyika Plateau, Manyenjere Forest, young fr. xii.1981, *Dowsett-Lemaire* 314 (K)?
  Also in southern Tanzania. In mid stratum of evergreen forest, near streams; 2050 m.
  Because the specimen cited above is rather fragmentary the identification is only tentative. The taxon is otherwise only known from the type gathering.

9. **Vangueria randii** S. Moore in J. Bot. **40**: 252 (1902). —Robyns in Bull. Jard. Bot. État **11**: 281 (1928). —F. White, F.F.N.R.: 424 (1962). —Drummond in Kirkia **10**: 276 (1975). —K. Coates Palgrave, Trees Southern Africa, ed. 3, rev.: 873 (1988). —Verdcourt in Kew Bull. **36**: 552 (1981); in F.T.E.A., Rubiaceae: 855 (1991). Type: Zimbabwe, Bulawayo, *Rand* 123 (BM, holotype).

Vangueria edulis var. *bainesii* Hiern in F.T.A. **3**: 148 (1877). Type: Zimbabwe, Lee's Farm, Mangwe R., 9.iii.1870, *Baines* (K, holotype).

Shrub or small tree, occasionally scandent, 1.2–7(9) m tall, with slender spreading branches; shoots brown, greyish to mid red-brown, dull usually with sparse to dense lenticels and sometimes rather peeling epidermis, glabrous; in some forms some side branches are modified into spines. Leaves thin or very thinly membranous, 2–15.5 × 0.8–6.6 cm, elliptic to oblong-elliptic, bluntly subacute to obtusely acuminate or distinctly narrowly acuminate at the apex, cuneate to almost rounded at the base, glabrous or with both bristly hairs and much shorter hairs on the upper and particularly lower surface and usually also bristly ciliate; lateral nerves in 5–6(7) main pairs; tertiary nerves rather coarsely reticulate; stipules with base 1–2 mm long, and filiform apex 2–11 mm long; petiole 3–7 mm long. Inflorescence graceful, 1–3.5 cm long, the rhachis, pedicels and calyx tube glabrescent to shortly pubescent; ultimate branches 3–8-flowered; peduncle 4–15 mm long with paired ovate-oblong bracts 1–2 mm long; pedicels 1–5 mm long. Calyx tube 1 mm long, subglobose, glabrous to pubescent; lobes 2–6(7) mm long, linear or oblong-linear. Corolla acuminate or acute, usually 5-tailed in bud, white, green or golden-green, glabrous; tube 2–3 mm long, not markedly hairy at throat; lobes 2.5–3 mm long, narrowly triangular, ± obtuse to distinctly appendaged and apiculate at the apex. Style slender, 3.5–4.5 mm long; pollen presenter 0.75 mm long, cylindrical, minutely 5-lobed. Fruit yellow, 1.4–2 cm in diameter, subglobose, on stalks 0.1–3 cm long; pyrenes 3–5, each 1.3–1.5 cm long.

Subsp. **randii**

Plants never spiny. Leaves glabrous. Calyx tube pubescent, the lobes short, up to c. 4 mm long. Peduncles short, up to 7 mm long.

Botswana. N: Ramokgwebane (Ramaquebane), fr. i.1952, *Miller* B1290 (K; PRE). **Zambia.** B: Masese, fl. 21.xi.1960, *Fanshawe* 5909 (K; NDO). C: 19.2 km S of Lusaka on Kafue road, fl. 5.i.1957, *Angus* 1481 (BM; BR; FHO; K; PRE). S: Mapanza Mission, fl. & fr. 20.xii.1953, *E.A. Robinson* 394 (K). **Zimbabwe.** N: Sebungwe, Lubu R., fl. 18.xi.1951, *Lovemore* 183 (K; SRGH). W: Bulawayo, 1908, *Chubb* 97 in *Herb. Rogers* (K). C: 12.8 km SE of Gweru (Gwelo), fl. 29.xii.1966, *Biegel* 1609 (K; SRGH). E: Mutare (Umtali), Commonage, fl. 1.xii.1953, *Chase* 5154 (BM; K; SRGH). S: Masvingo (Fort Victoria), fr. iv.1953, *Vincent* 152 (K; SRGH) (identif. not certain). **Malawi.** S: Mt. Mulanje, bottom of Likhubula (Likabula) Valley, fr. 6.iv.1989, *J.D. & E.G. Chapman* 9517 (K; MO) (probably this species - fruit red). **Mozambique.** Z: between Régulo Guja and Derre, fr. 10.vi.1949, *Barbosa & Carvalho* 3002 (LMA). MS: Manica, 15 km from Mavita on road to Chimoio (Vila Pery), fr. 18.vi.1942, *Torre* 4345 (C; COI; LISC; LMU; SRGH; WAG).

Known only from the Flora Zambesiaca area. Frequently on termite mounds, also on granite hills and other rocky places; dry grassland, *Julbernardia–Brachystegia* wooded grassland, plateau woodland, and *Acacia–Combretum* woodland on dambo margins; (500)1050–1890 m.

Torre & Paiva 11706 (BR; COI; LD; LISC; LMA; MO) [Mozambique N: 3 km from Montepuez towards Nantulo, fr. 7.iv.1964] falls outside the main distribution range of this species, but could perhaps belong to it.

Subsp. *vollesenii* Verdc. occurs in southern Tanzania, and differs in the following respects: sometimes spiny; peduncles c. 3 mm long; calyx tube glabrous; lobes 3–6(7) mm long.

Subsp. *chartacea* (Robyns) Verdc., distinguished by the longer peduncles, up to 8 mm long, and glabrous calyx tubes, occurs in South Africa (KwaZulu-Natal). It may well be present in Mozambique M: but no certain record has yet been found. An illustration can be found in Pooley, Trees of Natal, Zululand & Transkei: 470 (1993).

Subsp. *acuminata* Verdc. occurs further north in coastal Tanzania and Kenya. The leaf blades are very acuminate and the peduncles are up to 15 mm long.

10. **Vangueria sp.**

Shrub 2.5–5 m high; young stems glabrous or sparsely pubescent, covered with yellow-brown lenticellate bark. Leaves immature at time of flowering; blades up to

17 × 7 cm, broadly elliptic, acuminate at apex, acute at base, glabrous or sparsely pubescent below and with the midrib puberulus above, membranous to thinly papery; lateral nerves in 6–8 main pairs; tertiary nerves coarsely reticulate; stipules with triangular base c. 3 mm long, and linear appendage 4–6 mm long; petioles 3–7 mm long, glabrous or pubescent. Inflorescences many-flowered, shortly pedunculate; the branches bifariously puberulous. Calyx tube 1 mm long; lobes 1 × 0.8 mm, triangular-ovate. Corolla somewhat constricted at top of tube and not tailed in bud; tube 2 mm long; lobes c. 1.5 mm long sparsely puberulous. Fruit c. 3 cm across.

Malawi. S: Chikwawa Distr., Lengwe Game Reserve, young fr. 13.xii.1970, *Hall-Martin* 1085 (K; SRGH); fl. buds xii.1970, *Hall-Martin* 1252 (K; SRGH); fr. 5.iii.1970 *Brummitt* 8877 (K).
Thickets on termitaria in floodplain alluvium; 90 m.
The above cited specimens which are in different stages of growth have provisionally been assumed to be the same taxon. *Brummitt* 8877 is sparsely pubescent but does not seem to be a good match for *V. infausta*. The two glabrous *Hall-Martin* specimens are not a convincing match for either *V. esculenta* or *V. madagascariensis*.

42. TAPIPHYLLUM Robyns

Tapiphyllum Robyns in Bull. Jard. Bot. État **11**: 117 (1928); **32**: 133 (1962). — Verdcourt in Kew Bull. **36**: 533–539 (1981); **42**: 143–145 (1987). —Havard & Verdcourt in Kew Bull. **42**: 605–609 (1987).

Suffrutescent herbs with stems from a woody rhizome, subshrubs, shrubs or small trees, the foliage, inflorescences and young shoots characteristically covered with a dense velvety tomentum. Leaves opposite or in whorls of 3–4, subsessile or shortly petiolate, usually markedly discolorous; stipules connate in lower part, drawn out into a linear apex, eventually deciduous. Inflorescences axillary, dense, usually opposite with flowers few to many in ± globose cymes, or 1–few-flowered fascicles; bracts and bracteoles present, often fairly conspicuous. Flowers fairly small, but large in *T. rhodesiacum*, subsessile to shortly pedunculate. Calyx lobes 5, mostly erect and linear to linear-lanceolate, less often ovate to oblong-triangular. Corolla apiculate in bud, green, yellow or white, densely hairy outside; tube cylindrical, glabrous, occasionally hairy at the throat, with a ring of deflexed hairs within; lobes reflexed, usually apiculate. Stamens inserted at the throat, the anthers ± exserted. Ovary 4–5(6)-locular with one pendulous ovule per loculus; style slender, usually exserted; pollen presenter cylindrical or coroniform, sulcate and 4–5-lobulate. Fruit subglobose, usually crowned by persistent calyx, velvety or infrequently ± sparsely hairy, with 3–5 pyrenes, or fruit oblique and reduced to 1 pyrene.

An ill-defined genus characterised by little but the velvety tomentum and mostly elongate calyx lobes but certainly having a recognisable facies. The species are equally poorly defined; Robyns in his 1962 paper recognised 29 species but this could probably be reduced by a third. A more extensive reduction with most of the taxa reduced to variants of a very few species could certainly be made but would scarcely be useful. As with so many savanna and *Brachystegia* woodland genera the variation defies normal taxonomic treatment and a practical approach has to be made.
Seven species are recognised in the Flora Zambesiaca area. Species 1–3 have a habit reminiscent of *Pachystigma* and/or *Fadogia*; species 4 and 5 are characteristic of the central core of the genus, species 6 is similar except for the large corolla, while species 7 stands well apart. Species 4 has a plethora of distinct looking but connected variants.

1. Corolla tube 11–20 mm long · 2
– Corolla tube up to 6 mm long · 3
2. Corolla 11–13 mm long, slender and straight · · · · · · · · · · · · · · · · · · · 7. *burnettii*
– Corolla tube 18–20 mm long, slightly to distinctly curved · · · · · · · · · · · · 6. *rhodesiacum*
3. Upper surface of leaf glabrescent to sparsely pubescent or pilose but surface clearly visible; leaves very discolorous; subshrubs or suffrutescent herbs · 4
– Upper surface of leaf densely pubescent to distinctly velvety, the surface obscured; leaves less discolorous; subshrubs, suffrutescent herbs or small trees · · · · · · · · · · · · · · · · · 5
4. Venation impressed above; leaf blades up to 11 × 3.7 cm; subshrubs to 40 cm tall with several stems from a woody rootstock; calyx lobes up to 6 mm long, not densely hairy · · · · · · · ·
· 3. *cistifolium* var. *latifolium*

- Venation not impressed above; leaf blades up to 5 × 1.5 cm; subshrubs to 1.2 m tall; calyx lobes c. 5 mm long, densely hairy · 4. *cinerascens* var. *laetum*

5. Suffrutescent herbs with several apparently or truly unbranched stems 0.3–0.6(2.4) m tall from a woody rootstock · 6

- Subshrubs or small trees, the stems sparsely to densely branched · · · · · · · · · · · · · · · 7

6 Leaves in pairs or in whorls of 3, narrowly elliptic to oblong-ovate, 1.6–6.5(7.5) cm wide · 1. *discolor*

- Leaves appearing verticillate due to congestion of 3 lateral branches at each node, oblong-lanceolate, 0.3–1.2 cm wide (Zambia: B) · 2. *molle*

7. Shrub or small tree up to 3 m tall with mostly slender branches; leaves linear-lanceolate to oblong-elliptic, 2–11 × 0.5–3.3(5.5) cm, mostly not so reticulate beneath, often much less velvety above; calyx lobes 1.5–10 mm long · 4. *cinerascens*

- Shrub or small tree up to 4 m tall, usually larger than *T. cinerascens* and with mostly thicker more robust branches; leaves oblong to almost round, 2.7–7 × 1–6 cm, usually distinctly bullate with venation raised and reticulate beneath, always thickly velvety on both surfaces; calyx lobes 2–4 mm long · 5. *velutinum*

1. **Tapiphyllum discolor** (De Wild.) Robyns in Bull. Jard. Bot. État **11**: 105 (1928); in ibid. **32**: 138 (1962). —Verdcourt in F.T.E.A., Rubiaceae: 782 (1991). Type from Zaire (Dem. Rep. Congo).

   Fadogia discolor De Wild. in Repert. Spec. Nov. Regni Veg. **13**: 138 (1914); Notes Fl. Katanga **4**: 88 (1914); Contrib. Fl. Katanga: 213 (1921).

   Tapiphyllum confertiflorum Robyns in Bull. Jard. Bot. État **11**: 103 (1928). Type: Zambia, Batoka Distr., Magoye, 17.x.1911, *Rogers* 8943 (K, holotype & isotype).

   Tapiphyllum herbaceum Robyns in Bull. Jard. Bot. État **11**: 104 (1928). Type: Zambia, Mumbwa, *Macaulay* 1080 (K, holotype).

   Tapiphyllum grandiflorum Bullock in Bull. Misc. Inform., Kew **1933**: 148 (1933). Type: Zambia, Mumbwa, *Macaulay* 922 (K, holotype).

   Tapiphyllum fadogia Bullock in Bull. Misc. Inform., Kew **1933**: 147 (1933). —Verdcourt in Kew Bull. **36**: 537 (1981). Type from Tanzania.

   Tapiphyllum oblongifolium Robyns in Bull. Jard. Bot. État **32**: 137 (1962). Type: Zambia, Ndola, Lake Ishiku, 18.x.1953, *Fanshawe* 426 (K, holotype).

Shrub or subshrubby herb. Stems erect, 0.3–0.6(2.4) m tall, 1–several from one root, usually burnt back every year, densely ferruginous velvety, becoming pubescent or even glabrescent and purplish-brown beneath the indumentum. Leaves paired, or in whorls of 3 particularly further north in Tanzania; blades 3–11(14) × 1.6–6.5(7.5) cm, narrowly elliptic to oblong-ovate, subacute to shortly acuminate or rarely rounded at the apex, cuneate or rounded to slightly subcordate at the base, drying discolorous, densely softly velvety on both surfaces with pale ferruginous hairs, but said to be flannelly-white beneath (in life); petiole (2)5–6 mm long, velvety; stipules subtriangular and connate at the base, 3–4 mm long with a subulate apex 5–8 mm long. Flowers in dense, short, velvety several- to many-flowered cymes; cymes 1.5–2.5 cm long, including the 0.5–1.2 cm long peduncle; bracts 6–7(12) mm long, 1.5–3.5 mm wide, linear-lanceolate; pedicels 0–2 mm long. Calyx densely pale ferruginous-pubescent; tube 2 mm long; lobes 3–7 × 1 mm, linear-lanceolate to triangular. Corolla apiculate in bud, yellow to green, velvety outside; tube 4–5 mm long, cylindrical, glabrous at base outside; lobes 3.5–5 × 1.3–2 mm long, apiculate. Style 7 mm long; pollen presenter 1 mm long, cylindrical-coroniform, sulcate. Fruits green, turning yellow to orange but drying ferruginous, about 10–13 mm in diameter, subglobose, wrinkled in the dry state, ferruginous velvety and with sparser longer hairs, crowned with the persistent calyx, grooved between the 3–5 pyrenes in dry state; pyrenes 8 × 5 mm, rounded segment-shaped.

Zambia. B: Kaoma (Mankoya), near Luena R., fl. 20.xi.1959, *Drummond & Cookson* 6686 (K; LISC; SRGH). W: Chingola, fl. 8.x.1954, *Fanshawe* 1606 (K; NDO). C: Lusaka Distr., between Kasisi and Constantia, fr. 4.i.1973, *Kornaś* 2947 (K; KRA). S: Mazabuka, fr. 23.i.1960, *White* 6348 (FHO; K).

Also in southern Zaire (Dem. Rep. Congo) and Tanzania. Tall grassland with scattered trees, and in *Brachystegia, Monotes–Julbernardia–Burkea* woodlands and dambo margins, in sandy ground and laterite; 1130–1235 m.

A seedling said by Fanshawe to be of this species is almost glabrous. Confirmation of this extraordinary fact is needed.

2. **Tapiphyllum molle** Robyns in J. Bot. **69**: 186 (1931); in Bull. Jard. Bot. État **32**: 136 (1962). Type from Angola.

Suffrutex, unbranched; stems numerous about 1.2–1.5 m tall, densely pale ferruginous-velvety hairy, becoming less so with age. Leaves up to 12, appearing falsely verticillate due to congestion of 3 lateral branches at each node, but leaves actually paired, 1.5–8 × 0.3–1.2 cm, oblong-lanceolate, acute at the apex, cuneate at the base, discolorous, velvety-hairy with pale ferruginous hairs above, densely white velvety-woolly beneath; stipules connate, 3 mm long with appendix 5–6(10) mm long; petiole c. 3 mm long. Inflorescences congested in the axils forming contiguous verticils, densely velvety-hairy, many-flowered; peduncle up to 3 mm long; pedicels ± obsolete; bracts 6 mm long, linear. Calyx tube 2 mm long, subglobose, densely hairy; lobes 5 mm long, linear, ± obtuse. Corolla shortly apiculate in bud, pale greenish outside, white inside; tube 4.5–5 mm long, cylindrical, white-velvety outside and densely hairy at the throat and with a ring of deflexed hairs inside towards the base; lobes 3.5 × 1.5 mm, ± lanceolate with an appendage 2 mm long. Style scarcely exserted; pollen presenter cylindrical, ± sulcate, 5-lobulate at the apex. Fruit not seen (see below).

Zambia. B: Kaoma (Mankoya), 18.x.1964, *Fanshawe* 8980 (K; NDO).
Also in Angola. *Julbernardia–Brachystegia* woodland on Kalahari Sand, with *Gardenia, Lannea* and *Diplorhynchus*; 1120 m.
Fanshawe refers to an unbranched shrub but it is assumed that he means unbranched stems from a multiheaded woody rootstock, as described by Gossweiler in his collector's notes of the type specimen.
Several specimens from NW Zambia (e.g. *Milne-Redhead* 3899; 3929, *Brummitt et al.* 14069) have been provisionally included in this taxon pending a fuller investigation. The fruits of these specimens ripen to orange, are c. 10 mm in diameter and glabrescent.

3. **Tapiphyllum cistifolium** (Welw. ex Hiern) Robyns in Bull. Jard. Bot. État **11**: 108 (1928); in ibid. **32**: 136 (1962). Type from Angola.
 Ancylanthos cistifolius Welw. ex Hiern in F.T.A. **3**: 179 (1877); Cat. Afr. Pl. Welw. **1**: 484 (1898).

Suffrutex with several sparsely branched stems 10–40 cm tall from a woody rootstock, the lateral branches often short or more or less suppressed, youngest parts densely hairy with pale ferruginous hairs but soon glabrescent, dark purple-brown and with minutely flaky bark. Leaves paired or appearing in whorls due to abbreviated branches, 1.6–11(15) × 0.3–3.7(5.5) cm, linear-oblong to oblong or ± oblanceolate, ± rounded to mostly acute at the apex, cuneate at the base, rather thick, very discolorous, drying green to dark brown above with scattered bristly yellow hairs, densely grey-white to buff velvety tomentose beneath with woolly hairs; venation distinctly impressed above in the mature leaves; petiole 2–8 mm long; stipules 1–5 mm long, ± triangular, joined at the base, the filiform appendages 3–6 mm long. Inflorescences with 5–7 flowers in congested dichasial cymes; peduncle 0–5 mm long; pedicels 0–3 mm long; bracts up to 5 × 2 mm, ovate-oblong. Calyx tube c. 2 mm long, campanulate, pubescent to densely hairy; lobes 3.5–6 mm long, oblong to linear-lanceolate, slightly pubescent to densely hairy. Corolla tailed at the apex in bud; tube 6 mm long, cylindrical, yellow hairy outside and with a ring of deflexed hairs within; lobes about 4 mm long, distinctly apiculate, the appendage about 1 mm long. Fruits about 12 mm in diameter, subglobose, very sparsely pubescent; pyrenes c. 7 mm long.

Var. **latifolium** Verdc., var. nov.* Type: Zambia, Mwinilunga Distr., 20 km SSE of Salujinga on road to Kalene Hill, *Hooper & Townsend* 278 (K, holotype).

Leaf blades up to 11 × 3.7 cm. Calyx lobes c. 6 mm long, glabrous inside, slightly pubescent outside.

---

* A var. *cistifolio* laminis foliorum usque 11 × 3.7 cm, calycis lobis c. 6 mm longis intus glabris extra leviter pubescentibus differt.

Zambia. W: Mwinilunga Distr., 20 km SSE of Salujinga on road to Kalene Hill, corolla fallen 21.ii.1975, *Hooper & Townsend* 278 (K).
Known only from this locality. *Uapaca* woodland; 1250 m.
Bridson (Kew Bull. **51**: 349 (1996) commented that this species showed some affinity with both *Fadogia* and *Pachystigma*. See the note under *Pachystigma micropyren*, p. 244.
Var. *cistifolium* occurs only in Angola, and differs in its leaves and corolla lobes being smaller and more hairy.

4. **Tapiphyllum cinerascens** (Welw. ex Hiern) Robyns in Bull. Jard. Bot. État **11**: 107 (1928); ibid. **32**: 136 (1962). —Verdcourt in Kew Bull. **36**: 534 (1981). —Havard & Verdcourt in Kew Bull. **42**: 606, fig. 1F (1987). —Verdcourt in Kew Bull. **42**: 144 (1987); in F.T.E.A., Rubiaceae: 780 (1991). Type from Angola.
Ancylanthos cinerascens Welw. ex Hiern in F.T.A. **3**: 159 (1877); Cat. Afr. Pl. Welw. **1**: 484 (1898).

Erect shrub or small tree (0.15)0.9–3 m tall with ± spreading slender branches, or suffrutex with 1–several shoots from a woody rootstock; branchlets slender, densely velvety with appressed pale ferruginous hairs when young, later with sparser hairs or glabrescent. Leaves paired, or occasionally in whorls of 3; blades 2–11 × 0.5–3.3(4.6) cm, oblong-elliptic to ovate-lanceolate, or less often linear-lanceolate, acute or subacute at the apex, cuneate to subcordate at the base, occasionally somewhat unequal, slightly to distinctly discolorous, glabrescent, somewhat scabrid-pubescent or densely pubescent above, densely velvety tomentose beneath with appressed pale ferruginous or grey hairs; petiole 3–6 mm long; stipules joined into a sheath 1–2 mm long at base with linear apices 2–4(8) mm long. Flowers several to 15, in shortly pedunculate subglobose dense velvety hairy congested cymes; peduncles 2–7 mm long; pedicels 1–3 mm long; bracts 4–7 × 1–5 mm, elliptic to lanceolate. Calyx densely pale ferruginous hairy; tube 1–2 mm long; lobes 1.5–8(10) mm long, linear to linear-lanceolate. Corolla with distinct tails up to 2 mm long in bud, pale yellow or greenish-white, densely ferruginous-hairy or tomentose, tube 4–5 mm long, ± cylindrical; lobes 4–5 mm long (including caudate appendages c. 2 mm long) 1.2–2 mm wide, elliptic to lanceolate. Ovary 3–5-locular; style exserted c. 1.5 mm; pollen presenter 0.8 mm long, subcylindric to coroniform, sulcate. Fruits yellow or orange-brown to reddish when ripe, 8–9 mm in diameter, subglobose, crowned with the persistent calyx, ferruginous-velvety tomentose and with longer hairs; pyrenes 1–5.

This is an exceedingly difficult species and constitutes the intractable core of the genus. A key to the variants is scarcely feasible and the following is but a guide.

1. Upper surface of leaf blades glabrescent to sparsely pubescent · · · · · · · · · vi) var. *laetum*
– Upper surface of leaf blades densely pubescent to velvety · · · · · · · · · · · · · · · · · · · 2
2. Leaf blades up to about 4 × 1 cm, narrowly oblong to oblanceolate · · · · v) var. *richardsiae*
– Leaf blades relatively wider · 3
3. Leaf blades up to 10 × 5.5 cm, oblong-elliptic; petiole up to 9 mm long · · · · · · iii) var. *A*
– Leaf blades smaller and more shortly petiolate · 4
4. Calyx lobes usually much longer and narrower, 3–10 mm long; leaf blades 3.5–9.7 × 1–2.7 cm, oblong-elliptic to linear-lanceolate · iv) var. *inaequale*
– Calyx lobes 1.5–4(6) mm long; leaf blades relatively wider, 3.5–8 × 1–3 cm · · · · · · · · · · 5
5. Leaf blades mostly about 8 × 3 cm, oblong; calyx lobes 3–4(6) mm long · · i) var. *cinerascens*
– Leaf blades mostly smaller, up to 3.5(5) × 1–2 cm, oblong or elliptic; calyx lobes mostly very short, 1.5–2 mm long, less often up to 3(4) mm long · · · · · · · · · · · · · · · · ii) var. *laevius*

i) Var. **cinerascens** —Verdcourt in Kew Bull. **42**: 144 (1987); in F.T.E.A., Rubiaceae: 780 (1991).
Tapiphyllum cinerascens subsp. *cinerascens* Verdc. in Kew Bull. **36**: 535 (1981).

Leaf blades mostly about 8 × 3 cm, ± oblong, mostly densely pubescent above, often ± velvety. Calyx lobes mostly shorter, 3–4(6) mm long, glabrescent to usually densely pubescent.

Zambia. N: 16 km from Katibunga to the Great North Road, corolla fallen 30.iii.1961, *Angus* 2594 (FHO; K). W: Mwinilunga Distr., just S of Matonchi Farm, fl. 12.xi.1937, *Milne-Redhead* 3220 (K). C: Serenje Distr., Kundalila Falls, fr. 4.ii.1973, *Strid* 2783 (K). E: Lutembwe R., just

below Great East Road bridge, fr. 9.i.1959, *Robson* 1112 (BM; LISC; K). **Malawi**. N: Mzimba Distr., 22.5 km east of Mzambazi, corolla fallen 2.iii.1974, *Pawek* 8194 (K; MAL; MO; SRGH; UC). **Mozambique**. N: "Western Zone, Niassa, Luchimara R. Valley", corolla fallen 31.i.1942, *Hornby* 2522 (PRE).
Also in Zaire (Dem. Rep. Congo), Burundi, southern Tanzania and Angola. *Brachystegia–Julbernardia–Monotes–Uapaca* woodland, and *Cryptosepalum–Copaifera* woodland on Kalahari Sand, also bushland and scrub; 900–1500 m.
Some of the W Zambian specimens are close to *T. fulvum* Robyns, based on an Angolan specimen and probably only yet another variant of *T. cinerascens.*

ii) Var. **laevius** (K. Schum.) Verdc. in Kew Bull. **42**: 144 (1987); in F.T.E.A., Rubiaceae: 781 (1991). Type from Tanzania.
　　Vangueria velutina var. *laevior* K. Schum. in Bot. Jahrb. Syst. **28**: 194 (1900). —De Wildeman in Bull. Jard. Bot. État **8**: 66 (1922).
　　Tapiphyllum velutinum var. *laevius* (K. Schum.) Robyns in Bull. Jard. Bot. État **11**: 114 (1928). —Brenan, Check-list For. Trees Shrubs Tang. Terr.: 532 (1949).
　　Tapiphyllum tanganyikense Robyns in Bull. Jard. Bot. État **32**: 149 (1962). Type from Tanzania.
　　Tapiphyllum griseum Robyns in Bull. Jard. Bot. État **32**: 147(1962). Type: Malawi, Nyika Plateau, *E.A. Robinson* 3098 (K, holotype; PRE).
　　Tapiphyllum cinerascens subsp. *laevius* (K. Schum.) Verdc. in Kew Bull. **36**: 536 (1981).

Leaf blades mostly small, up to 3.5(5) × 1–2 cm, oblong or elliptic, not very discolorous, densely velvety on both surfaces. Calyx lobes mostly 1.5–2(3) mm long and not very narrow (but see note below).

Malawi. N: Chitipa Distr., Nyika Plateau, by Chisanga Falls, fr. 28.ii.1982, *Brummitt et al.* 16184 (K; MAL). **Mozambique**. N: Marrupa to Lichinga, 7 km from Marrupa, corolla fallen 19.ii.1982, *Jansen & Boane* 7893 (K; LMU).
Also in Tanzania. *Brachystegia* woodland; 800–1800 m.
Tapiphyllum griseum Robyns [type from Malawi, Nyika Plateau, N Rukuru waterfall, 1950 m, fl. 6.i.1959, *Robinson* 3098 (K, holotype; PRE)] is very similar but has slender calyx lobes to 4 mm long. It is considered here to be only a form of var. *laevius* connecting with var. *cinerascens* and var. *inaequale.*

iii) Var. **A**

Leaf blades up to 10 × 5.5 cm, broadly oblong-elliptic, very rounded at the base, very discolorous, sparsely pubescent above; petioles up to 9 mm long. Calyx lobes 6–7 mm long.

Zambia. W: Mwinilunga–Matonchi road, km 25, corolla fallen 21.xii.1969, *Simon & Williamson* 1913 (K; SRGH).
Known only from this locality. *Cryptosepalum* woodland; 1350 m.
Shows some similarity to *T. discolor*, and possibly of hybrid origin, but *T. discolor* has not been found in the Mwinilunga area; the long petioles are distinctive.

iv) Var. **inaequale** (Robyns) Verdc. in Kew Bull. **42**: 144 (1987); in F.T.E.A., Rubiaceae: 780 (1991). Type from Tanzania.
　　Vangueria kaessneri S. Moore in J. Bot. **48**: 221 (1910). —De Wildeman in Bull. Jard. Bot. État **8**: 57 (1922). Type from Zaire (Dem. Rep. Congo), Shaba Prov.
　　Tapiphyllum kaessneri (S. Moore) Robyns in Bull. Jard. Bot. État **11**: 106 (1928). —Brenan, Check-list For. Trees Shrubs Tang. Terr.: 532 (1949). —Robyns in Bull. Jard. Bot. État **32**: 135 (1962).
　　Tapiphyllum inaequale Robyns in Bull. Jard. Bot. État **32**: 145 (1962).
　　Tapiphyllum cinerascens subsp. *inaequale* (Robyns) Verdc. in Kew Bull. **36**: 535 (1981).

Leaf blades 3.5–9.5 × 1–2.7 cm, narrowly oblong, oblong-elliptic or linear-lanceolate, very discolorous, glabrescent to densely pubescent above. Calyx lobes 3–10 mm long, mostly very narrow.

Zambia. N: Mpika, fr. 9.ii.1955, *Fanshawe* 2033 (K; NDO). W: 2 km from Ndola, fr. 27.i.1963, *Symoens* 9998 (BR; K). C: Serenje Distr., 1 km from Kanona to Kundalila Falls, corolla fallen 12.iii.1975, *Hooper & Townsend* 669 (K). **Malawi**. N: Mzimba Distr., 17.4 km NE of Mzambazi, fr. 22.i.1978, *Pawek* 13678 (M; MAL; MO).
Also in Zaire (Dem. Rep. Congo), Tanzania and ?Angola. Woodland, and in grassland with scattered trees, on sandy soil; 1150–1650 m.

v) Var. **richardsiae** (Robyns) Verdc. in Kew Bull. **42**: 145 (1987) (as "*richardsii*"). Type: Zambia, Mbala, escarpment above Chilongowelo, *Richards* 36 (K, holotype).
   Tapiphyllum richardsii Robyns in Bull. Jard. Bot. État 32: 142 (1962).
   Tapiphyllum glomeratum Robyns in Bull. Jard. Bot. État **32**: 146 (1962). Type: Zambia, Mbala, 1.6 km E from Mpulungu, *Angus* 770 (K, holotype).

Leaf blades up to about 4 × 1 cm, narrowly oblong to oblanceolate, rather thicker, discolorous, pubescent to ± velvety above. Calyx lobes c. 4 mm long.

   Zambia. N: Luwingu Distr., Chishinga Ranch, fl. 28.x.1961, *Astle* 1003 (K). **Malawi**. C: Kasungu Distr., 38.4 km S of Champira, fl. 23.xii.1976, *Pawek* 12035 (K; MO).
   Known so far only from the above localities. Grassland, *Combretum–Grewia* thicket and *Brachystegia* woodland, on sandy soil, sometimes in rocky places; 1200–1500 m.

vi) Var. **laetum** (Robyns) Verdc. in Kew Bull. **42**: 145 (1987). Type: Zambia, Mbala, Inono Stream, *Richards* 3707 (K, holotype).
   Tapiphyllum laetum Robyns in Bull. Jard. Bot. État **32**: 143 (1962).

Leaf blades up to 5 × 1.5 cm, narrowly oblong, thin and very discolorous, green and very sparsely pubescent above. Calyx lobes c. 5 mm long, very narrow.

   Zambia. N: Mbala Distr., about 3.2 km from Kalambo Falls on path to Lake Tanganyika, corolla fallen 25.xii.1967, *Simon & Williamson* 1560 (K; LISC; SRGH). E: Lunkwakwa Valley, corolla fallen 23.iii.1955, *Exell, Mendonça & Wild* 1147 (BM) ('cf').
   Known so far only from the above localities. *Brachystegia* woodland, riverine forest, rocky places; 900–1350 m.

5. **Tapiphyllum velutinum** (Hiern) Robyns in Bull. Jard. Bot. État **11**: 111, figs. 15 & 16 (1928). — J.G. Garcia in Mem. Junta Invest. Ultramar **6** (sér. 2): 27 (1959) [Contrib. Conhec. Fl. Moçamb. IV (1959)]. —Robyns ibid. **32**: 137 (1962). —F. White, F.F.N.R.: 421 (1962). — Drummond in Kirkia **10**: 276 (1975). —A.E. Gonçalves in Garcia de Orta, Sér. Bot. **5**: 208 (1982). —K. Coates Palgrave, Trees Southern Africa, ed. 3, rev.: 875 (1988). Type: Mozambique, Morrumballa (Moramballa), *Kirk* (Jan. 1863) (K, lectotype).
   Vangueria velutina Hiern in F.T.A. **3**: 151 (1877). —K. Schumann in Engler, Pflanzenw. Ost-Afrikas **C**: 385 (1895). —S. Moore in J. Bot. **45**: 42 (1907). —Sim, For. Fl. Port. E. Africa: 75 (1909). —De Wildeman in Bull. Jard. Bot. État **8**: 66 (1922).
   Tapiphyllum vestitum Robyns in Bull. Jard. Bot. État **11**: 114 (1928). Type: Zimbabwe, Harare (Salisbury), *Eyles* 1924 (K, holotype; PRE; SRGH).

Shrub or small tree 1–4 m tall (see note), usually with rather robust shoots, divaricately branched and with ultimate branches often at right angles, or branchlets very reduced and cushion-like; young shoots densely velvety tomentose, later glabrous and with dark reddish-brown or black ± peeling bark, rugulose. Leaves 2.5–8 × 1–6 cm, narrowly oblong or elliptic to almost round, rounded or sometimes obtuse at apex, rounded or subcordate at base, ± thick, somewhat discolorous, densely thickly velvety tomentose, particularly beneath; venation impressed above and raised beneath, the leaf surface usually rugulose and distinctly bullate; petiole thick, 2–3.5 mm long; stipules shortly connate, 3–5 mm long with linear appendage 3–7 mm long, densely tomentose outside, villous inside, deciduous. Flowers in dense few- to many-flowered cymes in opposite axils, axillary, or sometimes appearing to be terminal on very short lateral branchlets with suppressed leaves; peduncles 2–4 mm long; pedicels c. 1 mm long; bracts 9 × 2–5 mm, ovate-oblong to lanceolate; all parts densely tomentose. Calyx densely velvety or woolly; tube 1–1.5 mm, subglobose; lobes 2–4 mm long, triangular to linear-subulate. Corolla tailed at apex in bud; green to yellow, densely velvety; tube 3.5 mm long, cylindrical, with a ring of deflexed hairs towards the base inside; lobes 2–4 × 2 mm, ovate to narrowly triangular, apiculate. Style exserted 0–2 mm; pollen presenter 0.5–1.2 mm long, cylindrical-sulcate, 4–5-lobed at the apex. Fruit orange-brown, 1–1.4 cm in diameter, subglobose, finely velvety tomentose, with 1–4(5) pyrenes, crowned by the persistent calyx. Pyrenes straw-coloured, 9 mm long.

   Zambia. N: Kawambwa, fr. 15.xi.1957, *Fanshawe* 3973 (K; NDO). C: Chilanga, Mt. Makulu, young fr. 26.xii.1959, *White* 6014 (FHO; K). S: Mazabuka Distr., Nachibanga Stream, Choma to Pemba, km 42, young fr. 9.ii.1960, *White* 6907 (FHO; K). **Zimbabwe**. N: Goromonzi Distr.,

Chindamora Reserve, Ngomakurira, fl. 19.x.1967, *Loveridge* 1758 (K; SRGH). W: Matobo Distr., Besna Kobila, fl. xii.1961, *Miller* 8074 (K; SRGH). C: Makoni Distr., 20 km from Rusape towards Nyanga (Inyanga), corolla fallen 2.xii.1930, *Fries et al.* 3370 (K; LD; PRE; SRGH). E: Mutare Distr., East Commonage, E ridge of Ngamagaree Peak, fl. 3.i.1960, *Corner* s.n. (E; K). S: Mberengwa Distr., Mt. Buhwa, below rock face on N slopes, sterile 28.iv.1973, *Pope* 979 (K; SRGH). **Malawi.** C: Lilongwe Distr., Dzalanyama Forest Reserve, valley NW of Kazuzu Hill, young fr. 24.ii.1982, *Brummitt et al.* 16089 (K). S: Chiradzulu Distr., above Kansisi Village, Lisau Hill, young fr. 4.i.1983, *Patel* 1069 (BR; K; MAL). **Mozambique.** N: Litunde, in dry places near Penedo da Saudade, fl. 15.xii.1934, *Torre* 542 (COI; LISC). MS: Barúè, Serra de Choa at km 28 from Catandica (Vila Gouveia), fl. 10.xii.1965, *Torre & Correia* 13501 (C; LISC; LMU). Z: near base of Morrumbala, fl. 30.xii.1858, *Kirk* (K). T: Cahora Bassa, Songo, fr. 21.i.1973, *Torre et al.* 18824 (COI; LD; LISC; LMA; MO).

Known so far only from the Flora Zambesiaca area. *Brachystegia* woodland and mixed deciduous woodlands, grassland with scattered trees, scrub particularly on quartzite and granite hills and steep rock outcrops, also recorded from "fringing forest"; 850–1800 m.

A field note to *Norrgrann* 477 claiming it to be a tree to 25 m cannot be correct. *Jacobsen* 2653 & 2605, both from Zimbabwe, Makonde Distr., lack the thick corrugated leaves of typical *T. velutinum*. The former was from highly copper-bearing rocky ground arkose and the latter from graphitic slates; both may be yet further distinct variants.

6. **Tapiphyllum rhodesiacum** (Tennant) Bridson in Kew Bull. **51**: 351 (1996). Type: Zambia, Serenje Distr., *Fanshawe* 1848 (K, holotype; NDO).
   Ancylanthos rhodesiacus Tennant in Kew Bull. **19**: 283 (1965).
   Ancylanthus sp. 1 of F. White, F.F.N.R.: 400 (1962).

Shrub or small tree up to 4.5 m tall; the internodes, particularly on the abbreviated lateral shoots, very short giving a gnarled appearance; youngest parts densely ± velvety ferruginous-hairy, but soon glabrous and with dark purplish-brown minutely scaling bark. Leaves 0.8–3.5 × 0.6–3 cm, broadly elliptic to almost round or oblate, rounded at both ends or ± subcordate at the base, markedly discolorous, pubescent above but not obscuring the bright green surface, densely grey velvety-tomentose beneath, bullate above due to deeply impressed venation; petiole 1–2 mm long; stipules 2 mm long, shortly triangular, densely hairy inside the bases, becoming persistently woody, the appendage 1–2 mm long, subulate. Flowers solitary, or in c. 3-flowered cymes; peduncles ± suppressed; pedicels c. 1 mm long; bracts and bracteoles up to 5 mm long, subulate. Calyx densely pubescent; tube 2 mm long, ovoid, drying brownish and ribbed; lobes 2–4 mm long, triangular at base, with subulate apex. Corolla shortly tailed in bud, creamy or lemon-yellow in colour, the tube more greenish; tube 1.8–2 cm long, slightly to distinctly curved, densely pubescent outside with a ring of deflexed hairs near the base inside; lobes 5–7 mm long, 2 mm wide at the base, triangular-lanceolate, with short appendage at apex, pubescent outside. Style 2.2–2.5 cm long; pollen presenter 2 mm long, subcylindric, shortly 4-lobed at apex. Immature fruit 7 mm wide, ± globose, densely tomentose, pale yellow-brown.

Zambia. N: Chinsali Distr., Ishiba Ngandu (Shiwa Ngandu), Machipara Hill, fl. 16.i.1959, *Richards* 10684 (K). C: Serenje, fl. 23.i.1955, *Fanshawe* 1848 (K; NDO).

Known so far only from the Flora Zambesiaca area. Escarpment vegetation, *Protea–Vellozia–Euphorbia–Tricalysia–Hymenodictyon* thicket, and in woodland on granite rocks and very stony ground; c. 1500 m.

7. **Tapiphyllum burnettii** Tennant in Kew Bull. **19**: 280, fig. 1 (1965). —Verdcourt in Kew Bull. **36**: 537 (1981). —Havard & Verdcourt in Kew Bull. **42**: 606, fig. 1C (1987). —Verdcourt in F.T.E.A., Rubiaceae: 784, figs. 132/7, 139 (1991). TAB. **52**. Type from Tanzania.

Shrub or small tree, 1.5–6 m tall, sometimes scrambling; branches slender, densely pale ferruginous-hairy on young parts but soon glabrescent or glabrous and revealing fissured ± purplish-brown bark. Leaves paired; blades 1.5–9.5 × 1–4.4 cm, oblong to ovate-lanceolate, rounded to acute at the apex, ± rounded at the base, discolorous, green with appressed ± ferruginous hairs above, densely velvety beneath with silvery to pale ferruginous hairs, the main nervation usually slightly darker ferruginous; petiole 1–2 mm long; stipules 8 mm long, linear-lanceolate, deciduous. Flowers in 3–several-flowered cymes, ferruginous pubescent; peduncles 4 mm long; lower bracts small, upper bracts ± leaf-like, up to 7 × 2 mm, lanceolate; pedicels

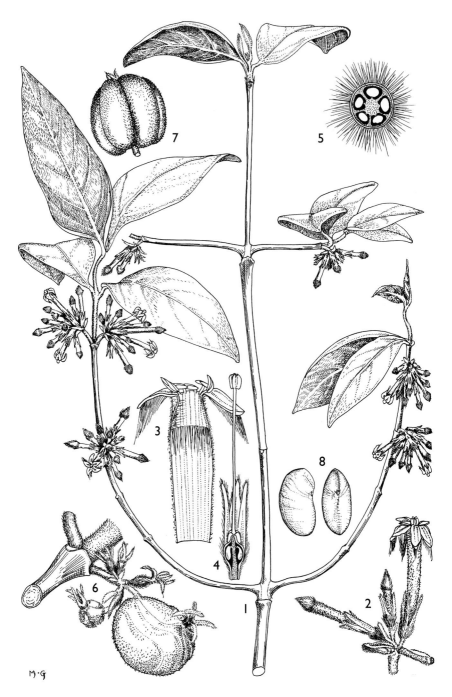

Tab. 52. TAPIPHYLLUM BURNETTII. 1, flowering branch (× 1); 2, flower and flower buds (× 2); 3, section through corolla (× 3); 4, section through ovary with calyx lobes, style and pollen presenter (× 3); 5, transverse section through ovary (× 6), 1–5 from *Richards* 7404; 6, fruiting node (× 2); 7, mature fruit (× 2); 8, pyrene, 2 views (× 2), 6–8 from *Haerdi* 445/0. Drawn by Mary Grierson, with 8 by Sally Dawson. From F.T.E.A.

obsolete. Calyx densely pale ferruginous-pilose; tube 1.2 mm long; lobes 2.5–6 × 0.4–1 mm, linear to lanceolate, sometimes slightly spathulate. Corolla capitate, acute or apiculate at the apex in bud, pale green, creamy-green or white, densely ferruginous hairy; tube 11–12.5 mm long, slender, cylindrical; lobes 3–4 × 1–2 mm, ovate-lanceolate, shortly apiculate. Anthers 1.1–1.5 mm long, shortly exserted. Style 11–13 mm long; pollen presenter 1.2 mm long, coroniform. Fruits orange-yellow, subglobose, 1–1.3 cm in diameter with 2–5 pyrenes, sulcate in dry state, densely ferruginous-tomentose with short and longer hairs; pyrenes reddish-brown, up to 9 × 5.5 mm, rugose.

Zambia. N: Isoka, fl. 22.xii.1962, *Fanshawe* 7205 (K; NDO).
Also in southern Tanzania. Dry rocky streambed; c. 1500 m.
This species stands apart from the rest of *Tapiphyllum* (see Bridson in Kew Bull. **51**: 349 (1996)) but nonetheless in the absence of flowers it can be difficult to distinguish from *T. cinerascens*. However, the cotyledons are shorter than the radicle in *T. burnettii* but longer than the radicle in *T. cinerascens*.
Brummitt et al. 15977 (Malawi, S: Mangochi Distr. fr. 19.ii.1982), a climber to 8 m, very likely belongs here, and one or two sterile specimens named *T. cinerascens* also resemble this.

43. FADOGIA Schweinf.

Fadogia Schweinf., Reliq. Kotschy: 47, t. 32 (1868). —Robyns in Bull. Jard. Bot. État **11**: 41 (1928). —Verdcourt in Kew Bull. **36**: 515 (1981).
Temnocalyx Robyns in Bull. Jard. Bot. État **11**: 317 (1928), pro parte.

Glabrous or hairy suffrutescent subshrubs, usually with several often annual stems from a woody rootstock; rarely shrubs or small trees. Leaves usually in whorls of 3–5(6), or paired, mostly subsessile or shortly petiolate; stipules small, often persistent, connate into a sheath, hairy inside at the base and produced into a subulate appendage at the apex. Inflorescences of axillary few-flowered cymes, usually shortly pedunculate, or in some species flowers solitary; flowers usually rather small; buds obtuse or shortly apiculate. Calyx tube campanulate or subglobose, the limb truncate, or 5–10-toothed or -lobed, the divisions obtuse to subulate. Corolla tube usually cylindrical, glabrous or hairy outside, with a ring of deflexed hairs inside, pubescent to densely hairy above it and densely hairy or less often glabrous at the throat; lobes 5(8), acute to apiculate at the apex, papillate on inner surface. Anthers usually oblong, shortly exserted; filaments short. Ovary 3–4(8)-locular; ovules solitary in each cell, pendulous. Style slender or rather thick, usually thicker towards the base, not or scarcely exceeding the anthers; pollen presenter coroniform or rarely cylindrical, sulcate, 3–5(8)-lobed. Disk usually depressed. Fruits small, often bearing the remains of the calyx limb, glabrous, containing 1–3(8) pyrenes.

A genus of about 40–50 closely related species, 15 of which occur in the Flora Zambesiaca area. It is rather well characterised in habit. Part of *Temnocalyx* Robyns has been merged with it; there is too much variation in the curvature of the corolla for this to be a useful character; nevertheless the Tanzanian species chosen as lectotype of Robyns's genus, *T. nodulosus*, differs markedly from the others he included in it by its distinctive habit, nodose stems, uniformly paired leaves, structure of the pollen presenter and characteristic corolla lacking deflexed hairs inside. The genus has been maintained for this species. The species with larger red flowers are probably bird pollinated. Several species are said to have edible fruits.

1. Leaves (only the youngest in *F. flaviflora* var. *calvescens*) densely velvety or papillate beneath, indumentum ± hiding the leaf lower surface which is markedly discolorous (concolorous in sp. 8) · 2
 – Leaves glabrous to pubescent beneath but not so densely velvety or papillate as to cover, or almost cover, the entire lower surface; the blades not markedly discolorous, any differences in colour between leaf upper and lower surfaces is due to the surfaces themselves (rarely flowering when leafless) · 8
2. Indumentum of leaf lower surface consisting entirely of very small closely spaced white papillae; leaf upper surface green and glabrous; corolla tube 3–4 mm long; calyx lobes triangular-lanceolate, 2–3.5 mm long · 2. *homblei*

– Indumentum of hairs, or a mixture of hairs and papillae · · · · · · · · · · · · · · · · · · 3

3. Indumentum of undersides of leaves consisting of a mixture of coarse hairs and an understorey of papillae*; corolla tube 2.5–3.5 mm long; calyx lobes triangular-subulate 1–2.5 mm long; leaves mostly narrow and tapering acute in the Flora Zambesiaca area · 1. *cienkowskii*

– Indumentum of undersides of leaves consisting of hairs only, or with only very few papillae · 4

4. Lower leaves with velvety indumentum wearing off, sometimes only the venation with hairs · 6. *tomentosa* var. *calvescens*

– Lower leaves with velvety indumentum persistent · 5

5. Leaves mostly distinctly narrowly acute; stems longer and more robust, to 90 cm tall; indumentum rather coarse and less matted particularly on costa · · · · · · · · · · 3. *elskensii*

– Leaves ± rounded at apex, or if somewhat acute then plants either small, 15–25 cm tall, with indumentum very fine and greyish-white, or plants taller but with a thick densely matted very softly velvety indumentum covering the entire lower leaf surface and venation · 6

6. Stems robust, 45–60 cm tall, either unbranched or branched; leaves larger 2–12 × 1.2–6 cm with dense thick velvety indumentum of very soft tangled hairs · · · · · · · · · · 6. *tomentosa*

– Stems slender and 15–25 cm tall; leaves small 1.2–6 × 0.5–2.5 cm with a very fine greyish-white indumentum more or less hiding the leaf lower surface · · · · · · · · · · · · · · · · · 7

7. Leaves markedly discolorous, densely grey pubescent above, grey velvety beneath, 1.7–6 × 0.8–2.5 cm · 7. *arenicola*

– Leaves concolorous, appressed grey velvety on both surfaces, mostly smaller and narrower, 1.2–3 × 0.5–0.9 cm · 8. *luangwae*

8. Calyx limb truncate, or with minute teeth less than 0.5(0.7) mm long · · · · · · · · · · · · 9

– Calyx limb with distinct teeth · 14

9. Corolla tube 14–28 mm long, sometimes slightly to distinctly curved · · · · · · · · · · · · 10

– Corolla tube shorter, straight · 12

10. Corolla tube usually more slender, 1.8–2.8 cm long, distinctly curved; flowers 5-merous · 14. *ancylantha*

– Corolla tube robust 1.4–3 cm long, 4–9 mm wide at apex, straight or curved; flowers 5-merous or 6–9-merous · 11

11. Corolla tube straight, 1.4–2.7 cm long; leaf blades 4.5–15 × 1.8–9 cm; flowers usually 6–9-merous; corolla with tube dark red, and lobes cream to buff within · · · · · · 15. *fuchsioides*

– Corolla tube usually distinctly curved, 2–3 cm long; leaf blades 1–6 × 0.2–1.6 cm; flowers usually 5-merous; corolla tube not red · 16. *verdickii*

12. Plant with stems and leaves ± hairy · 10. *triphylla* vars.

– Plant glabrous save for insides of stipules, and inside of corolla tube, or if rarely pubescent then inflorescences produced from lowermost nodes with reduced leaves · · · · · · · · · 13

13. Inflorescences borne at the lower nodes of shoots mostly with reduced leaves; calyx limb mostly erose or with short lobes to 0.7 mm long, but sometimes ± truncate; corolla tube 3–4 mm long; leaves mostly rounded at the apex; plants usually short geophytes 6–40 cm tall with numerous shoots from a woody stock · 9. *stenophylla*

– Inflorescences more evenly spaced, with some from at least the middle and often the upper nodes; calyx limb truncate; leaves mostly acute to acuminate at the apex; plants variable (10)22–45(120) cm tall · 10. *triphylla*

14. Inflorescences borne at the lower nodes, often subtended by leaves · · · · · · 9. *stenophylla*

– Inflorescences not restricted to lower nodes, usually some from the middle and often the upper nodes · 15

15. Small geophilous subshrub 5–15 cm tall with congested leaves, the apical internodes very short; leaves oblanceolate to narrowly oblong, 2–6.5 × 0.5–1.4 cm · · · · · · · 11. *chlorantha*

– Plant not so short, or if so then leaves more broadly elliptic, 1.8–8.3 × 0.8–2.7 cm · · · 16

16. Leaves very narrow, linear-oblong to linear-oblanceolate, 2–12 × 0.25–1.2 cm, mostly on 3 young slender branchlets at each node · 12. *gossweileri*

– Leaves much wider · 17

17. Flowers numerous, mostly at leafless nodes; shoots rather stout with bark finally peeling at base to reveal reddish underbark; leaves restricted to apex of stem, rather coriaceous, drying with venation rather prominently raised · 13. *schmitzii*

---

* Best seen under low power of a student's compound microscope, after scraping the surface in cases where hairs are also present.

– Flowers fewer, at leafy nodes; shoots more slender, the bark not peeling; leaves not so
thick · 18
18. Leaves ± densely hairy beneath · 3. *elskensii*
– Leaves glabrous and ± glaucous beneath, or with very sparse hairs, the surface with a
characteristic pattern of stomata* · 19
19. Stems 20–120 cm tall; leaves ± spaced, 4–11 × 1.5–4.8 cm, glabrous to pubescent;
inflorescences 3–9-flowered · 4. *tetraquetra*
– Stems 10–20 cm long; leaves rather closely spaced, 1.8–8.3 × 0.8–2.7 cm, glabrous; flowers
solitary, or in 2-flowered cymes ·5. *variifolia*

1. **Fadogia cienkowskii** Schweinf., Reliq. Kotschy: 47, t. 32 (1868). —Hiern in F.T.A. **3**: 154
(1877). —Robyns in Bull. Jard. Bot. État **11**: 79 (1928). —Aubréville, Fl. For. Soud.-
Guin.: 480, t. 106/4–5 (1950). —Hepper in F.W.T.A. ed. 2, **2**: 178 (1963). —Verdcourt in
Kew Bull. **36**: 517 (1981); in ibid. **42**, fig. 1L, fig. 5B (1987); in F.T.E.A., Rubiaceae: 788
(1991). Type from Ethiopia.

Suffrutex 0.3–1.2 m tall; stems simple, few to several (15) from the apical parts of
a branching woody rhizome, reddish-brown or yellowish-green, covered with dense
pale ferruginous hairs, or less often glabrous. Leaves 3–4-whorled, markedly
discolorous, 2–8.5 × 0.5–4.5 cm, narrowly to broadly elliptic or oblong-elliptic, to
narrowly lanceolate, ± rounded to acute or very shortly acuminate at the apex,
broadly to narrowly cuneate or ± rounded at the base, often ± bullate, sparsely to
fairly densely hairy above, the hairs not obscuring the ± glossy upper surface, very
densely (rarely ± sparsely) continuously velvety-tomentose beneath with grey to pale
ferruginous matted hairs, beneath which is a dense covering of short white papilla-
like hairs; petioles short and stout, 1–1.5 mm long, densely pubescent; stipules 6–9
mm long, shortly joined at the base, subulate from an oblong-triangular base,
pubescent. Inflorescences 2–6-flowered, or flowers sometimes solitary; peduncles
3–9(12) mm long; pedicels 1.5–9(12) mm long, pubescent. Calyx tube 1–1.5 mm
long, 7–10-toothed; teeth (1)1.5–2.5 mm long, triangular to triangular-lanceolate or
subulate, often membranous usually pubescent. Corolla distinctly apiculate with
subulate tails in bud, yellow or cream-coloured; tube cylindrical, 2.5–3.5 mm long,
glabrescent to pubescent outside; lobes narrowly lanceolate, about 3.5–5.5 mm long
including the subulate appendage (0.5)1.5–2 mm long. Style up to 5–7.5 mm long;
pollen presenter yellow-green or whitish, coroniform, 0.75 mm long, sulcate, 3–4-
lobed. Ovary 3–4-locular. Fruit dark green becoming black and shining, crowned
with calyx lobes, up to 10 mm in diameter and subglobose with 1–3 pyrenes, or when
pyrenes reduced to 1 then fruit oblique.

Var. **cienkowskii** —Verdcourt in F.T.E.A., Rubiaceae: 789 (1991).
Fadogia katangensis De Wild. in Repert. Spec. Nov. Regni Veg. **12**: 295 (1913). —Robyns
in Bull. Jard. Bot. État **11**: 77 (1928). Type from Zaire (Dem. Rep. Congo).
Vangueria tristis K. Schum. in De Wildeman, Études Fl. Katanga [Ann. Mus. Congo, Sér. IV,
Bot.] **1**: 227 (1903); Contrib. Fl. Katanga: 211 (1921). Type from Zaire (Dem. Rep. Congo).
Fadogia tristis (K. Schum.) Robyns in Bull. Jard. Bot. État **11**: 76 (1928).
Canthium cienkowskii (Schweinf.) Roberty in Bull. Inst. Fondam. Afrique Noire, Sér. A.,
Sci. Nat. **16**: 61 (1954).

Leaf blades up to 4.5 cm wide, mostly densely velvety beneath.

Zambia. B: c. 51 km NE of Mongu, fl. 10.xi.1959, *Drummond & Cookson* 6317 (K; SRGH). W:
Mwinilunga, slope east of R. Lunga, fl. 24.xi.1937, *Milne-Redhead* 3374 (BM; K). C: near
Chilanga, Mt. Makulu Research Station, fl. 8.x.1957, *Angus* 1763 (FHO; K; PRE). S: Mumbwa
Distr., Nambala Mission, fl. 16.ix.1947, *Greenway & Brenan* 8101 (EA; K; PRE). **Zimbabwe**. N:
Mazowe Distr., Mvurwi (Umvukwes), Ruorka Ranch, fr. 16.xii.1952, *Wild* 3921 (K; SRGH). E:
Chimanimani Distr., Glencoe Forest Reserve, fr. 22.xi.1955, *Drummond* 4948 (K; SRGH).
Malawi. N: Mzimba Distr., Viphya, 61 km SW of Mzuzu, fl. 9.ix.1969, *Pawek* 2981 (K). C: Lilongwe
Distr., Chisepo, fl. 20.xi.1963, *Salubeni* 152 (K; SRGH).
    Also from Mali to Nigeria in West Africa, and Cameroon, Cabinda, Zaire (Dem. Rep. Congo),
Sudan, Ethiopia, Uganda, Kenya, Tanzania and Angola. Grassland and grassland with scattered

* Seen under low power of a student's compound microscope.

trees, also in deciduous woodlands including; *Brachystegia–Julbernardia–Uapaca* woodland, *Cryptosepalum* woodland, *Albizia–Marquesia–Anisophyllea* woodland and *Combretum–Acacia* woodlands, often by dambos and on rock outcrops, and in cultivated land; 600–1740 m.

Fadogia fragrans links *F. homblei* with *F. cienkowskii* which are nevertheless retained here as species. It has been usual to maintain *F. katangensis* distinct from *F. cienkowskii*, and on the whole the southern material does have more tapering leaves with less tendency to be slightly bullate on drying, but occasional specimens are indistinguishable from those of northern populations. *Johnson* 46 (Manica, Choa Valley, fl. 24.xi.1906) was annotated and cited by Robyns as *Fadogia fragrans* Robyns, a species occurring in South Africa (North Prov., Gauteng, and Swaziland). It is undoubtably intermediate between *F. cienkowskii* and *F. fragrans*. Some extreme variants occur, e.g. *Fanshawe* 5337 [Zambia, Mufulira 6.i.1960] which had been named *F. tetraquetra*, but the leaf blades have very sparse hairs and papillae beneath, appearing glabrous to the naked eye. It could be of hybrid origin. Similar variants occur in West Africa.

Var. *lanceolata* K. Schum. ex Robyns, from Cameroon, Zaire (Dem. Rep. Congo) and Tanzania, has narrow leaf blades up to 10 mm wide.

2. **Fadogia homblei** De Wild. in Repert. Spec. Nov. Regni Veg. **12**: 295 (1913); Notes Fl. Katanga **3**: 27 (1913); Contrib. Fl. Katanga: 213 (1921). —Verdcourt in Kew Bull. **36**: 518 (1981); in F.T.E.A., Rubiaceae: 789 (1991). Type from Zaire (Dem. Rep. Congo) (Shaba Prov.).

   Fadogia monticola Robyns in Bull. Jard. Bot. État **11**: 70, figs. 11 & 12 (1928). —Codd & Voorendyk in Bothalia **8** (Suppl. 1): 57, figs. 60, 61, map 7 (1965). —A.E. Gonçalves in Garcia de Orta, Sér. Bot. **5**: 191 (1982). Type from South Africa.

Erect suffrutex 30–120 cm tall, with 3–4 or many stems from a slender rhizome; stems often reddish, 3–4-angled, glabrous or rarely very sparsely pubescent. Leaves in whorls of 3–5, markedly discolorous; 2.2–12 × 0.3–3.8 cm, narrowly elliptic or oblong-lanceolate to lanceolate, tapering to a fine acute apex, cuneate at the base, green and glabrous above, drying whitish-tomentose beneath due to a very dense covering of minute papilla-like hairs, pale green in life; petiole 1–2.5 mm long; stipules joined to form a sheath 1.5–3 mm long with 3 filiform appendages 5–8 mm long. Inflorescences 2–5-flowered, glabrous; peduncle 3–4 mm long; pedicels 1.5–6 mm long. Calyx tube 1.2–1.5 mm long; limb with triangular to lanceolate teeth 1–3.5 mm long. Corolla distinctly 5–6-apiculate at the apex in bud, the appendages diverging, yellowish or bright yellow; tube 3–4(5) mm long, with ring of deflexed hairs inside; lobes 4–5.5 × 2 mm, narrowly triangular or oblong-triangular, drawn out at the apex into a subulate appendage, 1–1.5 mm long. Style rather stout, about 7 mm long; pollen presenter coroniform, 1–1.5 mm long, 3–4-lobed. Ovary 3–4-locular. Fruit black, crowned with the persistent calyx limb, often oblique, 6–10(13) mm long, usually with 2–3 pyrenes.

Zambia. B: Zambezi (Balovale), Katuva (Katuba) Valley, fr. 25.x.1953, *Gilges* 76 (K; PRE). N: Kasama Distr., Mungwi, fl. 5.xi.1960, *E.A. Robinson* 4038 (K). W: Mufulira, fl. & fr. 1.i.1948, *Cruse* 138 (K). C: Lusaka Distr., Chakwenga Headwaters, fl. 16.xi.1963, *E.A. Robinson* 5833 (K). S: Livingstone Distr., Situmpa Forest Reserve, Machili, fr. 24.i.1952, *Fewdays* 7 (FHO; K). **Zimbabwe**. E: Mutare Distr., Dickers' Farm, Himalaya Mts., fl. 27.xi.1966, *Dale* 409 (K; SRGH). **Malawi**. N: Rumphi Distr., Kaziwiziwi R., fl. & fr. 8.i.1959, *Richards* 10557 (K). C: Lilongwe Distr., Dzalanyama Forest Reserve, valley NW of Kazuzu Hill, fl. 24.ii.1982, *Brummitt* 16076 (K; MAL). **Mozambique**. T: E Mt. Furancungo, fr. 17.iii.1966, *Pereira et al.* 1831 (BR; LMU) (fide Gonçalves). MS: Manica, Rotanda R. Valley near the Falls, fl. 21.x.1945, *Pedro* 420 (LMA; PRE).

Also in southern Tanzania, Angola and South Africa (North Prov. and Gauteng). *Protea* grassland, scrub, *Brachystegia* woodland, *Cryptosepalum* woodland, *Lannea*–bamboo association, often at dambo edges and on termite mounds; 1050–1800 m.

Reference to a shrub 4 m tall (*Fewdays* s.n.) from Zambia, Livingstone, is presumably an error for 4 ft. *F. oleoides* Robyns described from South Africa (Gauteng), and having longer peduncles, is merely a variant.

3. **Fadogia elskensii** De Wild., Pl. Bequaert. **3**: 201 (1925). —Verdcourt in Kew Bull. **36**: 518 (1981); in F.T.E.A., Rubiaceae: 790 (1991). Type from Burundi.

   Vangueria katangensis K. Schum. in De Wildeman, Études Fl. Katanga [Ann. Mus. Congo, Sér. IV, Bot.] **1**: 227 (1903) non *Fadogia katangensis* De Wild. Type from Zaire (Dem. Rep. Congo) (Shaba Prov.).

   Fadogia cienkowskii sensu Robyns in Bull. Jard. Bot. État **11**: 80 (1928), pro parte non Schweinf.

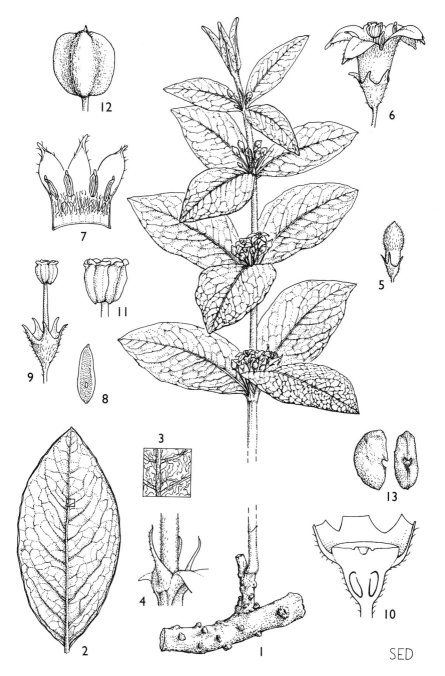

Tab. 53. FADOGIA ELSKENSII var. ELSKENSII. 1, habit (× ²/₃), from *Newbould & Jefford* 1707; 2, leaf lower surface (× ²/₃); 3, detail from lower surface of leaf (× 4); 4, stipule (× 2); 5, flower bud (× 2); 6, flower (× 3), 2–6 from *Milne-Redhead & Taylor* 7875; 7, part of corolla opened out (× 4); 8, dorsal view of anther (× 8); 9, calyx, with style and pollen presenter (× 4), 7–9 from *Newbould & Jefford* 1707; 10, longitudinal section through ovary (× 8), from *Milne-Redhead & Taylor* 7875; 11, pollen presenter (× 8), from *Newbould & Jefford* 1707; 12, fruit (× 2), from *Lovett* 1269; 13, pyrene, 2 views (× 2), from *Milne-Redhead & Taylor* 7875. Drawn by Sally Dawson. From F.T.E.A.

Suffrutex with several unbranched stems 25–90 cm tall from a woody rootstock; stems ± 4-angled, densely hairy, particularly above. Leaves in whorls of 3–4, very discolorous; blades 4–9 × 1.2–4.6 cm, elliptic, oblong-elliptic or, less often, oblong-lanceolate, rounded to acute or even retuse at the apex, usually apiculate, rounded to broadly cuneate at the base, with venation impressed above and raised beneath, the leaf upper surface almost bullate, pubescent above but not in any way obscuring the surface, densely woolly-velvety beneath with grey, whitish, ochraceous or ± ferruginous hairs usually completely obscuring the lower surface, occasionally thinner and not obscuring it, without an understorey of papillae but with a pattern of stomata; petiole c. 2 mm long; stipules with base c. 2 mm long and with linear appendage 4–11 mm long. Inflorescences 1–5-flowered, with flowers in sessile or pedunculate fascicles; peduncles 1–5 mm long; pedicels 1–6 mm long. Calyx tube 2 mm long, glabrous to pubescent; limb-tube up to 0.5 mm long; lobes 1–2.5 mm long, ovate to linear-lanceolate. Corolla apiculate in bud, yellow, greenish-yellow or cream-coloured; tube c. 4 mm long, glabrous outside; lobes 3–5 × 1.5–2.5 mm, triangular, apiculate, ± hairy outside above, the apiculae up to 2 mm long. Style pale green, exserted 2 mm; pollen presenter whitish, 1.5 mm long, oblong or obconic. Fruit black, glossy, about 10 mm in diameter, subglobose, with up to 5 pyrenes. Pyrenes 8 mm long and 4.5 mm wide, segment-shaped, reticulate-rugose.

Var. **elskensii** —Verdcourt in F.T.E.A., Rubiaceae: 790, fig. 141 (1991). TAB. **53**.

Leaves with a dense velvety indumentum beneath, obscuring the surface; shoots to 90 cm tall.

Malawi. N: Mzimba Distr., Mzuzu, Marymount, fl. 8.xi.1969, *Pawek* 2963 (K). S: Mt. Zomba, grasslands, fl. 6.xii.1957, *Banda* 359 (K; SRGH). **Mozambique**. N: about 30 km from Imala towards Mecubúri, fr. 16.i.1964, *Torre & Paiva* 10006 (LISC) (forma). Z: Gurué Mt summit, 5 km from Licungo waterfall, fr. 21.ii.1966, *Torre & Correia* 14758 (C; COI; LISC; LMU; WAG).

Also in Zaire (Dem. Rep. Congo), Burundi and Tanzania. Upland grassland, *Protea*-grassland and *Brachystegia* woodland; (450)540–1900 m.

This has always been identified as *F. cienkowskii*, and is closely related to that species, but the absence of an understorey of papillae beneath the indumentum of the leaf lower surface renders it easily recognisable. It is closer to *F. tetraquetra* having a similar stomatal pattern. A few intermediate specimens have been seen from northern Zambia, e.g. *Richards* 10730 (Shiwa Ngandu, Lake Young, fl. 17.i.1959) and *Robertson* 209 (between Mbala and the Kalambo Falls, fl. 2.xi.1952).

In Malawi some specimens with and without papillae are so similar that it is not easy to believe them distinct - the character needs field investigation.

Var. *ufipaensis* Verdc., from Zaire (Dem. Rep. Congo) and Tanzania, has shorter stems, up to 25 cm tall, and the leaf lower surface is not obscured by the indumentum.

4. **Fadogia tetraquetra** K. Krause in Bot. Jahrb. Syst. **39**: 544 (1907). —Robyns in Bull. Jard. Bot. État **11**: 68 (1928). —J.G. Garcia in Mem. Junta Invest. Ultramar **6** (sér. 2): 27 (1959) [Contrib. Conhec. Fl. Moçamb. IV (1959)]. —Verdcourt in Kew Bull. **36**: 519 (1981); in F.T.E.A., Rubiaceae: 792 (1991). Type: Zimbabwe, Mutare (Umtali), *Engler* (1905) 3139 (B†, holotype).

Suffrutex with up to 7 basically unbranched stems, 20–90(120) cm tall, from the crowns of a fairly stout rhizome; stems brown glabrous, or very sparsely pubescent to densely hairy, mostly 3–4-angled. Leaves in whorls of 3–5(6), drying distinctly discolorous; blades 4–11 × 1.5–4.8 cm, broadly elliptic to narrowly oblong-elliptic, acuminate to a fine point at the apex, cuneate to rounded at the base, glabrous to pubescent, paler and often glaucous beneath, the actual surface (when viewed under high magnification) with a distinct appearance due to denseness of the stomata, the main nerves often rather crinkly and undulate (perhaps due to slight fleshiness of leaf in life), the leaf upper surface slightly bullate above in some specimens; petioles 1–2.5 mm long; stipules connate at the base, the base 3–5 mm long, triangular, pubescent within, prolonged into a subulate appendage (2)4–10 mm long. Inflorescences 3–9-flowered; peduncles 2–5(16) mm long; pedicels 1.5–5 mm long; bracts and bracteoles 3–4 mm long, subulate. Calyx tube 1.3–1.5 mm long, often wrinkled when dry; limb with tubular part 0.5–1 mm long, bearing narrowly triangular teeth 1.5–3.5 mm long. Corolla apiculate in bud, greenish-yellow or honey-coloured, glabrous to ± pubescent outside; tube 2.5–4 mm long, densely hairy at the throat and with a row of deflexed hairs inside; lobes oblong-triangular, apiculate at the apex, 3.5–5(7) mm long. Style exserted for up to 1 mm; pollen presenter coroniform, up to 1.5 mm long, ribbed and 3–4-lobed. Ovary 3–4-locular. Fruit 6–7 mm long, subglobose or oblique, usually with 3 pyrenes but sometimes reduced to 1 by abortion.

Var. **tetraquetra** —Verdcourt in F.T.E.A., Rubiaceae: 792 (1991).
    Fadogia mucronulata Robyns in Bull. Jard. Bot. État **11**: 67 (1928). Type from South Africa (Mpumalanga).

Stems glabrous or sparsely hairy; leaves glabrous. Corolla glabrous in bud.

Zambia. N: Mbala Distr., Kawimbe, near leper settlement, fl. 12.x.1956, *Richards* 6431 (K). **Zimbabwe**. C: Makoni Distr., Rusape Road, fr. 13.ii.1961, *Rutherford-Smith* 530 (K; SRGH) (atypical). E: Mutare Distr., Shila Estate, near Stapleford Forest Reserve, fl. 25.xi.1955, *Chase* 5874 (BM; K; LISC; SRGH). **Mozambique**. MS: near Catandica (Vila Gouveia), fl. 3.xi.1941, *Torre* 3750 (C; COI; LD; LISC; LMU; MO; SRGH; WAG).
    Also in Zaire (Dem. Rep. Congo), western Tanzania, Swaziland and South Africa (North Prov. and Mpumalanga). Open grassland, burnt hillslopes, old cultivations and *Brachystegia* woodland, on sandy soil; 1200–1740 m.
    Fanshawe 5544 (Zambia S: Tara, fr. 19.iii.1960) may belong here, but has fruits about 1.5 cm diameter and deeply strongly pitted pyrenes.

Var. **grandiflora** (Robyns) Verdc. in Kew Bull. **36**: 520 (1981); in F.T.E.A., Rubiaceae: 792 (1991). Type from Tanzania.
    Fadogia grandiflora Robyns in Bull. Jard. Bot. État **11**: 65 (1928).
    Fadogia variabilis Robyns in Bull. Jard. Bot. État **11**: 63 (1928). — J.G. Garcia in Mem. Junta Invest. Ultramar **6** (sér 2): 27 (1959) [Contrib. Conhec. Fl. Moçamb. IV (1959)]. Type: Zimbabwe, Gazaland, Chirinda, *Swynnerton* 6502 (BM; K, holotype).
    Fadogia dalzielii Robyns in Bull. Jard. Bot. État **11**: 66 (1928). Type from Nigeria.
    Fadogia glauca Robyns in Bull. Jard. Bot. État **11**: 62 (1928). Type from Angola.
    Fadogia pobeguinii sensu Hutch. & Dalz. in F.W.T.A. **2**: 110 (1931). —Hepper in F.W.T.A. ed. 2, **2**: 178 (1963) non Pobéguin.

Stems covered with dense crisped hairs. Leaf upper surface with scattered short seta-like hairs; midrib beneath with dense hairs similar to those on the stems, or with pubescence ± all over the venation beneath. Corolla glabrous or hairy in bud.

Zambia. N: Lake Bangweulu, near Samfya, fl. 26.viii.1952, *Angus* 316 (FHO; K). C: Kasanka National Park, Luwombwa River, fr. 18.xi.1993, *Harder et al.* 1934 (K; MO). **Zimbabwe**. E: Gazaland, near Chirinda, fl. x.1905, *Swynnerton* 241 (K). **Mozambique**. MS: Espungabera (Spungabera), fl. 13.xi.1943, *Torre* 6188 (LISC).
    Also in Guinea, Ghana, Nigeria, Cameroon, Tanzania, Angola, Swaziland and South Africa (Mpumalanga). Open grassland, burnt "long grass – Chipya" vegetation, forest edges; 1140–1800 m.
    This variety links *F. tetraquetra* with *F. elskensii*.

5. **Fadogia variifolia** Robyns in J. Bot. **69**: 167 (1931). Type from Angola.

Suffrutex 10–20 cm tall, with several stems from a woody rootstock; stems yellow, drying yellowish or reddish-brown, sometimes ± 3-angular, ridged, mostly covered with a short bristly pubescence, later glabrescent. Leaves in whorls of 3; blades 1.8–8.3 × 0.8–2.7 cm, elliptic, ± obovate or obovate-elliptic, rounded to acute or shortly acuminate at the apex, rounded to cuneate at the base, often ± undulate on the margins, ± subcoriaceous, drying brownish, densely pale spotted on both surfaces and somewhat glaucous beneath, the venation visible as a darker reticulation, glabrous; petiole obsolete; stipules very shortly connate, c. 1 mm long, attenuate into a subulate appendage 3–7 mm long, mostly glabrous. Flowers solitary, or in 2-flowered cymes; peduncles of solitary flowers up to 6 mm long; joint peduncles 2.5–3 mm long; pedicels 1–2 mm long; bracts 1–2 mm long, filiform. Calyx tube 1–1.5 mm long, ovoid; limb-tube 0.5 mm long, the teeth 1–2 mm long, triangular to subulate, ± unequal, sometimes with additional ones between. Corolla apiculate in bud, yellow, glabrous; tube 2–2.5 mm long, cylindrical; lobes 2.5–3 × 1.5–2 mm, ± triangular with a short appendage. Style 3–3.5 mm long; pollen presenter coroniform, 1 mm long, sulcate, ± 3-lobulate at the apex. Ovary ± 3-locular. Fruit c. 10 mm in diameter (dry), ± globose, crowned by the calyx limb, glabrous, the stalk up to c. 10 mm long.

Zambia. W: Mwinilunga Distr., Cha Mwana (Chibara's) Plain, fl. 17.x.1937, *Milne-Redhead* 2765 (K).

Also in Angola. Open sandy ground on Plain and woodland edges; c. 1200 m.

6. **Fadogia tomentosa** De Wild., Études Fl. Bas-Moyen-Congo 2: 78 (1907); 348 (1908); 3: 296 (1910); 849 (1912). —T. Durand, Syll. Fl. Congol.: 271 (1909). Type from Zaire (Dem. Rep. Congo) (Kinshasa Prov.).

Suffrutex 20–60 cm tall with several stems from a woody rootstock, or less often a small tree 1.5 m tall; stems unbranched or branched, velvety-pubescent. Leaves in whorls of 3–4, very discolorous; blades 2–12 × 1.2–6 cm, elliptic or oblong, rounded to acute at the apex, cuneate to rounded at the base, finely densely pubescent to velvety above not obscuring the upper surface, softly grey-velvety beneath, the indumentum totally obscuring the lower surface, or in one variety the velvety indumentum soon disappearing, the leaves finally sparsely to densely pubescent or with hairs only on the venation beneath; venation raised beneath; petioles 0–6 mm long; stipule bases tomentose, 1–1.5 mm long, triangular, with a subulate appendage 3–10 mm long. Cymes 3–7-flowered, congested, the common peduncle suppressed or up to c. 1.5–4 mm long; pedicels 2–4 mm long; bracts 1–2 mm long, oblong. Calyx pubescent; tube 1–2 mm long, subglobose; limb scarcely 1 mm long, the teeth narrowly triangular, c. 1 mm long. Corolla acuminate, tailed in bud; tube cream-greenish or yellow, 4–7 mm long, cylindrical or narrowly funnel shaped, ± densely spreading bristly-pubescent outside, with a ring of deflexed hairs inside; lobes yellow or bright orange, 3.5–4.5 × 1.5 mm, lanceolate, shortly appendaged, the appendage scarcely 1 mm long. Anthers dark red. Ovary 3–5-locular. Style 5.5–8 mm long; pollen presenter 0.75–1.25 mm long, coroniform, sulcate, 3–5-lobulate. Fruit black, 5–9 × 8–12 mm, subglobose, distinctly lobed when dry, usually with (1)3–4 pyrenes 5–8 mm long.

1. Stems branched, woody (at least brown); leaves only reaching 6.5 × 2.4 cm · iii) var. *flaviflora*
   – Stems unbranched, ± green; leaves may reach 12 × 6 cm · 2
2. Indumentum of short hairs, dense and persistent on leaf blades beneath · · i) var. *tomentosa*
   – Indumentum of longer hairs, at first dense, but later glabrescent (save for the nerves) on leaf blades beneath · ii) var. *calvescens*

i) Var. **tomentosa**

Fadogia velutina De Wild. in Repert. Spec. Nov. Regni Veg. 13: 90 (1914); Notes Fl. Katanga 4: 90 (1914); Contrib. Fl. Katanga: 214 (1921). Type from Zaire (Dem. Rep. Congo) (Shaba Prov.).

Stems unbranched. Leaves often towards the upper limits of measurements given for the species, with the dense short velvety indumentum persisting, especially on the lower surface, save on the lowest leaves which are smaller and always narrowly elliptic.

Zambia. W: Mwinilunga Distr., 27 km west of Mwinilunga on the Matonchi road, west of Musangila R., fl. 22.i.1975, *Brummitt et al.* 13973 (K).

Also in Zaire (Dem. Rep. Congo) and Angola. *Brachystegia* woodland and *Cryptosepalum–Copaifera* thicket on Kalahari Sand; 1350 m.

Lewis 6158 from Mwinilunga (4–6 km SE of Angola-Zambia Border) is an odd form with narrower leaves, but they are perhaps not fully mature.

ii) Var. **calvescens** (Verdc.) Verdc., comb. nov.

Fadogia flaviflora var. *calvescens* Verdc. in Kew Bull. 36: 528 (1981). Type: Zambia, Machili, *Fanshawe* 6110 (BR; K, holotype).

Stems unbranched, or occasionally branched. Leaves at first densely grey-velvety tomentose but soon only ± densely pubescent or glabrescent beneath, save for hairs on the main veins; the indumentum usually distinctly longer than in var. *tomentosa*.

Caprivi Strip. Mashi, fl. 5.xi.1962, *Fanshawe* 7141 (K; NDO). **Zambia**. B: Barotseland, near confluence of Lungwebungu (Lungibungu) and Litapi Rivers, fl. 1.xii.1961, *Holmes* 1383 (K; NDO). W: Kasempa Distr., 48 km from Mufumbwe (Chizera) on road to Solwezi, fl. 26.i.1975, *Brummitt et al.* 14145 (K; NDO). S: Machili, fl. 10.i.1961, *Fanshawe* 6110 (K; NDO).

As yet known only from the Flora Zambesiaca area, but would be expected in adjacent Angola and Namibia. *Brachystegia* woodland and *Burkea–Cryptosepalum* woodland on Kalahari Sand, also in scrub transition and dambos on Kalahari Sand; 1060–1230 m.

The Kew specimen of *Bingham* 9903/1, from Mongu District, Looma, Nanganda in Zambia, has one unbranched and one branched portion.

The Caprivi Strip specimen is somewhat intermediate with var. *tomentosa*.

iii) Var. **flaviflora** (Robyns) Verdc., comb. et stat. nov.

   Fadogia flaviflora Robyns in J. Bot. **69**: 169 (1931). —Verdcourt in Kew Bull. **36**: 527 (1981). Type from Angola.

Stems branched, the main one ± woody, brown. Leaves mostly towards the lower limits of measurements given for the species, up to 6.5 × 2.4 cm, with the dense short velvety indumentum persisting, especially on the lower surface.

Zambia. B: Zambezi (Balovale), fr. 25.ii.1964, *Fanshawe* 8342 (K; NDO).
Also in Angola. *Brachystegia* woodland on Kalahari Sand; 930 m.
Luwiika et al. 247 [Kasama–Lukulu road, 57.2 km MW of Lusaka–Mongu road junction (K; MO)] is said to be a tree 1.5 m tall, but is otherwise a close match for this taxon.

7. **Fadogia arenicola** K. Schum. & K. Krause in Bot. Jahrb. Syst. **39**: 544 (1907). —Robyns in Bull. Jard. Bot. État **11**: 84 (1928). —Verdcourt in Kew Bull. **36**: 520 (1981); in F.T.E.A., Rubiaceae: 792 (1991). Type from Tanzania.

Suffrutex 10–25 cm tall, with several rather slender unbranched stems from a creeping slender woody rhizome; stems pale to dark reddish-brown, densely pubescent with short white spreading hairs. Leaves in whorls of 3, markedly discolorous; blades 1.7–6 × 0.8–2.5 cm, elliptic to oblong-obovate, obtuse to bluntly acute or even slightly retuse at the apex, cuneate at the base, the upper surface with a dense grey pubescence which does not hide the leaf surface, sometimes only the venation pubescent, the lower surface usually densely grey velvety; lateral nerves 4–6; petioles 1.5–3.5 mm long; stipules c. 1.5–2 mm long. Inflorescences 1–4-flowered; peduncle short, up to 2 mm long; pedicels 2–6 mm long, densely grey pubescent, mostly reflexed in fruit. Calyx tube and limb together about 1.5 mm long, pubescent; teeth 0.5–1 mm long, triangular. Corolla obtuse in bud; tube greenish-yellow or yellow, about 3.5–6 mm long, hairy outside near apex with a ring of deflexed hairs inside; lobes bright yellow, spreading, 4–5 × 1.3–2 mm, narrowly oblong, acute, densely hairy outside. Ovary 3-locular. Style 5–8 mm long; pollen presenter green, coroniform, 2–3-lobed, the lobes 4–6-ribbed, the ribs with transverse grooves. Fruits 5–7 mm in diameter, globose, pubescent, with only 1 pyrene developed (in material seen).

Zambia. N: Mbala Distr., Chinakila, path to Kaniya Flats, fl. 15.i.1965, *Richards* 19549 (K). W: Kitwe, fr. 26.ii.1956, *Fanshawe* 2804 (K; NDO). C: Serenje Distr., 70 km NE of Serenje, Nsalu Cave Nat. Mon., fl. 5.ii.1973, *Kornaś* 3218 (K; KRA). **Malawi**. N: Rumphi Distr., outskirts of Katumbi's Village, fr. iv.1952, *White* 2545 (K; FHO).

Also in Zaire (Dem. Rep. Congo) and Tanzania. *Monotes* and *Brachystegia–Julbernardia* woodlands, mostly on yellow laterite; 1200–1500 m.

In Zaire (Dem. Rep. Congo) (Shaba Prov.) specimens occur in which the leaf blades are glabrous or glabrescent beneath, but which seem to be only a variety of this species. *F. schumanniana* Robyns [Type: Zaire (Dem. Rep. Congo), Shaba Prov., *Verdick* 350 (BR, holotype) (=*Vangueria verdickii* K. Schum. in De Wildeman, 1903, non *Fadogia verdickii* De Wild. & T. Durand)] is probably the same species differing only in the more pointed buds resulting from the longer corolla lobe appendages).

8. **Fadogia luangwae** Verdc., sp. nov.* Type: Zambia, North Luangwa National Park, Muchinga Mts., 11°44'S, 32°05'E, *P.P. Smith* 0220 (K, holotype).

Suffrutescent herb 10–16 cm tall, with 3–4 slender hairy stems from a woody rootstock. Leaves in whorls of 3, concolorous, blades 1.2–3 × 0.5–0.9 cm, narrowly obovate (basal whorl) to oblanceolate or elliptic-oblanceolate, rounded to subacute at the apex, gradually narrowed to the base, appressed velvety grey pubescent on both sides so that surface is not visible save in lowest whorl where pubescence is sparser; lateral nerves in 3–4 pairs, obscure; petiole 2–3 mm long; stipules 1.5 mm long, acuminate. Cymes 1–3-flowered, sessile; pedicels 3–4 mm long, grey pubescent. Calyx densely pubescent; tube 2.5 mm long, very shortly toothed. Corolla yellow; tube 6 mm long, 2.5 mm wide, narrowed at both ends, glabrous save for pubescence near the apex; lobes 5 × 1.8 mm, narrowly lanceolate-triangular, spreading pubescent outside. Anthers exserted 1.5 mm; style exserted 1 mm; pollen presenter 1.5 mm long, obconic. Fruit not known.

Zambia. N: North Luangwa National Park, Muchinga Mts., 11°44'S, 32°05'E, fl. 5.ii.1994, *P.P. Smith* 0220 (K).
Known only from the above specimen. Hill miombo woodland; 800 m.

9. **Fadogia stenophylla** Welw. ex Hiern in F.T.A. **3**: 155 (1877); Cat. Afr. Pl. Welw. **1**: 483 (1898). —K. Schumann in Warburg, Kunene-Samb.-Exped. Baum: 390 (1903). —Robyns in Bull. Jard. Bot. État **11**: 60 (1928). —Verdcourt in Kew Bull. **36**: 521 (1981); in F.T.E.A., Rubiaceae: 793 (1991). Type from Angola.

A pyrophytic, usually glabrous, suffrutescent herb, 6–40 cm tall, with a single or usually several rather short unbranched stems from a slender creeping woody rhizome; stems glabrous, or with rather sparse bristly hairs in a very few specimens; lowest nodes usually leafless. Leaves in whorls of 3–4; blades 1–6.5 × 0.4–2.8 cm, rounded elliptic to narrowly elliptic, narrowly oblong or oblanceolate, rounded to bluntly acute at the apex, cuneate to almost rounded at the base, glabrous or rarely pubescent, usually drying pale olive-green; petioles 1–2 mm long; stipules connate at the base, oblong-triangular, 3–4 mm long including a shortly to distinctly produced apex. Inflorescences (1)3-flowered, sweet-scented, borne at the lower nodes, sometimes flowering before the leaves are produced; peduncles 5–20(30) mm long and pedicels 1.5–6.5(10) mm long, both glabrous or sparsely pubescent, sometimes flowers solitary on stalks up to 16 mm long. Calyx tube 2–2.5 mm long, the narrow limb almost membranous, truncate, erose or slightly to distinctly narrowly toothed; teeth 0.7–1.2 mm long. Corolla obtuse in bud, white or pale to bright yellow or greenish-yellow, glabrous; tube 3–4 mm long, cylindrical-campanulate, with ring of deflexed hairs inside; lobes 4–6(7), 3–5 × 2.5–2.8 mm, triangular-oblong or ovate, cucullate but scarcely apiculate at the apex, glabrous save for papillate margins and inner surface. Ovary 3–4-locular. Style 5.5–6 mm long; pollen presenter green, coroniform, 3–4-lobed. Fruit red at maturity, 7–9 × 9–10 mm, globose, or often drying very distinctly 2–3-lobed.

Subsp. **odorata** (Krause) Verdc. in Kew Bull. **36**: 521 (1981); in F.T.E.A., Rubiaceae: 794, figs. 132/9, 142 (1991). TABS. **47/B5 & 54**. Type from Angola.
   Fadogia stenophylla var. *rhodesiana* S. Moore in J. Bot. **40**: 253 (1902). Type: Zimbabwe, Harare, *Rand* 629 (BM, holotype).
   Fadogia lateritica Krause in Bot. Jahrb. Syst. **43**: 143 (1909). —Robyns in Bull. Jard. Bot. État **11**: 51 (1928). Type: Zimbabwe, Harare (Salisbury), *Engler* (1905) 3037 (B†, holotype).
   Fadogia odorata Krause in Bot. Jahrb. Syst. **43**: 144 (1909). —Robyns in Bull. Jard. Bot. État **11**: 52 (1928).
   Fadogia stolzii Krause in Bot. Jahrb. Syst. **48**: 415 (1912). —Robyns in Bull. Jard. Bot. État **11**: 51 (1928). Type from southern Tanzania.

---

* Affinis *Fadogia arenicolae* K. Schum. et K. Krause habitu minore, foliis minoribus angustioribus haud valde discoloribus utrinque appresse griseovelutinis differt.

Tab. 54. FADOGIA STENOPHYLLA subsp. ODORATA. 1, habit (× ²/₃); 2, stipule (× 4), 1 & 2 from *Stolz* 108; 3, inflorescence (× 3); 4, corolla opened out, with style and pollen presenter (× 6); 5, pollen presenter (× 8); 6, calyx (× 6); 7, longitudinal section through ovary (× 8); 8, transverse section through ovary (× 8), 3–8 from *Watermeyer* 138; 9, fruiting node (× 1); 10, pyrene (× 4), 9 & 10 from *Paulo* 235. Drawn by M.E. Church. From F.T.E.A.

Leaf blades rounded elliptic, narrowly elliptic, narrowly oblong or oblanceolate, up to 2.7 cm wide.

Zambia. B: Mongu, fl. 29.ix.1962, *Fanshawe* 7057 (K; NDO). N: Mporokoso, fl. 11.ix.1958, *Fanshawe* 4799 (K; NDO). W: Ndola, opposite Mwekera Rest House near Nkana, fl. 27.ix.1947, *Brenan & Greenway* 7969 (EA; FHO; K). E: Nyika, corolla fallen 31.xii.1962, *Fanshawe* 7367 (K; NDO). **Zimbabwe**. C: Harare, fl. 22.x.1924, *Eyles* 4417 (K; SRGH). E: Nyanga (Inyanga), fl. 27.x.1930, fl. *Fries, Norlindh & Weimarck* 2349 (K; LD; SRGH). **Malawi**. N: Mzimba Distr., 6.4 km SW of Chikangawa, fl. 6.ix.1978, *E. Phillips* 3875 (K; MO). C: Ntcheu–Dedza road, fl. 16.x.1949, *Wiehe* N/258 (K). S: Mt. Malosa, fl. xii.1896, *Whyte* (K). **Mozambique**. T: Angónia, near Ulónguè, fr. 21.xi.1980, *Macuácua* 1284 (LMA).

Also in Zaire (Dem. Rep. Congo), Burundi and Angola. Mainly open grassland and dambos, and grassland with *Protea, Annona, Lannea* etc., also on *Brachystegia* woodland edges, in areas subjected to annual burning, usually on laterite or Kalahari Sand; 1250–2150 m.

Mostly showing little variation but pubescent variants occur in N Malawi; *E. Phillips* 4149 from Nkhata Bay Distr. is not only pubescent but has the leaves more acute with venation reticulate beneath and flowers in axils above the middle, and is probably of hybrid origin.

Subsp. *stenophylla* occurs in Angola, and may be distinguished by its narrowly lanceolate to elliptic-lanceolate leaf blades.

10. **Fadogia triphylla** Baker in Bull. Misc. Inform., Kew **1895**: 68 (1895). —Robyns in Bull. Jard. Bot. État **11**: 57 (1928). —Verdcourt in Kew Bull. **36**: 522 (1981); in F.T.E.A., Rubiaceae: 796 (1991). Type: Zambia, Fwambo, *Carson* 43 (K, holotype).

Very variable herb or suffrutex (10)22–45(120) cm tall, from a thick or fairly slender usually horizontal woody rootstock; stems several, usually erect, branched or unbranched, yellowish, ± quadrate, grooved, glabrous to densely shortly pubescent. Leaves in whorls of 3–4, or opposite above; blades (1.7)3–8.5 × (0.6)1.2–4.8(6.7) cm, broadly to narrowly elliptic or ovate to ± obovate, acute to obtuse at the apex or very shortly acuminate, cuneate at the base, not markedly discolorous, but usually paler beneath and often drying with a reticulate darker pattern, glabrous to shortly densely pubescent; lowermost leaves often very reduced, 10 × 6 mm; petioles obsolete or very short; stipules about 2.5–3.5(4.5) mm long, the subulate cusp (1.5)2–3 mm long from a broad base. Inflorescences (1)2- but mostly 3–7-flowered; peduncles (2)5–18(32) mm long; pedicels 2–14 mm long, glabrous or shortly pubescent. Flowers fragrant. Calyx oblong-ovoid or obconic, 2.5–4 mm long including the narrow truncate or almost truncate limb; teeth if present minute, glabrous or shortly pubescent. Corolla apiculate and usually tailed in bud, white to greenish-yellow or yellow outside, creamy-green inside, glabrous; tube 5–6 mm long with ring of deflexed hairs inside; lobes 4.5–7 × 1.5–2.5(3.5) mm, oblong-triangular, appendaged, papillate inside. Anthers half-exserted with a papillate apical appendage, purple, brown or orange. Style green or white, thickened below, 7–8 mm long; pollen presenter green, 1–1.5 mm long, coroniform or truncate-obconic. Fruits dark bluish-green turning black when dry, about 10 mm in diameter, usually with 3 pyrenes.

1. Plant glabrous save for inside of corolla and stipule sheath · 2
− Stems and/or leaves pubescent · 3
2. Leaves broadly to narrowly elliptic or ovate to ± obovate, up to 8.5 × 5.7 cm, usually over 2 cm wide · i) var. *triphylla*
− Leaves narrowly elliptic, mostly smaller and narrower, 1.7–5 × 0.8–1.8 cm · · · · · · · · · · ·
· iii) var. *gracilifolia*
3. Stems, leaves and inflorescences shortly pubescent · · · · · · · · · · · · · · · · · ii) var. *giorgii*
− Stems and sometimes inflorescences sparsely to densely pubescent; leaves glabrous · · · · ·
· iv) var. *pubicaulis*

i) Var. **triphylla** —Verdcourt in Kew Bull. **36**: 522 (1981); in F.T.E.A., Rubiaceae: 796 (1991).
Fadogia kaessneri S. Moore in J. Bot. **49**: 152 (1911). —Robyns in Bull. Jard. Bot. État **11**: 57 (1928). Type: Zambia, Mulungushi (Malangashi) R., *Kassner* 2064 (BM, holotype).
Fadogia hockii De Wild. in Repert. Spec. Nov. Regni Veg. **11**: 535 (1913); Études Fl. Katanga [Ann. Mus. Congo, Sér. IV, Bot.] **2**: 152 (1913); Contrib. Fl. Katanga: 213 (1921); Pl. Bequaert. **3**: 204 (1925). Type from Zaire (Dem. Rep. Congo).

Fadogia hockii var. *rotundifolia* De Wild. in Repert. Spec. Nov. Regni Veg. **11**: 535 (1913). Type from Zaire (Dem. Rep. Congo).

Fadogia coriacea Robyns in Bull. Jard. Bot. État **11**: 49 (1928). Type from Zaire (Dem. Rep. Congo).

Fadogia buarica Robyns in Bull. Jard. Bot. État **11**: 48 (1928). Type from Cameroon.

Fadogia kaessneri var. *rotundifolia* (De Wild.) Robyns in Bull. Jard. Bot. État **11**: 58 (1928).

Plant glabrous save for inside of corolla tube and stipule sheath. Leaves often larger, up to 8.5 × 5.7 cm. Stems mostly with more than 4 leafy nodes. Inflorescences usually composed of pedunculate, several-flowered cymes.

Zambia. N: Old Isanya road to Mbala, fl. 18.i.1955, *Richards* 4128 (K). W: Mwinilunga Distr., about 1 km SW of Dobeka Bridge, fl. 14.xii.1937, *Milne-Redhead* (K). C: 100–109 km east of Lusaka, Chakwenga Headwaters, fl. 16.xi.1963, *E.A.Robinson* 5851 (K). S: Mumbwa, fl. no date, *Macaulay* 893 (K). **Malawi**. N: Nyika Plateau, Makanga, fl. 2.ii.1976, *E. Phillips* 857 (K; MO). S: Machinga Distr., Munde Hill, fr. 30.iv.1982, *Baloha* 84 (BR). **Mozambique**. N: near Mandimba, fl. 18.xii.1941, *Hornby* 3523 (K).

Also in Cameroon, Zaire (Dem. Rep. Congo), Tanzania and Angola. *Brachystegia, Uapaca, Protea, Marquesia* and *Cryptosepalum* woodland often of an open type, bushland, grassland with scattered trees, road edges, laterite pans; 540–1900 m.

A wide range of habit and leaf variation are included under this name; more study of the habit is needed in the field. Certain specimens are undoubtedly branched shrubs over 1 m tall and seem very different from the pyrophytes with several stems from a woody rootstock. Some may be due to protection from burning but specimens from Zambia from *Brachystegia* woodland are similar but scarcely protected from burning. *Greenway & Brenan* 8091 (Zambia, Mumbwa Distr., Nambala to Chenobi, fl. 15.ix.1947), has the inflorescences longer than the short ovate leaves (2.3 × 1.5 cm) which are crinkly and perhaps abnormally developed. It is unlikely to be anything but a form of *F. triphylla*. A specimen mentioned by Verdcourt & Trump, Common Pois. Pl. E. Africa: 127 (1969) as *Temnocalyx obovatus* (N.E. Br.) Robyns is in fact a shrubby form of *F. triphylla*; it has been used as an antidote to *Acokanthera* arrow poison.

ii) Var. **giorgii** (De Wild.) Verdc. in Kew Bull. **36**: 523 (1981); in F.T.E.A., Rubiaceae: 797 (1991). Type from Zaire (Dem. Rep. Congo) (Shaba Prov.).

Vangueria brachytricha K. Schum. in De Wildeman, Études Fl. Katanga [Ann. Mus. Congo, Sér. IV, Bot.] **1**: 227 (1903). —T. Durand, Syll. Fl. Congol.: 269 (1909). —De Wildeman, Contrib. Fl. Katanga: 211 (1921). Type from Zaire (Dem. Rep. Congo) (Shaba Prov.).

Fadogia brachytricha (K. Schum.) Robyns in Bull. Jard. Bot. État **11**: 54 (1928).

Fadogia viridescens De Wild. in Repert. Spec. Nov. Regni Veg. **13**: 138 (1914); Notes Fl. Katanga **4**: 91 (1914); Contrib. Fl. Katanga: 214 (1921). Type from Zaire (Dem. Rep. Congo) (Shaba Prov.).

Fadogia giorgii De Wild., Pl. Bequaert. **3**: 203 (1925). —Robyns in Bull. Jard. Bot. État **11**: 56 (1928).

Plant with stems, leaves, inflorescences, etc. shortly pubescent.

Zambia. N: S shore of Lake Tanganyika, Mpulungu, fl. 12.i.1960, *van Zinderen Bakker* 891 (BLFU; K; PRE). W: Ndola, fl. 12.iv.1957, *Fanshawe* 3173 (K; NDO). C: Serenje Distr., Kundalila Falls, fl. 4.ii.1973, *Kornaś* 3179 (K; KRA). S: Mazabuka Distr., Kafue Gorge, camping site, fl. 6.i.1973, *Kornaś* 2931 (K; KRA). **Malawi**. S: Zomba Rock, fr. 1892, *Whyte* (K).

Also in Zaire (Dem. Rep. Congo) and Tanzania. Bushland and open *Brachystegia* woodland, rocky hills; 750–1500 m.

Richards 201 (Lake Tanganyika, Mbulu Island, fl. 21.xii.1951) is probably only a form of this with small truly ovate leaves about 3 × 2 cm, very different from the narrowly elliptic, oblong or oblanceolate leaves of most specimens of the variety.

iii) Var. **gracilifolia** Verdc. in Kew Bull. **36**: 523 (1981); in F.T.E.A., Rubiaceae: 797 (1991). Type from Tanzania.

Plant glabrous save for inside of corolla tube and stipule sheath. Leaves mostly much smaller and narrower, 1.7–5 × 0.8–1.8 cm. Stems mostly with more than 4 leafy nodes, up to 1 m tall. Inflorescences usually composed of pedunculate, several-flowered cymes.

Zambia. N: 16 km from Katibunga to Great North Road, fl. 30.iii.1961, *Angus* 2595 (FHO; K).

Also in SW Tanzania. *Brachystegia–Julbernardia–Monotes–Uapaca* etc. woodland, also old cultivations; 1500–1830 m.

St. Clair-Thompson (1124) refers to 'stem on cutting yields white milky latex' but this seems unlikely; possibly a confusion of labels. No other collector mentions this for any *Fadogia* sp. Although extremes are distinctive there is no clear dividing line between this and var. *triphylla*.

iv) Var. **pubicaulis** Verdc. in Kew Bull. **36**: 524 (1981); in F.T.E.A., Rubiaceae: 797 (1991). Type from Tanzania.

Stems and sometimes inflorescence parts sparsely to densely pubescent. Leaves glabrous, mostly narrow as in var. *gracilifolia*.

Zambia. N: Mpika, fl. 26.i.1955, *Fanshawe* 1859 (BR; K; NDO). W: Mwinilunga, source of R. Zambezi, fl. 13.xii.1963, *E.A. Robinson* 5984 (K).
Also in SW Tanzania. *Brachystegia–Marquesia–Monotes–Julbernardia* woodland; 1350 m.
Numerous intermediates link this with var. *giorgii*, e.g. *Verdcourt* 3388 from Tanzania, Kigoma in which some leaves are slightly ciliate.

11. **Fadogia chlorantha** K. Schum. in Warburg, Kunene-Samb.-Exped. Baum: 388 (1903). — Robyns in Bull. Jard. Bot. État **11**: 65 (1928). Type from Angola.

Dwarf subshrub or suffrutex from a creeping woody rhizome; stems several short 5–15 cm* tall, erect thinly woody and glabrous to sparsely or densely appressed setulose with white or yellowish** hairs. Leaves in whorls of 3, congested, the internodes mostly very short; blades 2–6.5 × 0.5–1.4 cm (leaves in lowest whorl very small), oblanceolate or narrowly oblong to ± elliptic, subacute to acute at the apex or less often quite rounded, cuneate at the base, glabrous or sparsely to densely covered with appressed bristly yellow hairs, the costa sometimes very pale; petiole obsolete or 1–3 mm long; stipules about 1–1.5 mm long, ± triangular, keeled, ± obtuse or the appendage scarcely developed. Flowers solitary or in 2-flowered cymes; common peduncle suppressed, pedicels or peduncles of solitary flowers 1–6 mm long; bracts 2 mm long, lanceolate. Calyx glabrous to densely covered with yellowish appressed bristly hairs; tube campanulate 1.5 mm long; lobes 2–3.5 mm long, lanceolate. Corolla 5-tailed in bud, dark green, glabrous to densely appressed bristly-pubescent save at base of tube; tube 6–7 mm long, hairy at the throat and with a ring of deflexed bristly hairs inside; lobes greenish-white inside 8 × 1.5 mm including the 2–3 mm long subulate appendage. Style 7–9 mm long; the pollen presenter 1.2 mm long, cylindrical, 5-lobed at apex. Ovary 5-locular. Fruit subglobose, c. 12 mm in diameter, mostly with 3 pyrenes and 3-lobed when dry, crowned by calyx limb.

Var. **chlorantha**

Stems, leaves and flowers densely covered with ± yellowish bristly hairs.

Zambia. W: Mwinilunga aerodrome, fl. 30.xi.1937, *Milne-Redhead* 3448 (K).
Also in Angola. Sandy plains at edges of *Cryptosepalum* woodland; c. 1200 m.

Var. **thamnus** (K. Schum.) Verdc., comb. nov. et stat. nov. Type from Angola.
    Fadogia thamnus K. Schum. in Warburg, Kunene-Samb.-Exped. Baum: 390 (1903). — Robyns in Bull. Jard. Bot. État **11**: 85 (1928).

Stems, leaves and flowers glabrous or with very sparse bristly hairs.

Zambia. B: Senanga Distr., Sioma, 66 km S of Senanga, fr. 1.ii.1975, *Brummitt et al.* 14203 (K).
S: Machili, corolla fallen 13.iii.1961, *Fanshawe* 6411 (K; NDO).
Also in Angola and Namibia. Grassy plains and *Baikiaea* woodland on Kalahari Sand; 1050 m.

---

* K. Schumann's and Robyns' reference to "1 m altus" must presumably include a length of rhizome or be based on faulty collector's data.
** Milne-Redhead specifically mentions white hairs but these have dried yellow in the material accompanying the note.

12. **Fadogia gossweileri** Robyns in J. Bot. **69**: 171 (1931). —Verdcourt in Kew Bull. **36**: 527 (1981). Type from Angola.

Small shrub or suffrutex with several weak stems from a woody rootstock; stems (0.35)0.6–1.2 m tall, slender but eventually distinctly woody, dull purplish-brown or greyish with peeling epidermis, glabrous save for insides of stipule-bases, mostly with main stem leaves falling early, and with young slender branchlets up to about 15 cm long in whorls of 2 or 3 at each node from the leaf axils. Leaves in whorls of 2 or 3; blades 2–12 × 0.25–1.2 cm, linear-oblong to linear-elliptic or -oblanceolate, subacute or minutely rounded but apiculate at the apex, cuneate at the base, often somewhat curved, glabrous, glaucous-green drying pale yellow-green; petiole 1–3 mm long; stipules shortly sheathing, c. 1–2 mm long with lobes c. 1 mm long, keeled, becoming ± corky and persistent, with yellowish hairs inside at base. Inflorescences compact, at nodes all up the stem, mostly in axils of the branchlets, each about 10-flowered but forming c. 3 cm wide clusters with the other pair at the same node; peduncles (1)4–5(10) mm long, minutely asperulous; secondary axes and pedicels (1)3–5 mm long; bracts 2 mm long, ovate; bracteoles 1–2 mm long, elliptic. Calyx tube subglobose, 1 mm long; lobes distinctly leaf-like, 2–3 × 0.5–1 mm, lanceolate, very narrowly rounded at apex. Corolla tailed at apex in bud, yellow or pale yellow-green, glabrous; tube (3)4(5) mm long, broadly funnel-shaped, with long dense white or creamy hairs around the throat and a ring of deflexed hairs inside at base; lobes green outside, creamy inside, 4 × 1.5–2 mm, ± oblong, with short appendage at the apex. Style 6–7 mm long; the pollen presenter cream, 1 mm long, cylindrical-obconic, sulcate, lobed at apex. Ovary 4-locular. Fruits yellow, 10–15 mm in diameter, subglobose, mostly with 1–4 pyrenes about 10 mm long.

Zambia. B: Kataba, fl. 12.xii.1960, *Fanshawe* 5967 (K; NDO). W: Mwinilunga Distr., NE of Dobeka Bridge, fl. 9.xi.1937, *Milne-Redhead* 3158 (BR; K; PRE).
Also in Angola. Fringing riverine bushland, dambos on open sandy ground, *Brachystegia* woodland including dambo edges, mostly on Kalahari Sand, also on lake shores; c. 1200–1540 m.

13. **Fadogia schmitzii** Verdc., sp. nov.* Type: Zambia, 45 km east of Mwinilunga, *Milne-Redhead* 1084 (K, holotype; PRE).
Fadogia schmitzii Verdc. in Kew Bull. **36**: 530, fig. 8 (1981) nom. invalid.

Suffrutex with 10–20 caespitose stems from a thick rootstock; stems 35–50 cm tall, rather thick, ± trigonous, glabrous save inside the stipule-bases, swollen at the nodes and with pale peeling bark. Leaves in whorls of 3; blades 4–9 × 2.5–5 cm, obovate, rounded to acute at the apex or emarginate, mucronulate, cuneate at the base, ± discolorous, glabrous, ± coriaceous; petioles 4–6 mm long; stipules 2 mm long, densely pilose inside, with subulate appendage 6 mm long, soon deciduous. Cymes fasciculate, up to 50-flowered at one node, globose, often in axils of fallen leaves; peduncles 2–10 mm long; pedicels 3.5–10 mm long; bracts 2 mm long, triangular. Calyx tube 2 mm long; limb 0.8 mm long, shortly toothed. Corolla ± acute at the apex in bud, white or ± yellow, glabrous; tube 3 mm long, cylindrical to funnel-shaped, hairy inside top half; lobes 4–4.2 × 1.8–2 mm, ovate-lanceolate, minutely apiculate. Ovary 4-locular. Style 4.5 mm long; the pollen presenter 1–1.3 mm long, subcylindric or globose-obconic, 8-sulcate, 4-lobed at apex. Fruit black, about 15 mm in diameter, with 1–4 reddish-brown reniform pyrenes 7.3 × 5 × 3.5 mm.

Zambia. W: Mwinilunga Distr., 16 km west of Kakoma, slope down to Muzera R., 28.ix.1952, *White* 3396 (BR; FHO; K).
Also in Zaire (Dem. Rep. Congo) (Shaba Prov.). *Cryptosepalum* and *Brachystegia* woodland on Kalahari Sand; c. 1200 m.

---

* **Fadogia schmitzii** Verdc. sp. nov. ob inflorescentias multifloras distincta; *F. erythrophloeae* (K. Schum. & K. Krause) Hutch. & Dalz. probabiliter affinis sed cortice haud rubro-pulverulento, inflorescentiis majoribus, fructibus haud valde lobatis differt. Typus: Zambia, *Milne-Redhead* 1084 (K, holotype; PRE).
This species name was invalidly published in 1981 since two different specimens were cited as the holotype, one on page 530 and the other on page 532.

14. **Fadogia ancylantha** Hiern in F.T.A. **3**: 155 (1877). —Verdcourt in Kew Bull. **36**: 526 (1981). —Bridson & Troupin in Fl. Pl. Lign. Rwanda: 558, fig. 208.2 (1982); in Fl. Rwanda **3**: 159, fig. 74.2 (1985). —Verdcourt in Kew Bull. **42**: 125, fig. 1M (1987); in F.T.E.A., Rubiaceae: 799 (1991). Type from the Sudan.

'*Rubiacea* 481' —Thomson in Speke, J. Discov. Source Nile, Append.: 636 (1863).

Fadogia fuchsioides Oliv. in Trans. Linn. Soc., London **29**: 85, t. 50* (1873). —Hiern in F.T.A. **3**: 155 (1877), pro parte quoad *Grant* 481 excl. lectotype.

Fadogia obovata N.E. Br. in Bull. Misc. Inform., Kew **1906**: 105 (1906). Type: Zimbabwe, near Harare (Salisbury), Six-mile Spruit, *Cecil* 141 (K, lectotype).

Temnocalyx obovatus (N.E. Br.) Robyns in Bull. Jard. Bot. État **11**: 320, figs. 31 & 32 (1928). —F. White, F.F.N.R.: 422, fig. 67, H (1962). —A.E. Gonçalves in Garcia de Orta, Sér. Bot. **5**: 209 (1982).

Temnocalyx fuchsioides sensu Robyns in Bull. Jard. Bot. État **11**: 319 (1928), pro parte quoad *Grant* 481 non (Welw.) Robyns.

Temnocalyx ancylanthus (Hiern) Robyns in Bull. Jard. Bot. État **11**: 323 (1928). —F.W. Andrews, Fl. Pl. Anglo-Egypt. Sudan **2**: 463 (1952).

Temnocalyx ancylanthus var. *puberulus* Robyns in Bull. Jard. Bot. État **11**: 323 (1928). —F.W. Andrews, Fl. Pl. Anglo-Egypt. Sudan **2**: 464 (1952). Type from the Sudan.

Suffrutex 0.4–1.8 m tall, from a woody rootstock up to 2 cm in diameter; stems 2–6, branched or unbranched, glabrescent to shortly pubescent, rounded or triangular, often quite woody, yellowish-cream. Leaves paired, or in whorls of 3, glaucous-green, 2.3–11 × 1–7 cm, elliptic to rounded-obovate or ± round, obtusely often abruptly acuminate at the apex, cuneate to almost truncate at the base, rather thin, glabrous to shortly pubescent; petioles absent or short, under 2 mm long; stipules with a broad base 1–2 mm tall and a filiform appendage 1.5–7 mm long, the base densely pilose within. Inflorescences shortly stalked, mostly 2–3(4)-flowered, or flowers sometimes solitary; peduncles 2–16 mm long, sometimes supra-axillary and decurrent; pedicels 6–22 mm long, glabrous to shortly pubescent. Calyx 2.5–3 mm long; the limb very short and rim-like, truncate. Corolla obtuse to shortly acuminate at the apex in bud, usually distinctly curved, but sometimes straight (at least when dry), greenish-yellow, the tube usually greener, and the lobes green outside, whitish or pale yellow inside; tube 17–28 mm long, 5.5–7 mm wide at apex, 2.5–3.5 mm wide at the base, ± cylindrical, usually very distinctly widened above, glabrous outside but with a ring of hairs inside near base; lobes 5–6.5 × 2.5 mm, triangular, with a short blunt appendage. Style pale green, up to 3.4 cm long; pollen presenter 2.5 mm long, cylindrical, 10-angled, roundly 5-lobed at the apex. Ovary 5-locular. Fruit dark slate-grey-green, about 12 mm in diameter, ± globose, 5-lobed when dry.

Zambia. N: Mbala Distr., Lumi Marsh, fl. 17.xi.1956, *Richards* 7005 (K). W: Ndola, fl. 30.i.1954, *Fanshawe* 741 (K; NDO). C: Kabwe (Broken Hill), fl. xi.1909, *Rogers* 8637 (K). E: Chadiza, fl. 28.xi.1958, *Robson* 761 (BM; K; SRGH). S: Tara P.F.A., Siachitema Chieftaincy, fl. 25.xi.1963, *Bainbridge* 907 (K; SRGH). **Zimbabwe**. N: Darwin, Kandeya C.L. (Native Reserve), fl. 17.i.1960, *Phipps* 2289 (K; SRGH). C: Makoni Distr., 8 km from Rusape towards Nyanga (Inyanga), fl. 29.xi.1930, *Fries, Norlindh & Weimarck* 3298 (K; LD; SRGH). E: Nyanga (Inyanga), Chiduku, fr. 19.ii.1944, *Timson* in GHS 11749 (K; SRGH). S: east of Great Zimbabwe, sterile 1.vii.1930, *Hutchinson & Gillett* 3369 (K). **Malawi**. N: Chitipa Distr., 10 km from Nthalire towards Mwenewenya, fl. 1.iii.1982, *Brummitt et al.* 16209 (K; MAL). C: Ntchisi Forest Reserve, above Rest House, fr. *Brummitt* 9414 (K; MAL). S: Blantyre Distr., Chileka, fl. 17.i.1957, *Jackson* 2107 (K; SRGH). **Mozambique**. N: Massangulo, fl. xii.1932, *Gomes e Sousa* 1007 (K). Z: Lugela Distr., road to Muobede (Moebede), fl. 25.i.1948, *Faulkner* Kew 172(b). T: Macanga, Mt. Furancungo (Elefanta), fr. 15.iii.1966, *Pereira et al.* 1767 (BR; LMU).

Also in Nigeria, Zaire (Dem. Rep. Congo), Burundi, Sudan, Uganda and Tanzania. *Colophospermum* and *Brachystegia* woodlands, and chipya woodland, also in wooded grasslands, often by dambos and on termite mounds; 850–1680 m.

Fadogia ancylantha and *Fadogia obovata* had, up to 1981, been retained as separate species on the grounds of leaf shape (ovate and basally rounded contrasting with obovate and basally cuneate) and between 1928 and 1981 had been referred to *Temnocalyx*, and although Uganda and Sudan populations have the leaves less obovate with more truncate bases and with more indumentum, the same combination of characters turns up again, but rarely, in Malawi; otherwise the leaves are predominantly obovate throughout the range. The variation is such that only one taxon can be recognised but nevertheless needs further study in the field. Robyns' var. *puberula* is connected by many intermediates and does not seem worth maintaining.

* Figure is based on *Welwitsch* and *Speke & Grant* specimens; only the fruiting part refers to the present species.

15. **Fadogia fuchsioides** Welw. ex Oliv. in Trans. Linn. Soc., London **29**: 85, t. 50 (flowering part only) (1873), pro parte. —Hiern in F.T.A. **3**: 155 (1877); Cat. Afr. Pl. Welw. **1**: 482 (1898), pro parte excl. *Grant* 481. —F. White, F.F.N.R.: 406, fig. 68, A (1962). —Verdcourt in Kew Bull. **36**: 527 (1981); in F.T.E.A., Rubiaceae: 800 (1991). Type from Angola.

Temnocalyx fuchsioides (Welw. ex Oliv.) Robyns in Bull. Jard. Bot. État **11**: 319 (1928), pro parte.

Suffrutex 0.3–1.5 m tall, from a thick woody rootstock; stems several, reddish or purplish, angular or rounded, finely ridged (when dry), glabrous. Leaves in whorls of 3–4 (rarely 5–6); blades 4.5–15 × 1.8–9 cm, elliptic to obovate, rounded to very obscurely rounded-acuminate or bluntly acute at the apex, cuneate at the base, rather thicker than in other species, glabrous, the venation sometimes reddish when dry and in life; petiole rather thick, 3–20 mm long; stipules with a short broad base 1.5–2 mm long, densely pilose within, and with a subulate appendage 3.5 mm long. Inflorescences glabrous, 1–3-flowered; peduncle red, 5–11 mm long, or up to 25 mm long in solitary flowers; pedicels 3–23 mm long. Calyx red; tube 4–5 mm long, the limb rim-like, truncate, about 1.5 mm long. Corolla distinctly acuminate in bud; tube deep red, 14–27 × (4)6–8 mm, cylindrical and straight, glabrous inside and out, or ± hairy inside; lobes 6–9, cream, buff or yellow within, 8–14 × 3–3.5 mm, oblong-lanceolate, with a thick apical appendage 1–2 mm long. Ovary 6–8-locular. Style 28–34 mm long; pollen presenter large, 4.5–6 × 3–4 mm, broadly obconic-coroniform, c. 6–9-ridged or lobed. Fruits black, fleshy, 12–15 mm in diameter, ± globose or obovoid, containing 6–9 pyrenes.

Zambia. B: Zambezi (Balovale), Chavuma, fl. 12.x.1952, *White* 3474 (BM; FHO; K). N: Kalambo Falls, fl. 19.xii.1949, *Bullock* 2110 (K). W: Kitwe, fl. 28.x.1953, *Fanshawe* 467 (K; NDO). C: Chakwenga Headwaters, fl. 7.i.1964, *E.A. Robinson* 6138 (K). S: Mumbwa, *Macaulay* 975 (K). **Malawi**. S: Blantyre Distr., 3 km north of Limbe, Ndirande Forest Reserve, fl. 22.ii.1970, *Brummitt* 8701 (K; MAL).

Also in Zaire (Dem. Rep. Congo), SW Tanzania and Angola. *Brachystegia* and *Julbernardia–Brachystegia–Cryptosepalum* woodlands, often on Kalahari Sand, also bushland, dambo-edges and rocky places; 1200–1500 m.

16. **Fadogia verdickii** De Wild. & T. Durand in Bull. Soc. Roy. Bot. Belgique **40**: 21 (1901). —De Wildeman in Ann. Soc. Sci. Bruxelles **41**: 73 (1921); Contrib. Fl. Katanga: 214 (1921). Type from Zaire (Dem. Rep. Congo).

Temnocalyx verdickii (De Wild. & T. Durand) Robyns in Bull. Jard. Bot. État **11**: 319 (1928).

Suffrutex, glabrous or finely pubescent, ± glaucous, from a thick woody rootstock; stems several, 30–90 cm tall, often trigonous. Leaves in whorls of 3 or 4, blades 1–6 × 0.2–1.6 cm, narrowly oblong or elliptic to oblanceolate, acute at the apex, narrowed to the base, ± sessile; stipule bases short, c. 1 mm long with subulate appendage 1–2.5 mm long, densely hairy inside. Flowers scented, solitary, or sometimes apparently in 3-flowered inflorescences but actually terminating in short leafless shoots. Calyx tube c. 2 mm long, campanulate, ± ribbed in dry state; the limb spreading, c. 1 mm long, undulate or with short teeth. Corolla clavate, acute or shortly acuminate-tailed at the apex in bud; tube green or white, 20–30 mm long, tubular, usually distinctly curved, widened towards the apex, always glabrous outside, 3–4 mm wide at the base, 5–9 mm wide at the apex, glabrous above inside, pubescent in the lower two thirds but without a ring of deflexed hairs; lobes cream or white, often green outside, 8–13 × 2.5–4 mm, triangular-lanceolate, shortly appendaged, reflexed. Anthers exserted, shortly appendaged at base and apex; filaments 3 mm long. Style 30–40 mm long, thickened at base; pollen presenter large and coroniform, 4 × 5 mm, conspicuously lobed. Ovary (3)5–6-locular. Fruit not seen.

Zambia. N: Mbala Distr., near Chinakila Village, fl. 5.x.1956, *Richards* 6358B (K). W: Mufulira, fl. 20.xii.1948, *Cruse* 455 (K).

Also in Zaire (Dem. Rep. Congo) (Shaba Prov.). Dambos, grassland with scattered trees, flood plains, sometimes on termite mounds; 780–1500 m.

Used as a relish and sold in markets on the Copperbelt.

44. FADOGIELLA Robyns

Fadogiella Robyns in Bull. Jard. Bot. État **11**: 94 (1928).

Subshrubby herbs (suffrutices) or shrubs, completely softly tomentose, mostly much branched. Leaves opposite, or less often in whorls of 3, shortly petioled; stipules with triangular or sheathing connate bases and subulate appendages. Flowers relatively large in subsessile or shortly pedunculate, dense, few- to many-flowered cymes. Calyx tube globose with a short ± subundulate limb or short, triangular teeth. Corolla thick, cylindrical or obovoid, obtuse or shortly acuminate at the apex in bud, mostly yellow or greenish, densely tomentose; tube cylindrical, glabrous or hairy at the throat, with a ring of deflexed hairs inside; lobes reflexed, obtuse or shortly apiculate. Stamens inserted at the throat, anthers exserted or partly exserted. Ovary (3)4–5-locular, each locule with a pendulous ovule; style slender, exserted; pollen presenter cylindrical or coroniform, sulcate, (3)4–5-lobed at the apex. Fruit drying back, globose, crowned with the calyx limb, sometimes ribbed, containing (3)4–5 pyrenes, ± glabrous. Pyrenes thinly woody, not crested around apex, point of attachment ± beaked, very slightly textured. Cotyledons very much shorter than the radicle.

A genus of 3 species, mostly in Zaire (Dem. Rep. Congo) (Shaba Prov.) and the Flora Zambesiaca area, two extending into Tanzania, the third known only from the type now destroyed.

Corolla tube 3–4 mm long; indumentum on calyx gradually becoming less dense towards the limb; leaf blades narrowly to broadly elliptic or lanceolate to obovate · · · · · 1. *stigmatoloba*
Corolla tube 15–20 mm long; calyx entirely hidden by indumentum, or occasionally the very tips of the lobes exposed; leaf blades broadly elliptic or ovate · · · · · · · · · · · · · 2. *rogersii*

1. **Fadogiella stigmatoloba** (K. Schum.) Robyns in Bull. Jard. Bot. État **11**: 96 (1928). —Brenan, Check-list For. Trees Shrubs Tang. Terr.: 494 (1949). —F. White, F.F.N.R.: 407, fig. 68, F (1962). —Verdcourt in Kew Bull. **42**: 125, fig. 1N (1987); in F.T.E.A., Rubiaceae: 786, figs. 132/8, 140 (1991). TAB. **55**. Type from Tanzania.
   Fadogia stigmatoloba K. Schum. in Bot. Jahrb. Syst. **30**: 414 (1901).
   Fadogia manikensis De Wild. in Repert. Spec. Nov. Regni Veg. **13**: 139 (1914); Contrib. Fl. Katanga: 213 (1921). Type from Zaire (Dem. Rep. Congo).
   Fadogiella verticillata Robyns in Bull. Jard. Bot. État **11**: 95 (1928). Type from Zaire (Dem. Rep. Congo).
   Fadogiella manikensis (De Wild.) Robyns in Bull. Jard. Bot. État **11**: 97, figs. 13 & 14 (1928).

Erect suffrutex, 0.5–1.2(1.8) m tall, from a thick horizontal or vertical woody rootstock; stems usually single, or 2–3, much branched often 3–4-angled; shoots slender, velvety yellowish-tomentose above but soon glabrescent and ultimately dull purplish and glabrous. Leaves predominantly paired, or in whorls of 3 at a few nodes bearing 3 side branches; blades 1.7–8(10) × 0.5–4.4(5.8) cm, narrowly to broadly elliptic or obovate, or lanceolate, narrowed to a rounded or subacute apex, rarely shortly acuminate, cuneate at the base, mostly very discolorous, densely covered with yellowish-grey, curled, short hairs above which do not obscure the surface, ultimately sometimes becoming ± glabrous, white or yellowish-grey velvety tomentose beneath, the surface entirely obscured save in old leaves; venation impressed and ± bullate above and raised and reticulate beneath in mature leaves; petiole 2–3(6) mm long; stipules with triangular base, 1–2 mm long, and subulate appendage, 2.5–3(6) mm long, at first tomentose but becoming glabrous. Inflorescences dense and subsessile, the axillary pairs forming dense nodal clusters; peduncles 0–2.5 mm long, rarely to 15 mm in fruit; pedicels 1–1.5(5) mm long, all densely tomentose; bracts and bracteoles c. 1 mm long, ovate or narrowly lanceolate, scarious, turning dark brown. Calyx tube 1–1.5 mm long, the teeth short, 0.5–1 mm long, ± triangular, ± scarious, brown, ± glabrous or pubescent. Corolla with 5 short but distinct appendages in bud, green or yellow, densely tomentose with appressed hairs; tube 3–5 mm long, densely tomentose outside, shortly hairy at the throat; lobes 4–5 mm long, c. 2 mm wide, linear-triangular, apiculate, with similar indumentum to tube outside. Anthers exserted, 1.5–2 mm long. Style 6–6.5 mm long; pollen presenter pale green, 0.75–1.5

Tab. 55. FADOGIELLA STIGMATOLOBA. 1, flowering branch (× ²/₃); 2, stipule (× 2); 3, young inflorescence (× 4); 4, flower (× 5); 5, corolla opened out, with style and pollen presenter (× 6); 6, pollen presenter (× 10); 7, calyx (× 6); 8, longitudinal section through ovary (× 8); 9, transverse section through ovary (× 8), 1–9 from *Bullock* 3485; 10, fruiting branch (× 1); 11, pyrene (× 4), 10 & 11 from *Fanshawe* 687. Drawn by M.E. Church. From F.T.E.A.

mm long, subcylindrical, 3–5-lobed at apex. Fruits globose, 7.5–12(15) mm in diameter, crowned by a tomentose calyx limb, shiny, glabrescent or glabrous, with (1)3–5 pyrenes.

Zambia. E: between waterfall on North Rukuru R. and Apoka Village, 6.5 km north of Rest House, fl. 29.x.1958, *Robson & Angus* 466 (BM; K; LISC; SRGH). W: Luano, fl. 6.xi.1969, *Mutimushi* 3828 (K; NDO). **Malawi**. N: Rumphi Distr., Nyika Plateau, 30.5 km north of M1, fl. & fr. 23.xii.1977, *Pawek* 13332 (K; MAL; MO).
Also in Zaire (Dem. Rep. Congo) and southern Tanzania; also a variant in Angola. *Brachystegia, Uapaca* and *Julbernardia–Swartzia–Parinari–Diplorhynchus–Lannea–Acacia* woodlands, sometimes on dambo margins and termite mounds; 900–1900 m.
Milne-Redhead 719 (Solwezi Distr., Mbulungu Stream, west of Mutanda Bridge, fr. 16.vii.1930) has fruiting peduncles to 1.5 cm long (unless they are abbreviated branchlets) and leaves in whorls of 3 with upper surface ± glabrous; it presumably represents a mature stage of the form described as *F. verticillata*.

2. **Fadogiella rogersii** (Wernham) Bridson in Kew Bull. **51**: 351 (1996). Type from Zaire (Dem. Rep. Congo) (Shaba Prov.).
   Fadogia rogersii Wernham in J. Bot. **51**: 208 (1913). —De Wildeman, Études Fl. Katanga [Ann. Mus. Congo, Sér. IV, Bot.] **2**: 153 (1913); Notes Fl. Katanga 7: 73 (1921); Contrib. Fl. Katanga: 213 (1921).
   Ancylanthos rogersii (Wernham) Robyns in Bull. Jard. Bot. État **11**: 326 (1928). —F. White, F.F.N.R.: 400 (1962). —Verdcourt in Kew Bull. **42**: 126, fig. 2F (1987); in F.T.E.A., Rubiaceae: 802 (1991).

Shrub 0.3–4 m tall; stems with a dense white, cream or grey cottony tomentum, at length greyish but persistently tomentose; older stems bearing the new year's shoots can be blackish and glabrous; bark on tall specimens smooth and brown. Leaves paired or in whorls of 3, blades 1.7–7 × 1–4 cm, ovate to broadly elliptic, rounded to acute at the apex, rounded to cuneate at the base, very discolorous, densely covered with a floccose white or grey cottony tomentum, becoming glabrous above and the upper surface scarcely obscured, persistently thickly white cottony tomentose beneath; venation deeply impressed above; petiole 2–5(7) mm long; stipules 1–2 mm long, triangular with a subulate appendage c. 6 mm long. Cymes 3–6-flowered but sometimes appearing as dense verticillasters in inflorescences occurring at leafless nodes bearing 3 abbreviated new shoots; peduncle 0–3(15) mm long; pedicels 3–4(8) mm long; bracts 3.5–5 mm long, lanceolate. Calyx tube 3–4 mm long, ± globose, densely tomentose; limb ± truncate but with thin teeth c. 1 mm long, lacking tomentum. Corolla creamy-white, pale green or distinctly yellow, densely tomentose outside; tube 15–20 mm long, straight or somewhat curved, with a ring of deflexed and erect hairs ± 2/3 of way down inside; lobes 3–6 × 2.2–2.5 mm, ovate-triangular, ± acute. Stigma green, exserted 1–2.5 mm beyond the throat; the pollen presenter green, coroniform, deeply 5-lobed at the apex. Fruit about 10 × 8 mm, subglobose, distinctly 4-lobed in dry state, with scattered cottony tomentum.

Zambia. N: Mbala Distr., Mwembeshi, Top Farm, fl. 20.x.1954, *Richards* 2090 (K). W: Ndola, near Nkana, Mindola Forest Reserve, fl. 26.ix.1947, *J.P.M. Brenan & R.A.F. Brenan* 7693 (FHO; K). C: Kundalila Falls picnic area, 13 km S of Serenje–Mpika highway, fl. 22.xi.1993, *Harder et al.* 2072 (K; MO).
Also in southern Zaire (Dem. Rep. Congo). Deciduous woodlands including miombo woodland, *Brachystegia–Julbernardia–Marquesia* woodland and *Hymenocardia–Pterocarpus– Combretum* woodland, also in derived bushland on sandy soil, sometimes in rocky places; 1200–1740 m.

45. RYTIGYNIA Blume

Rytigynia Blume, Mus. Bot. **1**: 178 (1850). —Robyns in Bull. Jard. Bot. État **11**: 132 (1928). —Verdcourt in Kew Bull. **42**: 145–186 (1987).

Shrubs or small trees, occasionally somewhat scrambling; stems spiny in a few species, usually distinctly lenticellate. Leaves opposite, or occasionally in whorls of 3 particularly in some forms of *R. celastroides*, petiolate, often with domatia in axils of nerves beneath; stipules often ± persistent, connate at the base, oblong or triangular, villous within, ending in a mostly linear or subulate ± deciduous appendage. Flowers axillary, solitary, or in 2–10(15)-flowered umbel-like cymes, mostly 5-merous, small, usually white, yellowish or greenish; peduncles and pedicels mostly well-developed; bracts and sometimes bracteoles present, small; sometimes inflorescence axes scarcely developed and flowers appearing fasciculate. Calyx tube ± subglobose; limb short, truncate or denticulate, distinctly lobed in a few species (mostly in distinctive subgenera), mostly persisting on the young fruit. Corolla obtuse, acute or conspicuously long-apiculate in bud; tube cylindrical, glabrous or pubescent outside, glabrous or hairy within for part or for most of the tube length, or corolla tube with a ring of deflexed hairs in the middle; lobes shorter than to longer than the tube, acute to very distinctly long apiculate, or with a filiform appendage. Stamens slightly to distinctly exserted; anthers often mucronate to slightly appendaged, often papillate; filaments very short. Ovary mostly 3–5(6)-locular, but 2-locular in some species. Style usually exserted, slightly swollen at the base. Pollen presenter coroniform, subglobose or occasionally cylindrical, sulcate beneath where in contact with anther cells, mostly distinctly 2–5-lobed at the apex. Disk depressed, glabrous. Fruit mostly globose, or asymmetrical and compressed if only 1–2 pyrenes developed, c. 10 mm in diameter, with 1–5 pyrenes. Pyrenes narrowly ± reniform or boat-shaped, the notch usually about one third from the apex, the dehiscence line on a ± marked keel, often pitted.

About 50–60 species in tropical Africa and Madagascar.

The central core of the genus is distinctive enough and easily recognisable, but other species connect to other genera, *Canthium*, *Fadogia*, *Vangueria* and *Tapiphyllum*. Species 1–15 are placed in subgen. *Rytigynia*. Bridson (Kew Bull. **47**: 360 (1992) suggested that *R. bugoyensis* (*species no.* 16) should be considered for transfer to *Vangueriella*, but this decision is still pending. Subgen. *Fadogiopsis* is represented by *species no.* 17. A third subgenus, Subgen. *Sali*, is known only from 2 Tanzanian species. *Species* 18–21 (spp. D–G) are poorly known and therefore placed at the end of the genus, they are not included in the key below.

Some *Rytigynia* species with 2-locular ovaries may resemble *Canthium* very closely, especially in fruit. *Canthium* never has appendaged corolla lobes and the inflorescence is mostly of a different structure.

* Outside the Flora Zambesiaca area unusual forms of *R. celastroides* and *R. monantha* occur which have leaves velvety beneath; the possibility that similar forms might occur in the Flora Zambesiaca area should be borne in mind.

11. Calyx lobes triangular; flowers in nodal clusters made up of 2 axillary 5–10-flowered fascicles · 9. *torrei*
  – Calyx lobes obsolete, the limb ± truncate; flowers usually in fewer-flowered inflorescences · 12
12. Inflorescences usually umbel-like with pedicels supported by a small involucre formed by the bracts at one level; leaves paired, mostly drying bright-green, 1.8–11 × 0.7–5 cm, often widest below the middle sometimes rounded at the base; populations in lowland Mozambique glabrous, others glabrous or hairy · · · · · · · · · · · · · · · · · · ·1. *umbellulata*
  – Inflorescences often with bracts rather scattered and less distinctly umbel-like; leaves paired or often in threes, mostly drying brownish above, 2–5.5(9) × 0.8–2.5(4) cm usually widest at the middle and narrowed at both ends; plant of lowland Mozambique and usually pubescent · 10. *celastroides**
13. Leaves 2–4 × 1–1.8 cm, obtuse or very shortly obtusely acuminate, quite glabrous even lacking domatia beneath; corolla lobes exceeding the tube, acute but not appendaged · 11. *sp. B*
  – Leaves more acuminate, or if not then narrower · 14
14. Leaves about 2–2.5 × 0.7–1 cm with very obvious hairy domatia beneath but otherwise glabrous or glabrescent, not drying with a dark reticulate venation beneath · 10. *celastroides* var. *australis*
  – Leaves larger, or if as small then with venation drying dark beneath · · · · · · · · · · · · · · 15
15. Stipule appendages laterally flattened and rather scimitar-like rather than subulate; leaves discolorous, paler beneath but with venation drying dark and closely reticulate; corolla usually with some hairs outside, particularly on lobes, the tube 3–6 mm long; corolla lobe appendages 0.5–1.5 mm long · 13. *monantha*
  – Stipule appendages not so flattened, more subulate; leaves mostly less discolorous with venation less evident; corolla lobes without appendages or up to c. 1 mm long · · · · · · 16
16. Leaves very distinctly acuminate, the acumen almost half the total leaf length, corolla lobes with appendage c. 1 mm long, together longer than the corolla tube; whole plant glabrous · 4. *lewisii*
  – Leaves acuminate but acumen not so markedly long; corolla lobes with no, or vestigial appendages, shorter than or longer than the tube; whole plant glabrous or pubescent · · 17
17. Corolla lobes distinctly longer than the 2.5 mm long tube; peduncle longer than pedicels · 3. *senegalensis*
  – Corolla lobes distinctly shorter than or rarely ± equalling the 4.5–6 mm long tube · · · 18
18. Leaves thinner drying greener; corolla thinner and slightly smaller drying paler with lobes shorter 2 mm long; calyx limb truncate; buds rounded · · · · · · · · · · · · · · · · · 6. *uhligii*
  – Leaves thicker and more discolorous, usually drying darker and often less acuminate; corolla more robust, mostly slightly larger and drying darker with longer lobes 3–3.5 mm long; calyx lobes sometimes more developed; buds usually slightly apiculate or acuminate · 7. *adenodonta*

1. **Rytigynia umbellulata** (Hiern) Robyns in Bull. Jard. Bot. État **11**: 184 (1928). —F. White, F.F.N.R.: 420 (1962). —Hepper in F.W.T.A. ed. 2, **2**: 186 (1963). —A.E. Gonçalves in Garcia de Orta, Sér. Bot. **5**: 206 (1982). —Verdcourt in Kew Bull. **42**: 152, fig. 1E, 8C (1987); in F.T.E.A., Rubiaceae: 808 (1991). Type from Ghana.
    Vangueria umbellulata Hiern in F.T.A. **3**: 150 (1877); Cat. Afr. Pl. Welw. **1**: 480 (1898).
    Vangueria concolor Hiern in F.T.A. **3**: 150 (1877). —De Wildeman in Bull. Jard. Bot. État **8**: 50 (1922). Type from Princes Island.
    Canthium lagoense Baill. in Adansonia **12**: 190 (1878). Type: Mozambique, Delagoa Bay, *Forbes* (K; P**, holotype).
    Vangueria junodii Schinz in Mém. Herb. Boissier, No. 10: 68 (1900). —De Wildeman in Bull. Jard. Bot. État **8**: 57 (1922). Type: Mozambique, Delagoa Bay, *Junod* 220 (BR; G; K; Z, holotype).

* Although distinguishable at a glance, and not closely related, it is not easy to give a key which covers all variants. Many herbarium specimens of *Rytigynia celastroides* do not show the spines and some plants may not be spiny, hence the need to include it here.
** Baillon includes information not mentioned in the note given by Hiern at the end of his description of *Vangueria euonymoides*, where he suggests the Forbes specimen is an aberrant form, so it is presumed there must be a specimen at Paris. Baillon gives no locality nor collector but his reference to Hiern and the specific name clearly indicate what he meant.

Vangueria euonymoides sensu Burtt Davy in Bull. Misc. Inform., Kew **1908**: 173 (1908); Sim, For. Fl. Port. E. Africa: 75 (1909) non Hiern (1877).

Vangueria sparsifolia S. Moore in J. Linn. Soc., Bot. **40**: 92 (1911). —De Wildeman in Bull. Jard. Bot. État **8**: 64 (1922). Type: Mozambique, Gazaland, Madanda Forest, *Swynnerton* 551 (BM, holotype; K).

Vangueria ituriensis De Wild. in Bull. Jard. Bot. État **8**: 56 (1922); Pl. Bequaert. **2**: 275 (1923). Type from Zaire (Dem. Rep. Congo).

Rytigynia concolor (Hiern) Robyns in Bull. Jard. Bot. État **11**: 190 (1928).

Rytigynia junodii (Schinz) Robyns in Bull. Jard. Bot. État **11**: 179 (1928).

Rytigynia sparsifolia (S. Moore) Robyns in Bull. Jard. Bot. État **11**: 180 (1928).

Rytigynia welwitschii Robyns in Bull. Jard. Bot. État **11**: 193 (1928). —J.G. Garcia in Mem. Junta Invest. Ultramar **6** (sér. 2): 28 (1959) [Contrib. Conhec. Fl. Moçamb. IV (1959)]. Type from Angola.

Rytigynia perlucidula Robyns in Bull. Jard. Bot. État **11**: 175 (1928). Type from Zaire (Dem. Rep. Congo).

Scrambling or erect shrub or small tree 1–5(9) m tall, often deciduous, the branches usually slender; bark grey-black; branchlets grey to purple-brown, with or without lenticels, glabrous or pubescent, or setulose on young parts. Leaves 1.8–11 × 0.7–5 cm, oblong-elliptic to ovate or lanceolate, acuminate at the apex with the tip ± rounded, broadly cuneate to rounded at the base, glabrous save for barbellate domatia, or sparsely to fairly densely appressed pilose or setulose, particularly on the midrib, usually distinctly thin; petiole 2–7 mm long; stipules ± connate at the base, 1.5–3 mm long, with triangular-truncate bases, scarious but sometimes becoming woody, ± villous inside, with a subulate appendage 2–8 mm long or rarely lacking. Cymes umbel-like, 2–8-flowered, often produced together with young leaves; peduncle 0.5–4 mm long, or ± obsolete; pedicels 2–10(15) mm long; bracteoles 1–2 mm long, connate to form a scarious involucre. Calyx tube 1–1.5 mm long, the limb 0.3 mm long, truncate or obscurely toothed, scarious. Corolla rounded to sub-apiculate at the apex in bud, creamy, yellow or greenish-white; tube 2.5–4 mm long, with a ring of deflexed hairs within; lobes 1.8–3 mm long, triangular or oblong, subapiculate. Style 4–4.5 mm long. Stigma described as yellow below, green in middle and blue on top, coroniform, 5-lobed. Ovary (4)5–locular. Fruit black, 6–9 mm in diameter, subglobose, with (2)4–5 pyrenes; pedicels 7–12 (15) mm long. Pyrenes usually pitted.

Caprivi Strip. along northern border, about 67 km west of Katima Mulilo, fr. 17.ii.1969, *de Winter* 9214 (K; PRE). **Botswana**. N: Kwando R., island at 18°03'S, 23°19'E, fr. 29.i.1978, *P.A. Smith* 2306 (GAB; SRGH). **Zambia**. B: 14.4 km ESE of Kaoma (Mankoya), near Luena R., fl. 21.xi.1959, *Drummond & Cookson* 6696 (K; SRGH). N: Lake Kashiba, fr. 27.x.1957, *Fanshawe* 3784 (K). W: Ndola, Chichele Botanical Reserve, fl. 7.xii.1952, *White* 3826 (FHO; K). C: 64 km west of Lusaka, Mukulaikwa Agric. Station, fr. 6.i.1963, *Angus* 3477 (FHO; LISC; K). S: Mapanza, Choma, fl. 24.x.1953, *E.A. Robinson* 352 (K). **Zimbabwe**. E: Chimanimani Distr., Haroni Forest, east of Haroni River, fl. bud 18.xi.1994, *P. van Wyk* BSA 2843 (K; PRE). **Malawi**. S: Liwonde National Park, Namelembo Thicket, fl. 31.xii.1983, *Dudley* 831 (K). **Mozambique**. N: Nampula, Monapo, Mr Wolf's forest, fr. 13.ii.1984, *de Koning et al.* 9640 (LMU). Z: Quelimane, Lugela–Mocuba, Namagoa Estate, fl. & fr. Oct., *Faulkner* in PRE 323 (BR; EA; K; PRE). T: Mágoè, Serra de Songo near Cahora-Bassa, about 10 km on descent to Estima, fr. ii.1970, *Torre & Correia* 17853 (C; LISC; LMA; LMU; MO; WAG). MS: southern foothills of Chimanimani Mts., Makurupini Falls, fl. 25.xi.1967, *Simon & Ngoni* 1314 (K; LISC; SRGH). GI: Inhambane, Inharrime (Nhacongo), fl. 25.x.1947, *Barbosa* 532 (LISC; LMA). M: Maputo, Costa do Sol, fr. 14.xi.1963, *Balsinhas* 678 (K; LISC; LMA; PRE).

Also in West Africa from Guinea Bissau to Cameroon, and in Angola, Zaire (Dem. Rep. Congo), Uganda and Tanzania.

Deciduous forest and woodlands, and in riverine thicket and forest. Woodland types include: *Brachystegia–Julbernardia*, *Cryptosepalum–Copaifera*, *Maerua–Pteleopsis–Cassipourea–Entandrophragma caudatum*, and also *Baikaea–Fagara–Friesodielsia–Baphia* on Kalahari Sand, and *Afzelia–Garcinia–Bridelia–Albizia* on maritime sand dunes, sometimes in rocky places and on termite mounds; 10–1350 m.

As interpreted here *Rytigynia umbellulata* is very variable, but further work is needed to confirm the circumscription. Robyns recognised many species which are not separable, but possibly the above synonymy has gone too far and not recognised enough. The lowland Mozambique material is always glabrous and varies considerably in the peduncle length – if this is to be treated separately then Baillon's epithet *lagoensis* would be the earliest name. Some material from Zambia varies from glabrous to quite densely pubescent and corresponds with *R. welwitschii*. The pitting of the pyrenes varies considerably and needs investigation in the field.

The specimen *de Koning et al.* 9759 (BR; K; LMU) [Mozambique, N: Nampula, Mossuril, Serra de Mesa, fr. 19.ii.1984] differs in its slightly larger fruits with a more obvious persistent calyx limb, longer more slender pedicels, much longer peduncles and more finely reticulate leaves. *Patel & Seyani* 1330 (K; BR; MAL) [Malawi, S: Mulanje, Mpinda, Chingoa Hill, fl. 6.xi.1983] has yellowish-brown, almost shiny, shoots, the epidermis peeling to reveal a pale undersurface (stem described as cream-coloured in life). Both may prove to be distinct taxa when more material is available.

2. **Rytigynia sp. A**
   Rytigynia sp. 1 of F. White, F.F.N.R.: 420 (1962).

Small deciduous shrub 0.6–1.2 m tall, sparsely branched, the side shoots emerging almost at right angles before curving upwards, the main stems bearing short paired spines to 5 mm long, sometimes developed into a shoot on one side only. Stems brown, the epidermis ± glaucous and peeling, only the youngest shoots pubescent. Leaves not seen fully developed, 25 × 7 mm, lanceolate, acuminate to a narrowly rounded apex, cuneate at the base, membranaceous, with scattered short hairs above, glabrous beneath, minutely crinkly at the margin; lateral nerves about 3 on each side; petiole c. 1 mm long; stipules c. 2 mm long, ovate, glabrous or pubescent outside, densely hairy inside, with a subulate outwardly directed appendage 2 mm long. Inflorescences 1–5-flowered, borne on brachyblasts (contracted lateral spurs) in axils of the shoots; peduncle 0–1 mm long; pedicels 1–1.5 mm long; bracts scarcely 0.5 mm long, ± ovate. Calyx 0.6 mm long, the limb reduced to an entire rim. Corolla about 2 mm long, obtuse in bud; open corolla not seen. Ovary 3-locular.

Zambia. W: Mwinilunga Distr., 9.6 km north of crossing of Chikundulu Stream with Mwinilunga–Kabompo road, fl. bud 4.x.1952, *Angus* 604 (BM; FHO; K; WAG).
   Not known elsewhere. Dense undergrowth of *Cryptosepalum pseudotaxus* dry evergreen thicket (mavunda) on Kalahari Sand; c. 1350 m.
   This is very close to a plant which has been treated as a variety of *R. umbellulata* which exhibits hysteranthy but lacks spines.

3. **Rytigynia senegalensis** Blume, Mus. Bot. **1**: 178 (1850). —Robyns in Bull. Jard. Bot. État **11**: 163, figs. 19 & 20 (1928). —Verdcourt in Kew Bull. **42**: 155 (1987). Type from Senegal.
   Vangueria senegalensis (Blume) Hiern in F.T.A. **3**: 149 (1877).
   Vangueria euonymoides Hiern in F.T.A. **3**: 150 (1877). Lectotype from Sudan.
   Canthium euonymoides (Hiern) Baill. in Adansonia **12**: 189 (1878) (as "*C. euonymoides*").

Shrub or small tree 1.2–1.5 m tall, glabrous; bark on older shoots greyish, ridged and peeling revealing purplish-red beneath. Leaves 1.2–4 × 0.2–1.2 cm, oblong-lanceolate, long-acuminate at the apex with a narrowly rounded tip, cuneate at the base, with small pubescent domatia in axils of main nerves beneath; venation drying visible but not raised beneath, reticulate; petiole ± obsolete, appearing c. 2 mm long with the margins of the blade decurrent to the base; stipule sheath 1–2 mm long with a short narrowly flattened appendage 1–1.5 mm long. Flowers in 2-flowered cymes; peduncle 5–7 mm long, ± equalling or mostly longer than the 2–5 mm long pedicels. Calyx tube 1–1.2 mm long; limb c. 0.25 mm long, minutely denticulate. Corolla minutely apiculate in bud, greenish-white; tube 2.5 mm long, shortly cylindrical, with ring of deflexed hairs inside; lobes 3 × 1.3 mm, oblong, with scarcely any appendage. Style exserted about 2 mm; pollen presenter 1 mm long, obconic, obscurely sulcate, truncately 3-lobed at the apex. Ovary 3-locular. Fruit not seen from the Flora Zambesiaca area.

Zambia. W: Kitwe, Kafue R., fl. 25.x.1968, *Mutimushi* 2756 (K; NDO).
   Also in West Africa from Senegal to N Nigeria, and in the Sudan. Sandy places by river; c. 1260 m.
   The above description refers only to the specimen cited. Although very similar to many W African specimens I am not certain it is not an extreme case of convergence.

4. **Rytigynia lewisii** Tennant in Kew Bull. **22**: 441 (1968). —Verdcourt in Kew Bull. **42**: 155 (1987). Type: Zambia, Mwinilunga Distr., 24 km NNW of Kalene Mission, *Richards* 17165 (K, holotype).

A slender much-branched glabrous subshrub 40–60 cm tall, the stems with slightly peeling reddish-brown bark. Leaves 2.5–6.5 × 0.6–2.4 cm, lanceolate to ovate-lanceolate, or narrowly ovate, strikingly narrowly long-attenuate at the apex, rounded to cuneate at the base (or even subcordate in some Zaire (Dem. Rep. Congo) material fide *Robbrecht*), the venation beneath drying slightly darker, very evident and reticulate but not raised; petiole 2.5–6 mm long, slender; stipule sheath 0.5–1.5 mm long, hairy within, the appendage 3–5 mm long, subulate. Flowers in 1–2-flowered cymes; peduncle 5–10 mm long; pedicels 3–10 mm long; bracts small, 1.5 mm long joined to form a bifid structure ciliate within. Calyx tube 1.5 mm long, ovoid, the limb with short deltoid teeth about 0.25 mm long. Corolla markedly acuminate in bud, yellow-green, the lobes ± cream inside; tube 2.5 mm long, campanulate, with a ring of deflexed hairs below the throat inside; lobes 4 × 1.5–2 mm, oblong-triangular, including the apical 1 mm long appendage. Ovary 2-locular. Style exserted about 1.8 mm; pollen presenter 1.6 mm long, ovoid, sulcate, 2-lobed at the apex. Immature fruits subglobose, 9 mm in diameter.

Zambia. W: Mwinilunga Distr., 50 km from Mwinilunga on road to Solwezi, fl. 22.xi.1972, *Strid* 2606 (K; S).

Also in Zaire (Dem. Rep. Congo). *Brachystegia–Julbernardia* woodland on sandy soil; 1290 m.

It was at first thought the Zaire (Dem. Rep. Congo) material might represent a separate subspecies but the differences are inadequate. Much of it had been identified with *Rytigynia senegalensis* var. *ledermannii* Robyns [Type: Cameroon, Garua near Schuari, *Ledermann* 3595 (B†, holotype)] but the long appendages of the corolla lobe suggest no relationship with *senegalensis*. Nevertheless the naming is based on a sheet so-labelled by Robyns, so the synonymy could be correct and the distribution thus extended to Cameroon.

5. **Rytigynia orbicularis** (K. Schum.) Robyns in Bull. Jard. Bot. État **11**: 151 (1928). —F. White, F.F.N.R.: 420 (1962). —Verdcourt in Kew Bull. **42**: 156 (1987). Type from Angola.
    Plectronia orbicularis K. Schum. in Warburg, Kunene-Samb.-Exped. Baum: 388 (1903).
    Canthium orbiculare (K. Schum.) Good in J. Bot. **64**, Suppl. 2: 22 (1926).

Glabrous subshrub or shrub 0.6–1.2 m tall, with rather slender somewhat glaucous much branched stems with reddish-brown eventually somewhat flaking bark. Leaves sometimes folded and curved downward, glaucous or blue-green, 1.2–5 × 0.8–4 cm, ovate to round, or elliptic, mostly shortly acutely acuminate at the apex but sometimes rounded or even slightly emarginate, rounded to usually distinctly cordate at the base, glaucous; petiole 1–5 mm long; stipular sheath c. 1 mm long, with a subulate thickish appendage 1–2.5 mm long. Flowers solitary; peduncle 5–10 mm long; pedicels 5–8 mm long; bracts 2, minute, scarcely 1 mm long, lanceolate. Calyx tube 1–2 mm long, obconic, the limb 0.5–0.75 mm long undulate or shallowly toothed. Corolla apiculate in bud, greenish-white or yellowish, the lobes often a more distinct yellow; tube 4–6 mm long, funnel-shaped, with a ring of deflexed hairs below the throat inside; lobes triangular-lanceolate, 4–5 × 2 mm at the base, including the very short appendages. Stigma exserted about 1 mm; pollen presenter 1.5 mm long, ellipsoid, later subcoroniform, strongly 3-lobed at the apex. Ovary 3-locular. Fruit up to 8 × 10 mm with 1–3 pyrenes; pyrenes c. 8 mm long, deeply lobed in dry state.

Zambia. B: Kalabo to Sikongo, 21 km, fr. 14.ii.1952, *White* 2073 (FHO; K). W: Mwinilunga Distr., just east of R. Kasompa, fl. 1.ii.1938, *Milne-Redhead* 4440 (K). S: Machili, fl. 20.i.1961, *Fanshawe* 6152 (K; LISC; NDO).

Also in Angola. Woodland and thicket on Kalahari Sand; in *Cryptosepalum* and *Baikiaea plurijuga–Burkea africana* woodlands, and *Brachystegia bakeriana* thicket; c. 1500 m.

6. **Rytigynia uhligii** (K. Schum. & K. Krause) Verdc. in Kew Bull. **42**: 165, fig. 1F, 1G, 5F & 8D (1987); in F.T.E.A., Rubiaceae: 818, fig. 132/11 (1991). —Beentje, Kenya Trees, Shrubs Lianas: 544 (1994). Type from Tanzania.
    Vangueria neglecta sensu K. Schum. in Engler, Pflanzenw. Ost-Afrikas C: 384 (1895) non Hiern.
    Vangueria uhligii K. Schum. & K. Krause in Bot. Jahrb. Syst. **39**: 534 (1907). —De Wildeman in Bull. Jard. Bot. État **8**: 65 (1922).
    Plectronia kidaria K. Schum. & K. Krause in Bot. Jahrb. Syst. **39**: 539 (1907). Type from Tanzania.
    Rytigynia euclioides Robyns in Bull. Jard. Bot. État **11**: 146 (1928). —Brenan, Check-list For. Trees Shrubs Tang. Terr.: 529 (1949). —Dale & Greenway, Kenya Trees & Shrubs: 472 (1961). Type from Tanzania.

Rytigynia schumannii Robyns in Bull. Jard. Bot. État **11**: 158 (1928). —Brenan, Check-list
For. Trees Shrubs Tang. Terr.: 530 (1949). —Dale & Greenway, Kenya Trees & Shrubs: 473
(1961). —Drummond in Kirkia **10**: 276 (1975). —K. Coates Palgrave, Trees Southern
Africa, ed. 3, rev.: 875 (1988). Type from Tanzania.
  Rytigynia schumannii var. *uhligii* (K. Schum. & K. Krause) Robyns in Bull. Jard. Bot. État
11: 159 (1928).
  Rytigynia undulata Robyns in Bull. Jard. Bot. État **11**: 159 (1928). —Brenan, Check-list
For. Trees Shrubs Tang. Terr.: 530 (1949). Type from Tanzania.
  Rytigynia lenticellata Robyns in Bull. Jard. Bot. État **11**: 181 (1928). —Brenan, Check-list
For. Trees Shrubs Tang. Terr.: 530 (1949). Type from Tanzania.
  Rytigynia kidaria (K. Schum. & K. Krause) Bullock in Bull. Misc. Inform., Kew **1932**: 389
(1932). —Brenan, Check-list For. Trees Shrubs Tang. Terr.: 529 (1949).
  Canthium sarogliae Chiov., Racc. Bot. Miss. Consolata Kenya: 55 (1935). Types from
Kenya.
  Rytigynia murifolia Gilli in Ann. Naturhist. Mus. Wien **77**: 26, abb. 4 (1973). Type from
Tanzania.

Shrub or tree 0.9–9 m tall, rarely scandent(?), laxly branched; bark smooth, pale
brown to blackish-grey.   Stems pale chestnut-brown or purplish, longitudinally
ridged, lenticellate, glabrous, the epidermis or bark wearing off to reveal pale
undersurface.   Leaves mostly not fully developed at flowering time, 1.5–11 ×
0.5–4.8(5.2) cm, elliptic to ovate, narrowly or abruptly acuminate at the apex, the tip
usually rounded, cuneate, rounded or ± truncate at the base, glabrous save for
obvious tufted domatia, or rarely sparsely pubescent, ± thin, not or only slightly
discolorous; petiole 2–6 mm long; stipule bases 3–5 mm long, triangular or ovate,
joined, densely pilose within, with a ± spreading often compressed subulate
appendage 1–4.5 mm long, decurrent or sometimes completely lacking.
Inflorescences 1–2-flowered, rarely more; peduncle either ± suppressed or up to
6(10) mm long particularly in fruit; pedicels (2)6–12 mm long, often 15–23(30) mm
long in fruit; bracts up to 1.5 mm long. Calyx tube 1.2–1.5 mm long, the limb a rim
about 0.2 mm long with obsolete or very small teeth. Corolla oblong, very obtuse in
bud, the limb part characteristically much shorter than the tube, white to greenish-
yellow; the limb usually remains ± closed; tube 4.5–5.5 mm long, mostly subglobose
or urceolate-campanulate or sometimes ± cylindrical, with a ring of deflexed hairs
inside; lobes mostly green, 2 × 1.5 mm, ovate, acute but not appendaged.   Ovary
(2)3–5-locular.   Style swollen at the base; pollen presenter just exserted or exserted
up to 2.5 mm, 2–5-lobed.   Fruit blue-black, (7)9–10(?13 in life) mm in diameter,
globose, containing 1–5 pyrenes; pyrenes 6–7 × 3–4 mm.

Zimbabwe. E: Chimanimani Distr., Mutsangazi R., west of Mt. Peni, fl. 14.xii.1972, *Müller &
Goldsmith* 2068 (K; SRGH).   **Malawi**. N: Nyika Plateau, Nyamkowa (Nyamkhowa) Forest, fl.
14.xii.1982, *Dowsett-Lemaire* 534 (K). S: Mulanje Mt., Litchenya (Lichenya) Crater, sterile
4.ix.1970, *Müller* 1541 (K; SRGH).   **Mozambique**. Z: Serra do Gurué, Namuli, R. Malema, fl.
6.xi.1967, *Torre & Correia* 15973 (C; COI; K; LMU).

Also in Kenya and Tanzania.   Evergreen forest understorey or sub-canopy; (700)900–2000 m.
Rytigynia uhligii occurs very widely in upland forest in East Africa and in common with so
many forest species is well distributed there. It has its southern limit in E Zimbabwe. In East
Africa it is distinct from *Rytigynia adenodonta*, in fact typical material is not remotely similar. In
Malawi and Mozambique, however, there is difficulty separating it from *Rytigynia adenodonta* var.
reticulata which usually has smaller thicker less acuminate leaves, a more nodular stem and
slightly apiculate buds. Unfortunately, well collected material is lacking from some key areas.
Several sterile specimens may belong to *R. uhligii*. Some inadequate material from Malawi may
represent hairy forms of this species. *J.D. & E.G. Chapman* 9091 [Mt. Mulanje, Michesi Mt.,
Nandiwo Valley, fr. 27.v.1988] has the fruits described as flattened-spherical, c. 35 × 30 mm.
One dry fruit on the specimen is 16 mm in diameter. If the Mulanje population proves to be
uniformly large-fruited it must be at least a distinct subspecies.

7. **Rytigynia adenodonta** (K. Schum.) Robyns in Bull. Jard. Bot. État **11**: 148 (1928). —Brenan,
    Check-list For. Trees Shrubs Tang. Terr.: 529 (1949). —Verdcourt in Kew Bull. **42**: 166
    (1987); in F.T.E.A., Rubiaceae: 819 (1991). Type from Tanzania.
      Vangueria adenodonta K. Schum. in Bot. Jahrb. Syst. **30**: 414 (1901). —De Wildeman in
    Bull. Jard. Bot. État **8**: 42 (1922).

Much-branched shrub or small tree 1.5–3.6 m tall, entirely glabrous (save for
domatia and insides of stipules), or with young shoots slightly to densely appressed

pubescent; older branches greyish, ± nodulose and verrucose, the bark often red-brown and fissured and flaky. Leaves 1.3–6 × 0.4–2.8 cm, elliptic to broadly ovate-elliptic, abruptly shortly obtusely acuminate at the apex, ± rounded or cuneate at the base, scarcely discolorous, glabrous save for small pubescent domatia which may be ± absent or, rarely (Malawi) pubescent beneath and on midrib above; petioles 3–4 mm long; stipules connate at the base, glabrous outside, densely long-villous inside, 2 mm long, abruptly subulate-cuspidate, the appendage 1–2 mm long. Flowers solitary, or in lax 2–4-flowered cymes; pedicels rather thick, 3–5(11) mm long. Calyx tube 1.5 mm long; limb obsoletely dentate, truncate or teeth broadly triangular, 0.8 mm long. Corolla ± rounded to acuminate at the apex in bud, creamy-white; tube 5–6 × 3.5–5 mm obconic, dilated from the base, with a ring of deflexed long hairs inside from the top of the tube; lobes oblong, reflexed, reaching about the middle of the tube, about 3–3.5 × 2–2.5 mm, ovate, obtuse or with a short appendage to 0.5 mm long. Ovary 3–4-locular. Pollen presenter coroniform, 1 mm wide, 3–4-lobed. Fruit purple-black, about 8 mm long.

Var. **adenodonta** —Verdcourt in Kew Bull. **42**: 166 (1987); in F.T.E.A., Rubiaceae: 819 (1991).

Young stems often hairy, with mostly abbreviated nodose lateral branches. Leaf blades small, about 2–2.5 × 1–1.5 cm, obtuse to acuminate.

Zambia. N: Isoka Distr., Mafinga Hills., near Chisenga (in Malawi), fl. 21.xi.1952, *Angus* 820 (FHO; K). **Malawi**. N: Rumphi Distr., Nyika Plateau, Chowo Rock, fl. bud 6.v.1977, *Pawek* 13193 (K; MAL; MO; SRGH; UC).
Also in Tanzania. Submontane evergreen forest and forest margins, usually towards the upper end of ravines, on rocky slopes; 2100 m.

Var. **reticulata** (Robyns) Verdc. in Kew Bull. **42**: 167 (1987); in F.T.E.A., Rubiaceae: 819 (1991). Type: Malawi, Mt. Zomba, *Whyte* (K, holotype).

Young stems mostly glabrous, with side branches developed and leafy. Leaf blades up to 7.5 × 4.5 cm, acuminate.

Zambia. E: Nyika Plateau, 8.8 km southwest of Rest House, fl. 25.x.1958, *Robson & Angus* 354 (BM; K; LISC) (atypical). **Malawi**. N: Nyika Plateau, Engineers' Forest, Nthalire forest crossroad, fl. 27.xii.1975, *E. Phillips* 747 (K; MO). C: Dedza Mountain Forest, fl. & fr. 10.xii.1962, *Banda* 474 (K; SRGH). S: Zomba Distr., Zomba Plateau, Chingwe's Hole Nature Trail, fl. & fr. 13.ii.1982, *Brummitt & Chapman* 15864 (K). **Mozambique**. N: Serra de Ribáuè, Mepáluè, fr. 25.i.1964, *Torre & Paiva* 10228 (LISC) (?). Z: Serra do Gurué, 3 km after the falls on R. Licungo, fl. 24.ii.1966, *Torre & Correia* 14851 (C; COI; LISC; LMU; MO; SRGH; WAG).
Also in southern Tanzania. Submontane evergreen forest and forest margins, often in rocky places; (1200)1770–2200 m.
Robyns separated *Rytigynia reticulata* and *Rytigynia adenodonta* on whether the buds were obtuse or apiculate, but there is much variation between them, most specimens having very short corolla lobe appendages. The Tanzanian population of var. *reticulata* from Iringa was at first thought to be a subspecies of *Rytigynia uhligii* differing in having rather longer wider corolla tubes and longer lobes, thicker more discolorous leaves and often more developed calyx teeth. There seems no doubt that these Iringa specimens should be included in *R. adenodonta* var. *reticulata*. Genuine *R. uhligii* also occurs in the Iringa area. In the Flora Zambesiaca area however, the relationship between *uhligii* (which is widespread and well defined throughout eastern Kenya and Tanzania) and *adenodonta* is much less clear and needs more study. *Pawek* 11157 (EA; K; MAL; MO; SRGH; UC) and *Dowsett-Lemaire* 658 (K; MAL) from Malawi [N: Misuku Hills] are a variant of var. *reticulata* with pubescent stems and leaves.

8. **Rytigynia pawekiae** Verdc. in Kew Bull. **42**: 168 (1987), as "*pawekae*". Type: Malawi, Misuku Hills, *Pawek* 12192 (PRE, holotype; K; MAL; MO; SRGH; UC).

Shrub up to 2 m tall, the younger branches pale but later ferruginous, densely covered with appressed brown hairs. Leaves paired, 3–6 × 1.2–3 cm, oblong, shortly subobtusely acuminate at the apex, cuneate to rounded at the base, densely grey pubescent on both surfaces; stipule sheath 2 mm long, pubescent; stipule appendage 3–6 mm long, subulate. Cymes dichasial, (1)7–13-flowered, the flowers congested; peduncle 0–4 mm long; pedicels 3–4 mm long; bracts subhyaline, 2 mm long, elliptic; bracteoles 1–1.5 mm long, narrowly ovate. Calyx tube 1.5–2 mm long, campanulate, rugulose, densely pubescent; limb short, 0.2–0.8 mm long, ± hyaline, glabrous, subdentate or slightly crenate-dentate. Corolla shortly acuminate in bud, yellow, pubescent above; tube 3.5 mm long, glabrous below outside, inside with a ring of deflexed hairs at the throat which reach to the middle of the tube; lobes 4.6 × 1.6–2 mm including the 1 mm long appendage, narrowly ovate-triangular, pubescent outside. Style 4.5 mm long; the pollen presenter 1.5 mm long, subglobose, 8-sulcate, 3-lobed at the apex. Ovary 3-locular. Fruit not known.

Malawi. N: Chitipa Distr., Misuku Hills, Iponjola, fl. 2.i.1977, *Pawek* 12192 (K; MAL; MO; PRE; SRGH; UC).
Known only from this gathering. Submontane grassland with shrubs on hillsides and slopes; 1650 m.
The leaves are reminiscent of *Tapiphyllum* but the calyx lobes are scarcely developed.

9. **Rytigynia torrei** Verdc. in Bull. Jard. Bot. Belg. **62**: 416–417 (1993). Type: Mozambique, 40 km from Malema (Entre Rios) towards Ribáuè, Serra Murripa, fl. 16.xii.1967, *Torre & Correia* 16555 (LISC, holotype).

Shrub up to 4 m tall; young stems mostly glabrous but sparsely setose-pubescent in places, particularly at the nodes; older stems grey, ridged, densely but not very conspicuously lenticellate. Leaves 3–7.5 × 1.5–3.5, oblong, oblong-ovate or ± elliptic, acute at the apex, cuneate at the base, discolorous, glabrous; petiole 1–2 mm long; stipular sheath 1.5 mm long, pubescent, with a subulate appendage 4–5 mm long. Flowers in 5–10-flowered axillary fascicles, forming nodal clusters; peduncles ± obsolete; pedicels 3.5–4.5 mm long. Calyx tube c. 1.5 × 2 mm, cup-shaped, glabrous but micro-papillate; lobes about 1–2 × 1–1.5 mm, triangular, slightly joined at the base, more or less spreading, glabrous. Corolla with very short tails in bud, glabrous outside, greenish; tube 3–4 mm long, cylindrical or slightly urceolate, hairy at the throat and with a ring of deflexed bristly hairs about the middle; lobes c. 2–2.5 × 1–1.8 mm including the short apiculum, ovate. Anthers apiculate, just exserted. Style exserted 1 mm; pollen presenter 1 mm long, narrow, 5-lobed. Ovary 5-locular. Fruit not known.

Mozambique. N: at km 40 from Malema (Entre Rios) on road to Ribáuè, Serra Murripa, fl. 16.xii.1967, *Torre & Correia* 16555 (LISC).
Known only from this locality. Wooded grassland on rocky hills, on edge of mist forest; 1100–1200 m.
This distinctive species differs from all others in its broad calyx lobes, glabrous foliage and 5–10-flowered inflorescences.

10. **Rytigynia celastroides** (Baill.) Verdc. in Kew Bull. **42**: 171, fig. 1H, 1J & 8A (1987); in F.T.E.A., Rubiaceae: 823 (1991). —Beentje, Kenya Trees, Shrubs Lianas: 543 (1994). Syntypes from Kenya.
Canthium celastroides Baill. in Adansonia **12**: 190 (1878)* .

Densely branched shrub or small tree 1.8–7.5 m tall, or sometimes scrambling; bark rather rough grey, brown or whitish-grey; branchlets sparsely to densely pubescent on young parts, often with ± yellowish hairs, persistent or becoming glabrous; branches often in whorls of 3; spines frequently present, solitary, 7–13 mm long. Leaves usually paired but sometimes in whorls of 3; blades 2–5.5(9) × 0.8–2.5(4) cm, narrowly elliptic, ovate-lanceolate or lanceolate, obtuse to narrowly acuminate at the apex, the tip itself usually obtuse, cuneate or rarely rounded at the base, mostly densely pubescent and often ± velvety beneath when young, or persistently so in one variant, or entirely glabrous save for domatia, or only sparsely pubescent; petiole 0.5–2 mm long; stipule bases 1–1.5 mm long, connate, truncate, subscarious, pubescent to glabrescent outside, densely villous inside, with a subfleshy deciduous often ± reflexed appendage 2.5–4(6) mm long. Inflorescences (1)2–4(7)-

---

* Index Kewensis erroneously gives "= *Plectronia celastroides* Baker", a very different Seychelles species.

flowered, not subumbellate but with secondary branches or rhachis usually shortly developed; peduncle glabrous to pubescent, obsolete or up to 3(7) mm long; bracts scarious, forming a cup c. 1 mm long, glabrous to pubescent; pedicels 2–9 mm long, glabrous to pubescent, sometimes up to 14–20 mm long in fruit. Calyx tube 1–1.5 mm long, the limb truncate to shortly toothed, 0.25–0.5 mm long or rarely the calyx lobes more developed, 0.5–2 mm long. Corolla obtuse to distinctly but shortly acuminate in bud, white, cream, yellowish or greenish, glabrous or very sparsely to densely pubescent; tube 1.5–2 mm long, ± campanulate with a ring of deflexed hairs inside; lobes (1.5)2.5–3.5 × 1–1.5 mm oblong, glabrous to densely pubescent outside, ± papillate inside, acute or with short subulate appendage 0.5–0.75 mm long. Ovary 2(5)-locular. Style 2–2.5(4) mm long; pollen presenter coroniform, 3(5)-lobed. Fruit black, 6–8.1(9) × 8.5–9.5 mm, subglobose, usually deeply lobed in dry state, with 1–3(5) pyrenes, glabrous. Pyrenes about 10 × 6 mm, rounded reniform or oblong-segmentoid with notch just above the middle.

Var. **celastroides** —Verdcourt in Kew Bull. **42**: 171, fig. 8A (1987); in F.T.E.A., Rubiaceae: 824 (1991).
    Vangueria glabra K. Schum. in Engler, Pflanzenw. Ost-Afrikas **C**: 384 (1895). —De Wildeman in Bull. Jard. Bot. État **8**: 52 (1922). Type from Tanzania.
    Vangueria microphylla K. Schum. in Engler, Pflanzenw. Ost-Afrikas **C**: 385 (1895); in Bot. Jahrb. Syst. **28**: 494 (1900). —De Wildeman in Bull. Jard. Bot. État **8**: 59 (1922), pro parte. Syntypes from Tanzania.
    Vangueria oligacantha K. Schum. in Bot. Jahrb. Syst. **34**: 334 (1904). —De Wildeman in Bull. Jard. Bot. État **8**: 60 (1922). Type from Tanzania.
    Plectronia amaniensis K. Krause in Bot. Jahrb. Syst. **43**: 142 (1909). Syntypes from Tanzania.
    Rytigynia microphylla (K. Schum.) Robyns in Bull. Jard. Bot. État **11**: 171 (1928). —Brenan, Check-list For. Trees Shrubs Tang. Terr.: 530 (1949). —Dale & Greenway, Kenya Trees & Shrubs: 472 (1961).
    Rytigynia oligacantha (K. Schum.) Robyns in Bull. Jard. Bot. État **11**: 172 (1928). —Brenan, Check-list For. Trees Shrubs Tang. Terr.: 531 (1949). —Dale & Greenway, Kenya Trees & Shrubs: 471 (1961).
    Rytigynia glabra (K. Schum.) Robyns in Bull. Jard. Bot. État **11**: 174 (1928). —Brenan, Check-list For. Trees Shrubs Tang. Terr.: 530 (1949).
    Rytigynia amaniensis (K. Krause) Bullock in Bull. Misc. Inform., Kew **1932**: 389 (1932). —Brenan, Check-list For. Trees Shrubs Tang. Terr.: 531 (1949). —Dale & Greenway, Kenya Trees & Shrubs: 471 (1961).
    Canthium frangula sensu J.G. Garcia in Mem. Junta Invest. Ultramar **6** (sér. 2): 30 (1959) [Contrib. Conhec. Fl. Moçamb. IV (1959)], pro parte, quoad *Torre* 4887, non Hiern.

At least some parts of leaves, stems, inflorescences and flowers sparsely to densely pubescent; leaves 2–5.5(9) × 0.8–2.5(4) cm, usually narrowly acuminate; inflorescences 2–7-flowered.

Mozambique. N: Mutuáli–Malema road, 5 km from Mutuáli, fr. 14.ii.1954, *Gomes e Sousa* 4194 (K; LMA; PRE). Z: between Milange and Mocuba, 3.iii.1943, *Torre* 4887 (LISC).
    Also in eastern Tanzania and Kenya. Open deciduous forest; 450 m.

Var. **australis** Verdc. in Kew Bull. **42**: 172 (1987). Type: Mozambique, Nhachengue (Inhachengo), *Exell, Mendonça & Wild* 618 (BM; LISC, holotype).

Plant glabrous or glabrescent save for large hairy domatia on the leaf lower surface, or sparsely setulose; leaves 2–2.5 × 0.7–1 cm, shortly acuminate to a rounded tip; inflorescences 1(2)-flowered.

Mozambique. GI: Nhachengue (Inhachengo), fl. & fr. 26.ii.1955, *Exell, Mendonça & Wild* 618 (BM; LISC). M: Mangulane, fr. iv.1931, *Gomes e Sousa* 522 (K).
    Also in South Africa (KwaZulu-Natal). Deciduous woodland with scattered trees and shrubs, in sandy places.
    Recognising this variety is merely a convenience to cope with a few specimens which are remarkably similar to specimens from the opposite end of the geographic range (Kenya and Somalia) of *Rytigynia celastroides*, referred to *Rytigynia parvifolia* Verdc.
    Rytigynia celastroides var. *nuda* Verdc. (from Tanzania) is similar to var. *celastroides* but completely glabrous. *Torre & Paiva* 10026 (LISC) [Mozambique N: 36 km from Imala to Mecubúri] has the foliage of var. *australis* and spines and indumentum of more typical *R. celastroides*.

11. Rytigynia sp. B

Shrub up to c. 3 m tall; young shoots glabrous, with a purplish-brown longitudinally fissured flaking epidermis; older stems with grey wrinkled bark. Leaves 2–4 × 1–1.8 cm, elliptic, obtuse to very shortly but obtusely acuminate, the apex always rounded, cuneate at the base into a petiole c. 1 mm long, glabrous, ± discolorous; lateral nerves ± visible, but tertiary venation not visible; stipular sheath 1–1.5 mm long with a slender subulate appendage c. 3 mm long. Flowers solitary or in 2-flowered cymes; peduncles 2–4 mm long; pedicels 5–7 mm long; bracteoles 1 mm long, all glabrous. Calyx tube 1.2–1.5 mm long, cup-shaped, the limb reduced to a very short minutely denticulate rim or occasionally developed on one side into a ± triangular lobe 1.5 mm long. Corolla slightly acuminate but not tailed in bud, greenish, glabrous outside; tube shortly funnel-shaped or sub-urceolate, 2 mm long, inside with a prominent ring of deflexed hairs at the throat, extending for ± two thirds of the tube; lobes 3.5–4.2 × 1.3–1.8 mm, ovate-lanceolate, acute. Anthers just exserted. Ovary 4-locular. Style swollen below, exserted 0–2 mm; the pollen presenter 1.2 mm long, 3-lobed. Fruits not seen.

Mozambique. Z: Alto Molócuè, 10 km from Gilé, Mt. Gilé, fl. 21.xii.1967, *Torre & Correia* 16684 (LISC).
Known only from this locality. Rocky slopes, 'rupideserta', in clefts in the rock or on the ground; 300 m.
Perhaps allied to the very variable *R. celastroides* but even glabrous-leaved variants of that usually have domatia and different inflorescences. Nevertheless the facies suggests such an affinity and it would be foolish to describe it on a single specimen.

12. Rytigynia sp. C

Shrub up to c. 2 m tall; young shoots densely spreading yellowish pubescent; older shoots with epidermis peeling to show purplish-brown bark, glabrescent. Leaves 1.4–2.7 × 0.7–2 cm, elliptic or broadly elliptic, very shortly obtusely acuminate at the apex, the tip quite rounded, cuneate at the base, discolorous, ± appressed velvety-hairy on both sides but the actual surfaces not obscured; domatia present beneath; stipular sheath c. 1 mm long with a subulate appendage c. 3 mm long, densely pubescent. Cymes 1–2(3)-flowered; peduncles 0–3 mm long, pedicels 5–7 mm long; bracteoles 1 mm long, all parts densely spreading pubescent. Calyx limb truncate, minutely toothed. Corolla obtuse, limb densely pubescent in bud. Immature fruits 7 mm wide, subglobose, pubescent.

Mozambique. N: Serra de Ribáuè, Mepáluè, buds, young fr. 23.i.1964, *Torre & Paiva* 10152 (LISC).
Brachystegia woodland on red sandy soil; 800 m.
This, like *Rytigynia sp. B*, is closely related to *R. celastroides*, and at least one specimen of that species from Tanzania (*Greenway* 4589 from the Pare Mts.) is similarly velvety. Unusual forms of widespread species in other groups have been found on Ribáuè.

13. **Rytigynia monantha** (K. Schum.) Robyns in Bull. Jard. Bot. État **11**: 153 (1928). —Brenan, Check-list For. Trees Shrubs Tang. Terr.: 529 (1949), pro parte. —Bridson & Troupin in Fl. Rwanda **3**: 217, fig. 68 (1985). —Verdcourt in Kew Bull. **42**: 173 (1987); in F.T.E.A., Rubiaceae: 825 (1991). Type from Tanzania.
   Vangueria monantha K. Schum. in Bot. Jahrb. Syst. **28**: 493 (1900). —De Wildeman in Bull. Jard. Bot. État **8**: 59 (1922).

Much-branched shrub or small tree 0.3–4.5 m tall; older stems brown, purple-brown or reddish, the bark very finely fissured smooth, or sometimes in exposed areas becoming rough and nodose, the dark very rough bark flaking to reveal a reddish or pale brownish powder; branching often ± horizontal; young shoots green in life, densely shortly spreading pubescent with sometimes bright rusty hairs, or in some areas (e.g. Malawi) ± glabrous. Leaves 1–6 × 0.5–4 cm, ovate, oblong or elliptic, narrowly acuminate at the apex, rounded to cuneate or sometimes subcordate at the base, typically markedly discolorous with a dense appressed yellowish or whitish pubescence above and a dense ± coarsely velvety pubescence beneath, the pubescence not entirely obscuring the ± reticulate venation, or sometimes the

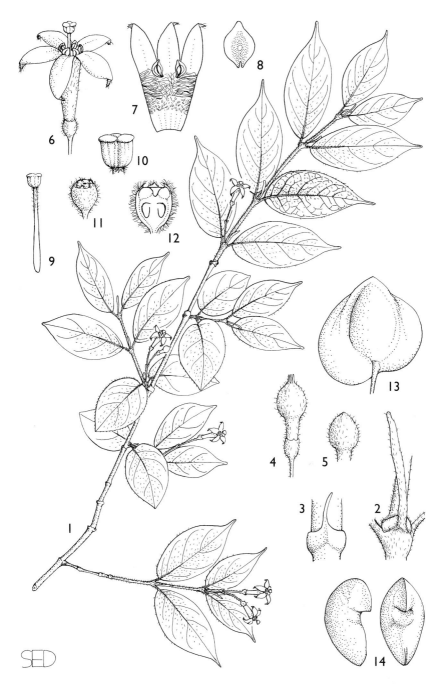

Tab. 56. RYTIGYNIA MONANTHA var. MONANTHA. 1, flowering branch (× ²/₃); 2, young stipule (× 4), 1 & 2 from *Polhill & Paulo* 1599; 3, mature stipule (× 4), from *Procter* 940; 4, bud (× 4), from *Procter* 381; 5, top of untailed bud (× 4), from *Bridson* 280; 6, flower (× 3), from *Polhill & Paulo* 1599; 7, part of opened corolla (× 3); 8, dorsal view of anther (× 8); 9, style and pollen presenter (× 3); 10, pollen presenter (× 8); 11, calyx (× 4); 12, section through ovary (× 6), 7–12 from *Bridson* 280; 13, fruit (× 3), from *Peter* 37734; 14, pyrene, 2 views (× 4), from *Peter* 37611. Drawn by Sally Dawson. From F.T.E.A.

indumentum is much sparser and some Malawi specimens have almost glabrous leaves; petiole 1–4 mm long; stipules with the hairy basal hyaline parts joined to form a sheath 1.5–2 mm long; stipule appendages compressed, 3–10 mm long, lanceolate or narrowly falcate, drying dark, or rarely the appendages pale and filiform. Inflorescences 1–2-flowered; peduncle 4–16 mm long; pedicels (2)3–9 mm long, lengthening to 19 mm in fruit; bracts ± connate, 1.5–3.5 mm long, ovate, acuminate. Calyx tube 1–1.5 mm long, densely pubescent outside, densely hairy inside all over; lobes 5–8, 0.5–1.5 mm long, linear to triangular, (long and linear in Malawi). Corolla clavate, acuminate, apiculate or with 5 distinct tails in bud, but very variable and sometimes (in Rwanda) ± blunt, white, pale yellow or greenish, usually spreading bristly pubescent but occasionally glabrous; tube 3–6 mm long, ± glabrous inside, with few deflexed hairs outside or pubescent in upper half; lobes 2–4 × 1–2 mm, triangular to lanceolate, acute, with an apiculus or short appendage 0.5–1.5 mm long, often creamy-white inside, usually bristly pubescent outside or less often glabrous (particularly in Malawi material). Style white, usually exserted 1–3 mm, sometimes with few hairs just beneath the pollen presenter, ± dilated at base; pollen presenter whitish, 1 mm long, depressed subglobose or oblong-obconic, sulcate beneath, 3–4 lobed at apex. Ovary 3–4-locular. Fruit 6–8 × 7–8 mm, subglobose, pubescent, with 1–4 pyrenes.

Var. **monantha**   —Verdcourt in Kew Bull. **42**: 174 (1987); in F.T.E.A., Rubiaceae: 826, figs. 132/12, 145 (1991). TAB. **56**.
  Rytigynia castanea Lebrun, Taton & Toussaint, Contr. Fl. Parc Nat. Kagera **1**: 139 (1948). —Bridson & Troupin in Fl. Pl. Lign. Rwanda: 598, fig. 203.1 (1982); in Fl. Rwanda **3**: 217 (1985). Type from Rwanda.
  Rytigynia sp. of Brenan in Mem. N.Y. Bot. Gard. **8**: 452 (1954).
  Rytigynia sp. 2 & 3 of F. White, F.F.N.R.: 420 (1962).

Leaves ovate, oblong or elliptic, mostly densely pubescent, usually velvety beneath. Corolla usually with dense spreading hairs outside.

  Zambia. N: 30 km SW of Mpika on Lusaka road, fl. 14.i.1975, *Brummitt & Polhill* 13778 (K). E: Lundazi Distr., Nyika Plateau, upper slopes of Kangampande, fr. 6.v.1952, *White* 2735 (FHO; K; WAG). C: near Mfuwe Camp, fr. 24.ii.1966, *Astle* 4598 (SRGH). **Malawi**. N: Nyika Plateau, N Rukuru waterfall, fl. 6.i.1959, *E.A. Robinson* 3095 (K; PRE). C: Ntchisi Mt., fl. & fr. 21.ii.1959, *Robson* 1708 (BM; K; LISC).
  Also in Rwanda, Burundi and Tanzania. Extending from *Brachystegia* woodland at lower altitudes to evergreen forest at higher altitudes, very often in rocky places; 585–2250 m.
  Flora Zambesiaca material often has the corolla glabrous or glabrescent.
  R. monantha var. *lusakati* Verdc., from Tanzania, may be distinguished by its leaves which are lanceolate or oblong-lanceolate and glabrescent save for pubescence along the main nerves, and by its flowers which are glabrous or with a very few scattered scarcely visible hairs.

14. **Rytigynia rubiginosa** (K. Schum.) Robyns in Bull. Jard. Bot. État **11**: 209 (1928). — Verdcourt in Kew Bull. **42**: 175 (1987). Type from Zaire (Dem. Rep. Congo).
  Vangueria rubiginosa K. Schum. in Bot. Jahrb. Syst. **23**: 457 (1897); **28**: 72 (1899). —T. Durand, Syll. Fl. Congol.: 270 (1909). —De Wildeman in Bull. Jard. Bot. État **8**: 61 (1922).

Shrub or tree 1.5–4 m tall (or climber fide *Nawa et al.* 140); branches dark reddish-brown or blackish, densely covered with ± spreading ferruginous hairs when young, later ± glabrous. Leaves held in one plane, 2–10.5 × 0.8–3.5 cm, oblong, narrowly long-acuminate at the apex, rounded and sometimes somewhat subcordate at the base, discolorous, thinly or densely pubescent with pale hairs on both surfaces, costa with ferruginous hairs on both surfaces and the c. 5 lateral nerves reddish beneath contrasting with glaucous-green blade; petiole 2–6 mm long, hairy like the stems; stipule-sheath 1.5 mm long, ferruginous-pubescent, ± persistent, with oblong-lanceolate glabrous appendage 3–4 mm long, soon deciduous. Flowers solitary, or in 2-flowered cymes; peduncles 1–1.5 mm long; pedicels c. 2 mm long; bracts up to 3 mm long, filiform, all parts densely ferruginous-pubescent. Calyx tube 1.5 mm long, obconic, ferruginous-pubescent; limb c. 0.5 mm long, with distant teeth under 0.5 mm long. Corolla acuminate and with distinct tails in bud; tube 2–2.5 mm long, with upper part ferruginous-pubescent, inside with a ring of deflexed hairs; lobes 3–4 mm long, triangular-lanceolate, with long filiform appendages 2–2.5 mm long. Style 4–5 mm long; pollen presenter coroniform, sulcate, deeply 5-lobed at the apex. Ovary 5-locular. Fruit 10–12 mm in diameter, globose but 3–6-lobed when dry, glabrous; pyrenes 8 mm long with a pitted-reticulate surface.

Subsp. **rubiginosa** —Verdcourt in Kew Bull. **42**: 175 (1987).
  Rytigynia vilhenae Cavaco in Bull. Mus. Hist. Nat. (Paris) sér. 2, **29**: 515 (1958); Diamang, Publicações Culturais, Museu do Dundo No. **42**: 149, pl. 7/4–5 (1959). Type from Angola.

Leaves rather thinly pubescent, the lower surface clearly visible, the ferruginous veins contrasting with the glaucous green blade.

  Zambia. W: Mwinilunga Distr., 6 km north of Kalene Hill, 12.xii.1963, *E.A. Robinson* 5932 (K). Also in Zaire (Dem. Rep. Congo) (Kasai Prov.) and Angola. Riverine forest; 1230–1300 m.
  Subsp. *cymigera* (Bremek.) Verdc. occurs in eastern Zaire (Dem. Rep. Congo), and may be distinguished from the typical subspecies by its leaf blades densely velvety pubescent beneath, the indumentum completely obscuring the leaf surface.

15. **Rytigynia macrura** Verdc. in Kew Bull. **42**: 178 (1987). Type: Zimbabwe, Mutare Distr., Vumba, *Drummond* 5076 (K, holotype; LISC; SRGH).
  Rytigynia sp. 1 of Drummond in Kirkia **10**: 276 (1975). —K. Coates Palgrave, Trees Southern Africa, ed. 3, rev.: 876 (1988).

A shrub 1.5–4.5 m tall, or small tree; older parts with a smooth grey bark; branches elongate with graceful horizontal branchlets, at first pubescent. Leaves paired, 1–6.5 × 0.4–3 cm, usually narrowly obovate or sometimes elliptic, or ± rounded, thin, narrowly acuminate at the apex, cuneate at the base, sparsely to densely appressed pilose on both sides and also with domatia beneath; petiole 0–3 mm long, pubescent; stipule bases ± triangular, subscarious, ± pubescent, joined into a sheath 1–3 mm long with a subulate laterally compressed appendage 1.5–5 mm long. Flowers solitary; pedicels 5–6 mm long, pubescent. Calyx tube pubescent, 1.2 mm long; limb 0.5 mm long with minute filiform lobes 0.5 mm long. Corolla produced at the apex into 5 tails in bud, white or greenish-cream, pubescent or glabrescent; tube 3.5–4 mm long, pubescent or glabrescent outside, with a ring of long deflexed hairs inside at the middle; lobes 4–4.5 × 1.7–2 mm, oblong, spreading, with linear-lanceolate greenish appendages 5–7 mm long. Ovary 2-locular; style exserted; pollen presenter 1.5 mm long and wide, distinctly bilobed. Fruit yellow or yellowish, or whitish-green, globose and 14 mm in diameter in live state but usually bilobed, 9–12 mm long and wide when dry, glabrous or sparsely pilose; pyrenes c. 8 mm long.

  Zambia. E: Nyika Plateau, Kasoma Forest, fl. 10.i.1983, *Dowsett-Lemaire* 562 (FHO).
  Zimbabwe. E: Mutare Distr., Vumba Mts., Bunga Forest, fl. 28.i.1979, *Müller* 3595 (K; SRGH).
  Malawi. N: Chitipa Distr., Misuku Hills, Mughesse, fl. 4.i.1974, *Pawek* 7775 (K; MA; MO; SRGH; UC). C: Nkhota Kota Distr., Ntchisi (Nchisi) Mt., fl. 14.i.1967, *Hilliard & Burtt* 4477 (E; K). S: Mulanje, W side of Mauzi Mt. (Maudzi Hill), fr. 6.ix.1983, *Seyani & Balaka* 1370 (K; MAL).
  Mozambique. N: Ribáuè, Serra de Chinga, fl. 12.xii.1967, *Torre & Correia* 16469 (K; LISC; LMA). Z: Milange, Serra Tumbine, fl. 19.i.1966, *Correia* 522 (LISC). MS: Chimoio Distr., Garuso (Garuzo) Mt., 14.xii.1943, *Torre* 6294 (COI; LISC; LMU).
  Known only from the Flora Zambesiaca area. Submontane mixed evergreen rain forest with *Aningeria–Chrysophyllum–Macaranga*, etc., also in *Brachystegia spiciformis–Uapaca* woodland below these forests; 1040–2000 m.
  Chapman 5836 [from woodland protected from burning for at least 20 years] has the fruiting pedicel about 10 mm long and a peduncle 6 mm long, with paired bracts forming a small involucre, and with dark purplish bark. However, it seems to belong to this species. *Fanshawe* 5601 [Zambia. N: Sunzu Hill] has the characteristic habit of this species, but has shorter blunter leaf apices; unfortunately it is in fruit only.

16. **Rytigynia bugoyensis** (K. Krause) Verdc. in Bull. Jard. Bot. Belg. **50**: 515 (1980); in Kew Bull. **42**: 184 (1987); in F.T.E.A., Rubiaceae: 833 (1991). Type from Zaire (Dem. Rep. Congo).
  Plectronia bugoyensis K. Krause in Mildbraed, Wiss. Ergebn. Deutsch. Zentr.-Afrika Exped., Bot., part 4: 327 (1911).
  Vangueria butaguensis De Wild. in Bull. Jard. Bot. État **8**: 47 (1922); Pl. Bequaert. **2**: 270 (1923). Type from Zaire (Dem. Rep. Congo) (Ruwenzori).

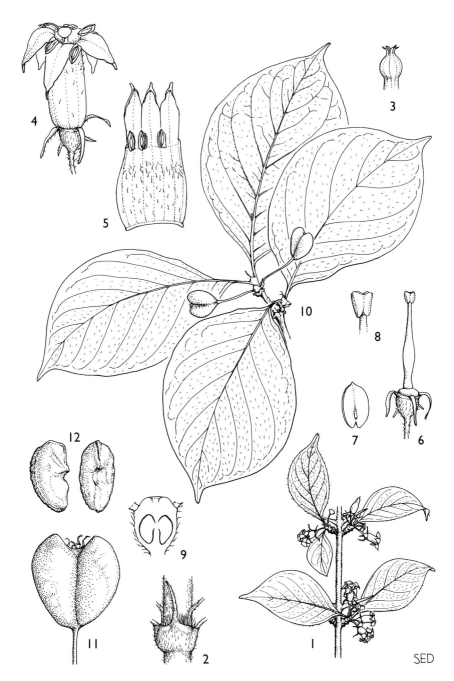

Tab. 57. A. —RYTIGYNIA BUGOYENSIS subsp. GLABRIFLORA. 1, flowering branch (× ⅔), from *Purseglove* 2576; 2, stipule (× 4), from *Paulo* 583; 3, top of flower bud (× 2), from *Snowden* 907/a; 4, flower (× 4); 5, portion of corolla opened out (× 4); 6, calyx with style and pollen presenter (× 4); 7, dorsal view of anther (× 8); 8, pollen presenter (× 8); 9, longitudinal section through ovary (× 8); 10, fruiting branch (× ⅔), 4–10 from *Purseglove* 2576; 11, fruit (× 2), from *Osmaston* 1189; 12, pyrene, 2 views (× 2), from *Snowden* 907/a. Drawn by Sally Dawson. From F.T.E.A.

Rytigynia butaguensis (De Wild.) Robyns in Bull. Jard. Bot. État **11**: 145 (1928); Fl. Sperm. Parc Nat. Alb. **2**: 347 (1947). —Dale & Greenway, Kenya Trees & Shrubs: 472 (1961).

Hutchinsonia bugoyensis (K. Krause) Bullock in Bull. Misc. Inform., Kew **1932**: 389 (1932).

Canthium urophyllum Chiov., Racc. Bot. Miss. Consolata Kenya: 54 (1935), as "*C. urophyllam*". Type from Kenya.

Shrub or small tree 1–6 m tall with slender spreading branches, sometimes ± scandent; shoots pubescent when young, becoming glabrous, with ± lenticellate grey-brown bark on older branches; older branches with supra-axillary ± opposite paired spines 6–25 mm long, and often bearing leaves and flowers on brachyblasts (short condensed nodular lateral shoots), the lateral shoots bearing only one pair of leaves or the apical portions of otherwise normal shoots can have the internodes much condensed and similarly bear a pair of leaves. Leaf blades 3–10.5 × 1.2–7 cm, ovate to elliptic or oblong-elliptic, long-acuminate at the apex, the actual tip narrowly rounded, rounded to cuneate at the base, very thin, rather sparsely pilose with appressed ± rough hairs on both surfaces, with ± obscure barbellate domatia, ciliate; petiole 3–10 mm long, bristly pubescent; stipules connate at the base, 4 mm long including the 1.5–2 mm long filiform apex, triangular, deciduous. Inflorescences 2–3-flowered axillary cymes, or flowers solitary; peduncles 4 mm long and pedicels 3–10 mm long, both ± densely pubescent. Calyx tube hemispherical, 1.2–1.5 mm long, ± glabrous or pilose; lobes 1–3.5 mm long, linear-lanceolate, glabrous. Corolla clavate, shortly tailed in bud, greenish to yellowish-cream; tube 3–6 mm long, glabrous to pilose outside, and pubescent with tangled hairs in most of inside; lobes 1.5–3.5 mm long, ovate-oblong. Ovary 2-locular. Style 4.5–5 mm long, thickened towards the base; pollen presenter 5–8 mm long, coroniform, 2-lobed. Fruit yellowish-green (apparently ripe), or black, 12–13 × 11–12 mm, compressed-obcordate in lateral view, broadly emarginate at the apex, glabrous or with a few bristly hairs at the apex, crowned by the calyx lobes; pedicels up to 15–30 mm long. Pyrenes 2.

Subsp. **glabriflora** Verdc. in Kew Bull. **42**: 184 (1987); in F.T.E.A., Rubiaceae: 835, fig. 147/13 (1991). TAB. **57**. Type from Tanzania.

Corolla glabrous outside; buds with distinct tails at the apex. Fruiting pedicels up to 30 mm long.

Malawi. N: Chitipa Distr., Misuku Hills, Chisasu–Itera road, fl. 27.xii.1977, *Pawek* 13398 (K; MAL; MO).

Also in southern Tanzania. Evergreen rainforest edges; 1600–1840 m.

Subsp. *bugoyensis* is restricted to Zaire (Dem. Rep. Congo), Rwanda, Burundi and Tanzania. The corolla lobes, and often the tube as well, are hairy outside, the buds have less distinct tails than the typical subspecies, and the pedicels reach 1.5 cm long.

17. **Rytigynia decussata** (K. Schum.) Robyns in Bull. Jard. Bot. État **11**: 195 (1928). —Brenan, Check-list For. Trees Shrubs Tang. Terr.: 531 (1949). —Verdcourt in Kew Bull. **42**: 185 (1987); in F.T.E.A., Rubiaceae: 835 (1991). Type from Tanzania.

Pachystigma decussatum K. Schum. in Engler, Pflanzenw. Ost-Afrikas **C**: 387 (1895).

Vangueria longisepala K. Krause in Bot. Jahrb. Syst. **39**: 534 (1907). —De Wildeman in Bull. Jard. Bot. État **8**: 58 (1922). Type from Tanzania.

Rytigynia sessilifolia Robyns in Bull. Jard. Bot. État **11**: 194 (1928). Type: Mozambique, Mecótia Mt. (Mt. Mkota), *Stocks* 138 (K, holotype).

Shrub or suffrutex 0.6–1.5* m tall, with slender stems from a woody rootstock, densely spreading-pubescent when young, becoming glabrescent; older stems covered with dark red-brown ± peeling bark. Leaves mostly paired but frequently in whorls of 3, 1.7–8.5(11) × 0.8–5(6) cm, elliptic, oblong or ± ovate, shortly pubescent above, velvety-woolly beneath with tangled appressed grey hairs which ± completely obscure the surface, but the main venation clearly evident when dry; petiole 4–7 mm long; stipules joined to form a short sheath 1.5–2.5 mm long, densely pilose within near the node, stipule appendages 2–3 mm long, subulate. Cymes short, 1–5-

---

* A report by Semsei (Tanzania) that it attains 4.5 m seems unlikely.

flowered; peduncles (1)2–4 mm long and pedicels (0.5)2–6 mm long, both ± spreading-pubescent; bracts and bracteoles ± obsolete. Calyx pubescent; tube 1–1.5 mm long, campanulate or globose; limb-tube very short; lobes 1.5–4 mm long, linear-lanceolate. Corolla acuminate in bud, white, cream or yellow; tube 2–4.5 mm long, glabrous or with a few scattered hairs outside, with a ring of deflexed hairs inside; lobes 2.5–4 × 1–2 mm, triangular, apiculate. Anthers purple-brown, half exserted with rather distinct connective appendages. Style exserted c. 1.5 mm; pollen presenter green, 1 mm long. Ovary 2-locular. Fruit black, didymous, usually only one per cyme developing, about 10 mm long and wide, 4 mm thick, crowned by calyx lobes; pedicel lengthening to 9 mm; pyrenes 2.

Mozambique. N: N slopes Serra de Ribáuè, Mepáluè, 25 km from Ribáuè, fl. 27.i.1964, *Torre & Paiva* 10269 (COI; EA; LISC; LMU; SRGH; WAG). Z: Mocuba, 90 km towards Alto Molócuè, fl. 27.xi.1967, *Torre & Correia* 16247 (C; LISC; LMU; WAG). M: Mecótia Mt. (Mt. Mkota), fl. 4.ii.1907, *Stocks* 138 (K).
Also in Tanzania. *Brachystegia* woodland; 'rupideserta a granito'; 300–850 m.
This is a peripheral species linking *Rytigynia*, *Fadogia* and *Tapiphyllum*, and resembling *Fadogia* in habit. Since its position is likely to remain doubtful a subgenus was erected within *Rytigynia* (*Fadogiopsis* Verdc.) to accommodate it, rather than transfer it to *Fadogia* which has a 3–4(9)-locular ovary. *Welch* 450 from Tanzania has leaves in whorls of 3 and a very *Fadogia*-like habit.

18. Rytigynia sp. D

A sparsely branched shrub with slender branches; branchlets very distinctly supra-axillary, densely covered with very short spreading stiff ± curved hairs. Leaves 1.5–4.7 × 1–1.7 cm, ovate to elliptic-lanceolate, very slenderly acuminate at the apex, rounded to cuneate at the base, glabrous save for traces of fine pubescence on the nerves beneath, and with distinctive pit-like domatia in the nerve axils beneath which are visible above as slightly raised areas in the dried state; petioles slender, 3–5 mm long, with same indumentum as the branchlets. Stipule-bases c. 1 mm long, puberulous with a subulate appendage c. 1 mm long. Flowers and fruits unknown.

Zimbabwe. E: Chimanimani Mts., below Mt. Peza, 26.xii.1957, *Goodier & Phipps* 149 (K; SRGH).
Dense evergreen rain forest in large kloof; 1440 m.
No other material of this has been seen but it should be easily recognised again by its rather distinctive branching and foliage.

19. Rytigynia sp. E

A slender shrub up to 1.3 m tall; bark smooth; stems with a dark brown epidermis, slightly longitudinally ridged, lenticellate, bristly with upwardly directed ferruginous hairs. Leaves 2.5–6.5 × 0.7–1.7 cm, narrowly oblong, acuminate to a very narrowly rounded tip at the apex, rounded at the base, tending to dry reddish-brown with spaced appressed ferruginous bristly hairs similar to the indumentum of the stem; the lateral leafy shoots are subtended by persistent shorter more ovate leaves; petiole short, 1–1.5 mm long, similarly hairy; stipule bases free, 2–3 mm long, oblong-triangular, drawn out into a filiform tip 1–2 mm long. Flowers and fruits unknown.

Mozambique. MS: Serra Macuta, 2.vi.1971, *Müller & Gordon* 1785 (K; LISC; SRGH).
Mixed evergreen rain forest, on slope in gully; 750 m.
The indumentum and leaf shape of this plant are characteristic, however it needs to be recollected in flower and fruit. The lateral leafy shoots are subtended by persistent shorter more ovate leaves.

20. Rytigynia sp. F

Shrub up to 1.2 m tall, with branchlets in whorls of 3 all in one plane; stems sparsely pubescent at first, soon glabrous, with a whitish peeling epidermis revealing a dark dull purplish underbark beneath. Leaves paired or in whorls of 3; blades 4–7 × 0.7–2.2 cm, oblong- or elliptic-lanceolate, very narrowly acuminate at the apex, cuneate at the base, glabrous save for some bristly hairs on the midrib beneath; petiole up to 2 mm long; stipule sheath 2–4 mm long with a subulate appendage 5 mm long, hairy within the base, subpersistent on the branches from which the leaves have fallen, the woolly inside showing. Inflorescences short, c. 3-flowered; peduncle 2 mm long; pedicels 3–5 mm long; bracts forming a narrow involucre 3 mm wide. Unripe fruits c. 6 mm in diameter, globose.

Zambia. W: Chingola, young fr. 5.ii.1957, *Fanshawe* 2998 (K).
"Granite boulder mateshi" – dry evergreen thicket associated with chipya woodland.
This closely approaches *Fadogia* but the habit is more that of *Rytigynia*.

21. **Rytigynia sp. G**

Small shrub to about 1 m tall; stems at first pale olive in colour, densely pubescent with short spreading hairs, the epidermis later peeling to reveal a reddish-brown underbark. Leaves 2.5–10.5 × 1.2–4.5 cm, elliptic to oblong-elliptic, obtusely acuminate to rounded at the apex, cuneate at the base, very discolorous, densely shortly appressed pubescent above but this indumentum not obscuring the surface, velvety beneath with tangled grey hairs forming a woolly tomentum covering the entire surface; petiole 1–5 mm long; stipules 1.5 mm long, triangular or oblong with a ± laterally compressed linear-lanceolate pubescent appendage c. 3 mm long. Cymes 2-flowered; peduncle 2 mm long; pedicels up to 10 mm long, glabrescent or pubescent. Calyx limb ± entire or with minute teeth. Ovary up to at least 3-locular. Fruits 9 × 12 mm, depressed globose, glabrous.

Malawi. S: Mulanje Mt., Machemba Hill, fr. 2.ii.1984, *Balaka & Nachamba* 347 (MAL).
Hill slope.
There are two loose corollas lodged in axils of the above specimen which might well belong to the plant, but they are in poor condition. These corollas have unappendaged triangular lanceolate lobes 3.5 × 1.5 mm from a short tube probably c. 2 mm long. More adequate material is needed for description.

UNIDENTIFIABLE SPECIES

Müller & Pope 3628 & 3629 [Zimbabwe, Mutasa Distr., S end of Nyazengu Ridge] are undoubtedly a species of *Rytigynia* but are completely sterile.

46. CUVIERA DC.

Cuviera DC. in Ann. Mus. Natl. Hist. Nat. **9**: 222, t. 15 (1807). —Hallé in Bull. Soc. Bot. France **106**: 342 (1960), nom. conserv.

Small trees or shrubs, unarmed or sometimes spinose. Leaves opposite, often large, mostly ± oblong or elliptic, petiolate, usually coriaceous and persistent; stipules small, basally ± connate, often acuminate, deciduous. Flowers hermaphrodite or sometimes infertile (polygamo-dioecious fide Benth. & Hook.f., Gen. Pl.), in subsessile or pedunculate many-flowered axillary cymes; bracts and bracteoles linear to lanceolate or elliptic, often leaf-like and accrescent. Calyx tube obconic or turbinate, sometimes 3–4-angled; limb-tube ± suppressed; lobes 3–6, linear to ovate, often leaf-like and accrescent, mostly longer than the corolla, persistent. Corolla funnel-shaped, campanulate or barrel-shaped, retrorsely hairy or bristly inside but throat glabrous, glabrous or pilose outside; lobes 5–6, spreading or reflexed, elongate, mostly caudate-acuminate, sometimes very markedly so. Stamens 5, inserted in the throat, the anthers exserted. Disk depressed, lobed. Style thick, narrowed at both ends, stiffly pubescent to glabrous, sometimes with a conspicuous globular swelling near the base (some West African species only). Pollen presenter cylindrical, mitriform or peltate, 2–10-grooved. Ovary (1)2–5-locular; ovules solitary in each locule, pendulous. Fruit drupaceous, often large, ovoid or subglobose, sometimes obscurely angled, with 1–5, 1-seeded pyrenes.

Tab. 58. CUVIERA SEMSEII. 1, flowering branch (× ²⁄₃), from *Vollesen* in *MRC* 1889; 2, branch with reduced shoots (× ²⁄₃), from *Schlieben* 5202; 3, domatium (× 9), from *Vollesen* in *MRC* 1889; 4, stipule (× 6), from *Vollesen* in *MRC* 3702; 5, bud with one calyx lobe folded back (× 2); 6, flower (× 3); 7, corolla opened out (× 3); 8, pollen presenter (× 9); 9, section through ovary (× 6), 5–9 from *Schlieben* 5202; 10, fruit, immature (× 1), from *Ludanga* in *MRC* 1280; 11, pyrenes, 3 views (× ²⁄₃), from *Bidgood et al.* 1760. Drawn by Sally Dawson, with 11 by Diane Bridson. From F.T.E.A.

A genus of about 20 species, mostly in western and central Africa. Three geographically disjunct species from Tanzania, Mozambique and southern Malawi have also been placed in this genus. Further investigation is needed to be sure that they are truly congeneric; *C. semseii* has very distinctive pyrenes, and both *C. schliebenii* and *C. tomentosa* are still only poorly known.

1. Leaf blades velvety pubescent on both surfaces · 3. *tomentosa*
 – Leaf blades glabrous save for hairy domatia in leaf axils beneath, or with sparse long setae
   beneath · 2
2. Calyx lobes ovate to lanceolate, 6–15 × (2)4.7–6 mm, ± ciliate; ovary 5-locular · · 1. *semseii*
 – Calyx lobes ± oblong, 2.3–2.6 × 0.7–0.9 mm, glabrous; ovary 2-locular · · · · · · 2. *schliebenii*

1. **Cuviera semseii** Verdc. in Kew Bull. **11**: 449 (1957); in F.T.E.A., Rubiaceae: 771, fig. 132/27, 136 (1991). TAB. **58**. Type from Tanzania.

Small tree or shrub 2–6 m tall, with grey-brown rugulose glabrous lenticellate branchlets. Leaves 4.4–8.5 × 1.6–4 cm, elliptic, acuminate at the apex, the actual tip sometimes narrowly rounded, cuneate at the base, thin, glabrous above, with hairy domatia in the main nerve axils and sparse long setae beneath; petioles 5–13 mm long, glabrous or sparsely setose; stipules 5 × 1 mm, linear-triangular to oblong, rather thick, ± acute; nodes bristly within. Inflorescences axillary, opposite, several- to many-flowered, the branches ± bifariously shortly pubescent; peduncles 5 mm long; secondary peduncles up to 10 mm long; bracts and bracteoles 2–5 × 0.5–1.5 mm, oblong-oblanceolate; true pedicels 1–2 mm long. Calyx tube 2 mm long and wide, subglobose to turbinate; lobes 4–5, 6–15 × (2)4.7–6 mm, ovate to lanceolate, acuminate, narrowed and often with colleters at the base, leaf-like, nervose, sparsely to densely pubescent and ciliate. Corolla yellow or greenish-yellow; tube 4.5 mm long, 3 mm wide at the base and apex, 3.75 mm wide at the middle, glabrous outside, with a ring of deflexed bristles 1 mm long inside; lobes 5, 5–6 × 1.5 mm, linear-triangular, acuminate, sparsely to densely pilose outside. Ovary 5-locular. Style 7.5–9 mm long, narrowed towards the apex; pollen presenter 1.1 mm long, cylindrical. Fruits about 3–3.5 × 2.5 cm, subglobose, glabrous and a little shining, angular and wrinkled when dry. Pyrenes about 2.5 cm long, sharply keeled.

Malawi. S: Liwonde National Park, Namelembo Thicket, fl. 3.xii.1983, *Dudley* 810 (K; MAL).
Mozambique. N: 27 km from Nampula on road to Muecate, fr. 2.iv.1964, *Torre & Paiva* 11586 (LISC).
Also in southern Tanzania. Dry forest derived thicket on sandy soil; 380–500 m.
The single specimen seen from Malawi has much more pubescent flowers than material from Tanzania, and shorter corolla lobes.

2. **Cuviera schliebenii** Verdc. in Kew Bull. **33**: 497 (1979); in Kew Bull. **42**: fig. 130/3E (1987); in F.T.E.A., Rubiaceae: 771 (1991). Type from Tanzania.

Shrub or small tree 2–18 m tall; youngest shoots drying black, sparsely to ± densely pubescent; older shoots covered with a pale brown rather corky bark, irregularly fissured. Leaves 5.2–15 × 1.8–7.4 cm, elliptic to oblong-elliptic, acuminate at the apex, cuneate at the base, drying dark, glabrous save for very small hairy domatia in the nerve axils beneath; petiole 2–3 mm long; stipules up to 15 mm long, linear-lanceolate, from broad bases which are slightly connate basally. Inflorescences appearing before, or together with, the leaves, cymose, much branched, 3–5 cm long, the ultimate parts spreading pubescent; peduncle 2–3 cm long; pedicels 1–3 mm long, pubescent; bracts and bracteoles c. 3 × 0.8 mm, linear-lanceolate. Calyx tube 1.2 mm long, top-shaped, densely covered with short spreading hairs; lobes 2.3–2.6 × 0.7–0.9 mm, oblong or linear-oblong or slightly elliptic, more or less glabrous. Corolla with 5 minute apiculae at the apex in bud, yellow, probably slightly fleshy; tube 3 mm long, shortly cylindrical, glabrous outside, inside with a ring of white strongly deflexed hairs 1.3 mm long affixed at about the upper third of the tube; lobes 2.2 × 1.2 mm, triangular-lanceolate, with a slender apical appendage c. 1 mm long. Anthers inserted at the throat, 1 mm long, oblong, ± acute, just exserted. Style slightly thickened at the base, 4.8 mm long, glabrous; pollen presenter cylindric-clavate, 12-ridged, slightly bifid at the apex.

Disk annular, thick. Ovary 2-celled. Fruit 23 × 13 × 9 mm, oblong, compressed, with 1–2 pyrenes. Pyrenes 19 mm long, c. 5 mm thick and wide, fusiform to canoe-shaped, the hilar notch close to the top of the seed.

Mozambique. N: Pemba (Porto Amélia), between Ancuabe and Metuge, fl. 7.ix.1948, *Barbosa* 2007 (LMA). Z: Namacurra, 28 km from Nicoadala, fr. 1.ii.1966, *Torre & Correia* 14340 (LISC).
Also in southern Tanzania. Open woodland on sandy-clay soil, with *Acacia nigrescens, Erythrophloeum africanum, Parinari, Uapaca* and *Milicia excelsa*; 40 m.

3. **Cuviera tomentosa** Verdc. in Kew Bull. **36**: 557 (1981); in F.T.E.A., Rubiaceae: 773 (1991). Type from Tanzania.

Shrub to 3 m, or tree 10–15 m tall; stems of young leafy shoots densely covered with spreading grey hairs; older stems glabrous or obscurely pubescent, with grey-brown finely flaking bark, minutely punctate with scars of hair bases; branchlets spreading at right angles. Leaves opposite; blades up to c. 11 × 6 cm, elliptic or oblong-elliptic, narrowly rounded to acuminate at the apex, obtuse or subcordate at the base, densely velvety on both surfaces with short ± spreading hairs; petioles short, up to 2 mm long; stipules with a ± triangular base 1.5 mm long and a linear-lanceolate apex 8–11 mm long. Flowers fairly numerous in pubescent branched dichasial cymes; peduncle 10–20 mm long; secondary branches 3–10 mm long; pedicels 1–3 mm long; bracts leaf-like, 4–5 × 1–2 mm, oblong-subspathulate, glabrous save for base and margins, or pubescent. Calyx tube 1 mm long, densely spreading pubescent; lobes 3–4 × 0.8–1.2 mm, narrowly oblong-spathulate, glabrous inside, ciliate at margins and pubescent outside. Corolla yellowish-green, acute and shortly 5-tailed in bud, olive-green, finely rather sparsely spreading puberulous; tube 2–3 mm long with ring of long deflexed hairs inside but throat glabrous; lobes 2.5–3.5 mm long including the tails, narrowly triangular. Ovary 2-locular. Style exserted 1.5 mm; pollen presenter 0.6 mm long. Fruit 15 × 9–11 mm (single-seeded only seen). Pyrenes 12 × 6 × 4.5 mm, ± ellipsoid with ventral face plane, not keeled, not beaked at point of attachment, ± smooth.

Mozambique. N: Niassa Prov., road from Nangade to Mueda, fl. 18.ix.1948, *Andrada* 1370 (COI; LISC; LMA).
Also in southern Tanzania. Tall forest with dense shrub layer.
Fruit said to be edible.

47. VANGUERIOPSIS Robyns

Vangueriopsis Robyns in Bull. Jard. Bot. État **11**: 248 (1928) (pro parte). — Verdcourt in Kew Bull. **42**: 187 (1987).
Vangueriopsis subgen. *Rostranthus* Robyns in Bull. Jard. Bot. État **11**: 248 (1928).

Shrubs or small trees, with mostly stiff thick branches. Leaves opposite, somewhat coriaceous, hairy or velvety tomentose; stipules thick, triangular, ± joined into a sheath at the base, long caudate at the apex. Flowers large and conspicuous, ± thick, in opposite simple to much branched many-flowered axillary cymes; bracts fairly conspicuous. Calyx limb-tube ± obsolete; lobes erect, triangular to linear-lanceolate or linear-oblong. Corolla elongate-lanceolate, beaked in bud; tube cylindrical, two thirds to one eighth the length of the linear-lanceolate lobes, with a ring of deflexed hairs or glabrous within save at base; throat glabrous. Stamens inserted at the throat; anthers exserted, linear. Style slender, long-exserted; pollen presenter cylindrical. Ovary 2-locular (5-locular in *V. gossweileri*), each locule with a solitary pendulous ovule. Fruit fairly large, didymous when 2 pyrenes develop, but unilateral and oblique when 1 is aborted, smooth or irregularly ribbed, presumably ± globose in *V. gossweileri*.

As circumscribed in Kew Bull. **42**: 187 (1987), and here, *Vangueriopsis* is restricted to the 4 species which comprised Robyns's subgenus *Rostranthus*; one occurs in the Flora Zambesiaca area.

Tab. 59. VANGUERIOPSIS LANCIFLORA. 1, leafy shoot (× ⅔), from *Rogers* 8588; 2, portion of large leaf (× ⅔); 3, detail from lower surface of large leaf (× 6), 2 & 3 from *Milne-Redhead* 2844; 4, stipule (× 1), from *Rogers* 8588; 5, flowering shoot (× ⅔); 6, flower (× 2), 5 & 6 from *Richards* 12870; 7, portion of corolla opened out (× 4); 8, dorsal view of anther (× 6); 9, pollen presenter (× 6); 10, longitudinal section through ovary (× 4), 7–10 from *Fanshawe* 274; 11, fruiting twig (× ⅔), from *Codd* 7222; 12, pyrene, 2 views (× 1), from *Stewart* 148; 13, embryo (× 2), from *Milne-Redhead* 2844. Drawn by Sally Dawson.

Vangueriopsis lanciflora (Hiern) Robyns in Bull. Jard. Bot. État **11**: 252 (1928)* . —Brenan, Check-list For. Trees Shrubs Tang. Terr.: 538 (1949). —Pardy in Rhod. Agric. J. **49**: 259 (1952). —K. Coates Palgrave, Trees Central Africa: 392, figs. (1957). —F. White, F.F.N.R.: 425, fig. 68, B,C (1962). —Launert in Merxmüller, Prodr. Fl. SW. Africa, fam. 115: 27 (1966). —Palmer & Pitman, Trees Southern Africa **3**: 2086 (1973). —Drummond in Kirkia **10**: 276 (1975). —Palmer, Field Guide Trees Southern Africa: 44 (1977). —K. Coates Palgrave, Trees Southern Africa, ed. 3, rev.: 876, pl. 302 (1988). —Verdcourt in Kew Bull. **42**: 187, fig. 3C (1987); in F.T.E.A., Rubiaceae: 774 (1991). TAB. **59**. Type: Zambia, near Victoria Falls, *Kirk* (K, holotype).

    Canthium lanciflorum Hiern in F.T.A. **3**: 146 (1877). —Oliver in Hooker's Icon. Pl.: t. 2252 (1893).

    Canthium platyphyllum Hiern, Cat. Afr. Pl. Welw. **1**: 479 (1898). Type from Angola.

    Vangueria lateritia Dinter in Repert. Spec. Nov. Regni Veg. **24**: 267 (1928). Type from Namibia.

Deciduous shrub or much branched small tree (0.9)1.5–6(13) m tall; branches thick and stiff; bark grey, flaking off to expose a brownish-pink or rusty-red underbark; young branches densely grey-pubescent. Leaves opposite; blades (1.5)5.3–21.5 × (0.5)1.6–12.5 cm, elliptic or oblong-elliptic, ± rounded or subacute at the apex, rounded to cuneate at the base, markedly discolorous, rather scabrid pubescent above, velvety grey or yellowish tomentose or pubescent beneath, very rarely glabrous; petiole 5–15 mm long; stipules thick, up to 9 mm long, triangular, joined to form a sheath at the base, grey tomentose outside, densely hairy inside, with a thick obtuse subulate apex 8–12 mm long, eventually deciduous. Inflorescences densely yellowish velvety pubescent, simply cymose or branched, axillary on the leafless nodes of older branches; cymes several-flowered; bracts 4–6 mm long, lanceolate or subulate; peduncles and secondary branches 10–20 mm long; pedicels very short, or up to 8 mm long. Calyx tube 4 × 4 mm, subcampanulate, densely tomentose; lobes erect, 1.5–5.5 mm long, narrowly lanceolate to triangular to oblong-lanceolate, densely tomentose. Corolla beaked, elongate, with divaricate tails at apex in bud, whitish or yellow-green, grey tomentose; tube 5 mm long, cylindric, with a ring of deflexed hairs inside; lobes reflexed, 20–25 mm long, linear-lanceolate, glabrous inside, tomentose outside, apiculate at the apex. Anthers 2.5–3 mm long, exserted for 5 mm. Style green, rather stout, exserted for up to 15 mm, constricted near the apex; pollen presenter 2–2.5 mm long, cylindric, smooth, 2-lobed at the apex. Fruit rather like a "medlar", 2–4 × 1.7–2 cm, or the size of a peach (ex Zimbabwe label), rounded, compressed, didymous or oblique, crowned by the calyx limb, sparsely pubescent, with 1–2 pyrenes; pyrenes up to 21 × 9 mm, narrowly ellipsoid, with a woody wall 1.7 mm thick.

    Zambia. B: near Senanga, fl. & fr. 29.vii.1952, *Codd* 7222 (BM; COI; K; PRE; SRGH). N: Mbala Distr., Kawimbe, fl. 12.vii.1960, *Richards* 12870 (K). W: between Mwinilunga and Matonchi Farm, fl. 6.ix.1930, *Milne-Redhead* 1060 (K). C: Chilanga, Quien Sabe, fl. & fr. ix.1929, *C. Sandwith* 26 (K). E: Chipata (Fort Jameson), young fr. 13.x.1967, *Mutimushi* 2302 (K; NDO). S: Mapanza, fl. 5.viii.1954, *E.A. Robinson* 864 (EA; K). **Zimbabwe**. N: Zvimba Distr., about 3 km west of Mutorashanga (Mtoroshanga) Village, fl. 15.ix.1963, *Leach & Müller* 11719 (K; SRGH). W: Shangani Distr., Gwampa Forest Reserve, fl. i.1958, *Goldsmith* 37/58 (K; SRGH). C: Marondera Distr., Lendy, fl. 30.ix.1949, *Corby* 493 (K; SRGH). **Malawi**. C: Lilongwe, Nature Sanctuary Zone B, Northern Trail, fr. 13.xi.1984, *Patel, Salubeni & Banda* 1700 (K; MAL). S: Shire Highlands, fl. *Buchanan* 475 (E; K). **Mozambique**. N: Mandimba, sterile 14.v.1948, *Pedro & Pedrogão* 3398 (EA; LMA).

    Also in Tanzania, Angola and Namibia. Common at medium to higher altitudes, in deciduous woodlands and open grassland with scattered trees, often on Kalahari Sand, sometimes on rocky ground, associated with *Albizia–Combretum, Brachystegia–Julbernardia, Parinari, Monotes, Burkea–Pterocarpus–Copaifera*, also retained in old cultivations and fields for its edible fruit; 900–1740 m.

    One glabrous specimen has been seen [Zambia W: Mwinilunga Distr., c. 3.5 km east of Ikelenge, Karunga Protected Forest Area, adjacent to Hillwood Farm's Nchila Wildlife Reserve, sterile 28.ii.1995, *Luwiika et al.* 103 (K; MO)].

---

* Launert, in Prodr. Fl. SW. Africa, gives the authority for this species as (Hiern) Robyns ex Good in J. Bot. **64**, Suppl. 2: 22 (1926), but the genus *Vangueriopsis* Robyns was not described at that time and the combination has no validity.

48. PYGMAEOTHAMNUS Robyns

Pygmaeothamnus Robyns in Bull. Jard. Bot. État **11**: 29 (1928).

Suffrutescent herbs, mostly under 30 cm tall, with several to many stems from a woody rhizome. Leaves opposite or in whorls of 3–4, mostly fairly shortly petiolate; stipules triangular, usually shortly subulate-caudate at the apex. Flowers in simple or branched cymes or fascicles, sometimes borne at lower often leafless nodes; bracts small. Calyx limb-tube ± obsolete; lobes 5, erect, linear-oblong, oblong, lanceolate or ± triangular, obtuse or subacute. Corolla shortly apiculate, usually 5-tailed in bud; tube cylindric, glabrous or pubescent, with a ring of deflexed hairs within, sparsely to densely pilose at the throat; lobes reflexed, usually shortly apiculate, glabrous within, about equalling the tube. Stamens inserted at the throat, the anthers oblong or lanceolate, just exserted. Ovary 2-locular, each locule with a solitary pendulous ovule; style slender, slightly exserted; pollen presenter cylindric or subcapitate, deeply bilobed. Fruit globose or obpyriform, didymous or asymmetrical, containing (1)2 pyrenes.

A genus of 4 species mostly confined to southern Africa, but 2 extending to central Africa and 1 occurring in the Flora Zambesiaca area. *P. concrescens* Bullock has been transferred to *Multidentia* Gilli.

Pygmaeothamnus zeyheri (Sond.) Robyns in Bull. Jard. Bot. État **11**: 30, figs. 7 & 8 (1928). — Miller in J. S. African Bot. **18**: 84 (1952). —F. White, F.F.N.R.: 418, fig. 67, J & K (1962). — Verdcourt in Kew Bull. **42**: 126, fig. 2C (1987); in F.T.E.A., Rubiaceae: 776 (1991). Type from South Africa.
    Pachystigma zeyheri Sond. in Linnaea **23**: 56 (1850).
    Vangueria zeyheri (Sond.) Sond. in F.C. **3**: 15 (1865). —De Wildeman in Bull. Jard. Bot. État **8**: 66 (1922).
    Fadogia zeyheri (Sond.) Hiern in F.T.A. **3**: 153 (1877).

Suffrutex with 3–6 slender stems 17–30 cm tall from a fairly slender ± horizontal rhizome, glabrous, or stems covered with dense short spreading rather bristly hairs and also some longer appressed ones. Leaves opposite, or mostly in whorls of 3 or 4, drying dark brownish-green or yellow-green (as in aluminium accumulators) usually ± subcoriaceous; blades 4.2–16 × 1.8–5.7 cm, obovate to obovate-oblong or elliptic, those of the lowest whorls rounded, others obtusely to subacutely shortly acuminate at the apex, cuneate at the base, glabrous or with rather sparse yellowish bristly hairs on both surfaces, which in no way obscure the surface; petioles distinct, 5–20 mm long, bristly pubescent; stipules 4–5 mm long, triangular at the base, drawn out at the apex, glabrous or bristly pubescent. Inflorescences borne in the lower axils, including the lowest from which the leaves have fallen or were never fully developed, few to c. 40-flowered, much branched, up to 4 cm long; peduncles (5)10–20 mm long; secondary cymose branches sometimes elongate, up to 30 mm long with the flowers spaced; pedicels 3–7 mm long; bracts and bracteoles 3–7 × 0.5 mm, narrowly elliptic-lanceolate, acuminate. Calyx tube globose, about 1 mm in diameter, bristly pubescent; lobes leaf-like, 0.8–2.5(5) × 0.3–0.8 mm, narrowly oblong, elliptic or lanceolate, glabrous or sparsely bristly pubescent. Corolla slender, acuminate in bud, red or greenish, bristly pubescent or glabrous; tube 3–4 mm long; lobes 4–5 mm long, oblong-lanceolate, reflexed or sometimes probably erect, apex acuminate to shortly tailed. Ovary 2-locular, filled with dark-coloured material. Fruits 25–30 × 15–20 mm, oblong, subglobose or pyriform, containing (1)2 pyrenes, each 12 × 6 mm, glabrous.

Var. **zeyheri** —Verdcourt in F.T.E.A., Rubiaceae: fig. 138 (1991). TAB. **60**.
    Canthium oatesii Rolfe in Oates, Matabele Land and the Victoria Falls, Appendix 5: 400 (1889). Type: Zimbabwe, Matabeleland, *Oates* (K, holotype).
    Fadogia welwitschii Hiern, Cat. Afr. Pl. Welw. **1**: 481 (1898). Type from Angola.
    Vangueria stenophylla K. Krause in Bot. Jahrb. Syst. **39**: 535 (1907). Type: Botswana, Mochudi, *Marloth* 3333 (B†, holotype).
    Plectronia abbreviata K. Schum. in Bot. Jahrb. Syst. **28**: 73 (1899); in Warburg, Kunene-Samb.-Exped. Baum: 387 (1903). Type from Zaire (Dem. Rep. Congo).

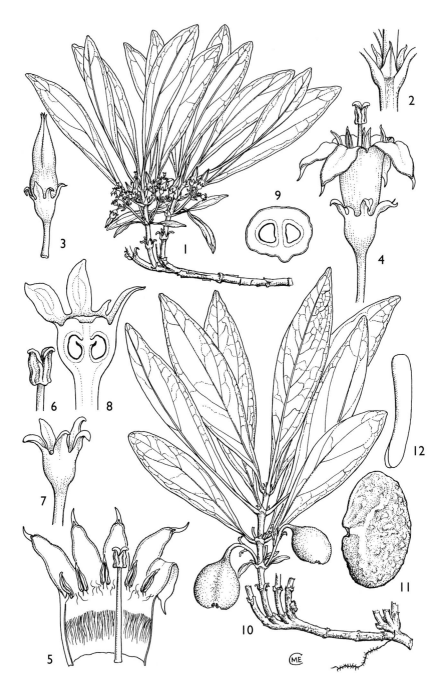

Tab. 60. PYGMAEOTHAMNUS ZEYHERI var. ZEYHERI. 1, habit, flowering (× ²⁄₃); 2, stipule (× 2); 3, flower bud (× 4); 4, flower (× 6); 5, corolla opened out, with style and pollen presenter (× 6); 6, pollen presenter (× 10); 7, calyx (× 6); 8, longitudinal section through ovary (× 10); 9, transverse section through ovary (× 12), 1–9 from *Holmes* 1224; 10, habit, fruiting (× ²⁄₃), from *Milne-Redhead* 3458; 11, pyrene (× 3); 12, embryo (× 3), 11 & 12 from *Brummitt et al.* 13879. Drawn by M.E. Church. From F.T.E.A.

Canthium abbreviatum (K. Schum.) S. Moore in J. Linn. Soc., Bot. **37**: 308 (1906).
Pygmaeothamnus zeyheri var. *oatesii* (Rolfe) Robyns in Bull. Jard. Bot. État **11**: 31 (1928).
—Miller in J. S. African Bot. **18**: 84 (1952).

Plant entirely glabrous, or inflorescence axes and calyx tubes spreading pilose.

Caprivi Strip. 16 km from Katima Mulilo on road to Singalamwe, fr. 30.xii.1958, *Killick &*
Leistner 3185 (K; SRGH). **Botswana**. SE: Mahalapye, v. young fr. 13.i.1958, *de Beer* 557 (K;
SRGH). **Zambia**. B: Sesheke, 3.2 km northwest of Masese Valley, fl. 11.viii.1947, *Brenan & Keay*
7672 (FHO; K). W: Solwezi, fl. 16.x.1953, *Fanshawe* 442 (K; NDO). C: north of Lusaka, 4.8 km
west of Karubwe, fl. 12.ix.1963, *Angus* 3744 (FHO; K). S: Namwala, corolla fallen 21.x.1963,
Lawton 1131 (K; NDO). **Zimbabwe**. N: Gokwe, fl. 31.x.1963, *Bingham* 889 (K; SRGH). W:
Bullima-Mangwe Distr., Dombodema Mission Station, fl. 24.x.1972, *Norrgrann* 255 (K; LISC;
SRGH). C: Charter Distr., 16 km from Chivhu (Enkeldoorn) on road to Buhera, fl. 3.xi.1971,
Pope 486 (K; PRE; SRGH). E: Mutare Distr., Stapleford Forest Reserve, fl. 21.xi.1955, *Chase* 5870
(BM; COI; LISC; SRGH).
Also in Zaire (Dem. Rep. Congo), Angola, Namibia and South Africa. Widespread in
grassland and open woodland, often on dambo margins where it usually occurs with *Parinari*
dwarf woodland, also in mixed woodlands with *Burkea* and *Terminalia*, and in *Cryptosepalum*
woodland on Kalahari Sand; 900–1500 m.
Most material of this species is entirely glabrous ('var. *oatesii*') but some specimens have hairy
inflorescence axes and ovaries, and many intermediates occur. The fruit is edible.

Var. **rogersii** Robyns in Bull. Jard. Bot. État **11**: 34 (1928). —Verdcourt in F.T.E.A., Rubiaceae:
778 (1991). Type: Zimbabwe, Victoria Falls, *Rogers* 5431 (?PRE, holotype; BOL; K).
Fadogia livingstoniana S. Moore in J. Bot. **57**: 88 (1919). Type: Zambia, Livingstone,
Rogers 7466 (BM, holotype; K).
Pygmaeothamnus zeyheri var. *livingstonianus* Robyns in Bull. Jard. Bot. État **11**: 34 (1928).
Type: Zambia, Livingstone, *Rogers* 7466 (?PRE, holotype; BM; K).

Stems, calyces, corolla and leaves sparsely to quite densely spreading hairy.

Zambia. B: Kalabo, fl. 15.x.1963, *Fanshawe* 8082 (K; NDO). N: Mporokoso Distr., about
16–32 km from Mporokoso by the Kawambwa road, fl. 16.x.1947, *Brenan & Greenway* 8129
(FHO; K). W: Kitwe, v. young fr. 26.iii.1966, *Mutimushi* 1351 (K; NDO). E: Nyika National Park,
between Zambian Resthouse and Manyenjere Forest, fl. 24.i.1992, *Goyder, Paton & Tawakali* 3561
(K; NDO). S: Livingstone, Dambwa Forest Reserve, fl. 3.x.1969, *Mutimushi* 3540 (K; NDO).
Also in Burundi, Zaire (Dem. Rep. Congo), Tanzania and South Africa. High rainfall
Brachystegia woodlands and submontane grassland, and in *Diospyros–Parinari* woodland on
dambo margins, also on Kalahari Sand; 915–2200 m.
Robyns's name "*livingstonianus*" is not a new combination but the same epithet based on the
same gathering. He appears to have overlooked S. Moore's name altogether and does not
mention it under *Fadogia* in his revision.
The fruits are eaten and were stated to be a staple food of some bushmen (fide Schoenfelder
in 1935).

49. MULTIDENTIA Gilli

Multidentia Gilli in Ann. Naturhist. Mus. Wien **77**: 21 (1973). —Bridson in Kew
Bull. **42**: 641–654 (1987); in F.T.E.A., Rubiaceae: 841–848 (1991).
Canthium sect. *Granditubum* Tennant in Kew Bull. **22**: 438 (1968).

Shrubs or small trees to 12 m tall, or sometimes pyrophytic subshrubs (suffrutices),
mostly glabrous, unarmed; stems often lenticellate. Leaves mostly restricted to new
growth, paired or rarely ternate, petiolate; blades chartaceous to coriaceous, typically
glaucous beneath, with a conspicuous network of finely, or less often, coarsely
reticulate tertiary nerves; domatia present as tufts of hair, or absent; stipules
sheathing at the base, pubescent within, provided with a linear somewhat keeled
lobe. Flowers (4)5(6)-merous, usually medium-sized, borne in pedunculate cymes,
usually at nodes from which the leaves have fallen, except in pyrophytic species;
bracteoles linear to lanceolate, small. Calyx chartaceous or coriaceous; limb cupular,
truncate, repand, dentate, or cupular below and lobed above. Corolla normal or
coriaceous and drying wrinkled; tube cylindrical, subequal to the lobes or
occasionally much longer, glabrous outside with a ring of deflexed hairs inside and

usually rather sparsely pubescent at the throat; lobes reflexed, rounded or obtuse, thickened towards the apex. Stamens set at throat; anthers partly or fully exserted, oblong-ovate or oblong with dark coloured connective tissue on dorsal face except for the margin, often apiculate. Ovary 2-locular. Style slightly longer, or less often up to twice as long as the corolla tube, slender; pollen presenter ± as broad as long or less, often elongate, ribbed, hollow to mid-point, apex cleft to about the mid-point when mature; disk glabrous. Fruit a 2-seeded drupe, large, subglobose, laterally compressed, often somewhat didymous, often lenticellate or rarely small and didymous (in the West African *Multidentia pobeguinii* (Hutch. & Dalz.) Bridson), crowned by a persistent calyx limb. Pyrenes thickly woody, broadly ellipsoid, sometimes curved, truncate at point of attachment, with a line of dehiscence extending from the point of attachment to the apex then arching on either side back towards point of attachment but stopping short of it, very strongly rugulose, (except in *Multidentia pobeguinii*); seeds with endosperm entire; testa finely reticulate; embryo slightly curved; radicle erect; cotyledons about one third the length of the embryo, set perpendicular to the ventral face of the seed.

A genus of 11 species (2 poorly known) restricted to tropical Africa, 5 of which occur in the Flora Zambesiaca area.

1   Pyrophytic subshrub up to 35 cm tall; inflorescences borne at leafy nodes, very often supra-axillary and forming an acute angle with stem; calyx limb with very distinct lobes up to 6 mm long · 5. *concrescens*
  − Shrubs or small trees more than 100 cm tall; inflorescences mostly borne at leafless nodes, axillary; calyx limb repand to dentate or sometimes with lobes up to 2 mm long · · · · · · 2
2.   Leaves coriaceous, often drying yellow-green, oblong, elliptic-oblong or rotund; stems frequently thick, covered with pale corky bark; corolla leathery and wrinkled when dry · 4. *crassa*
  − Leaves papery or subcoriaceous, usually drying blackish, brown or dark green above, elliptic or occasionally as above; stem not as above; corolla occasionally leathery when dry · · · · · 3
3.   Tertiary nerves finely reticulate; corolla only known from bud, 3 mm long; bark rusty-red, powdery · 1. *sp. A*
  − Tertiary nerves coarsely reticulate; corolla tube 3–14 mm long; bark light brown, dark red-brown or sometimes rusty-brown, smooth or powdery · 4
4.   Corolla tube 8–14 mm long, greatly exceeding the lobes; style somewhat longer than corolla tube; midrib drying cream-coloured · 2. *fanshawei*
  − Corolla tube 3–5 mm long, ± equalling the lobes; style up to twice as long as corolla tube; midrib not drying cream-coloured · 3. *exserta*

1. **Multidentia sp. A**

  Multidentia sp. B of Bridson in Kew Bull. **42**: 647 (1987).

Tree up to 7 m tall, glabrous; stems covered with rusty-red powdery to flaky, obscurely lenticellate bark. Leaves restricted to apices of branches; blades not fully mature up to 7 × 3.6 cm, elliptic, acute to subacuminate at apex, acute at base, papery, drying black-brown above and glaucous beneath; lateral nerves in 5–6 main pairs; tertiary nerves finely reticulate, clearly apparent; domatia present as hairy tufts, conspicuous; petioles up to 5 mm long; young stipule base 2 mm long, shortly sheathing, bearing a linear-subulate lobe 2.5–4 mm long; older stipules becoming bark-like and breaking down to reveal white hairs inside. Flowers only known in bud, 5-merous, borne in many-flowered pedunculate cymes; peduncle 1–14 mm long, sparsely puberulous; pedicels up to 2.5 mm long, puberulous; bracts and bracteoles linear. Calyx tube 1.75 mm long, sparsely pubescent; limb 0.75 mm long, dentate. Corolla bud 3 mm long, blunt; tube with ring of deflexed hairs inside. Fruit only known from young/galled examples, up to 15 mm long.

Malawi. N: Mzimba Distr., just north of Chikangawa, fl. bud & fr. (young/galled) 21.x.1978, *E. Phillips* 4124 (BR; K).
  Only known from the above specimen; growing off forest edge; 1700 m.
  The data states "fruit of 4087" but this gathering has not been traced.

2. **Multidentia fanshawei** (Tennant) Bridson in Kew Bull. **42**: 647, map 1 (1987); in F.T.E.A.,
    Rubiaceae: 843 (1991). Type: Malawi, Ntchisi Mt., *Robson* 1668 (K, holotype; EA; LISC;
    PRE; SRGH).
        Canthium fanshawei Tennant in Kew Bull. **22**: 438 (1968).

Shrub or tree 2–3(12) m tall, erect or sometimes scandent or a tree 14 m tall,
glabrous; stems covered with reddish-brown rather powdery bark, lenticellate.
Leaves tending to be restricted to apex of branches but not strictly so; blades 4–12
× 2.5–5(7) cm, elliptic to broadly elliptic or occasionally round or oblong-elliptic,
acute to subacuminate or acuminate at apex, acute to cuneate or sometimes obtuse
at base, stiffly papery, drying blackish-brown with nerves pale above and glaucous
beneath; lateral nerves in 4–5(6) main pairs; tertiary nerves coarsely reticulate,
either apparent or obscure; domatia present as hairy tufts; petioles 6–10 mm long;
stipule base 2–3 mm sheathing, bearing a linear-subulate lobe 2–3 mm long,
pubescent within. Flowers 5-merous, borne in many-flowered pedunculate cymes;
peduncle 10–25 mm long, glabrous to sparsely pubescent; pedicels 2–6 mm long,
sparsely pubescent to pubescent; bracts and bracteoles linear, 1–2 mm long. Calyx
tube 1–1.5 mm long, glabrous; limb 1–2 mm long, repand to slightly dentate.
Corolla white; tube 8–14 mm long, slightly pilose at throat, with a ring of deflexed
hairs set at about one third of the way from the top inside; lobes 5–9 × 2.2–5 mm,
oblong-elliptic to narrowly obovate, acute. Anthers erect or sometimes spreading,
but not reflexed. Pollen presenter 2.25–2.5 × 1.5–2 mm, well exserted, ovoid,
ridged. Fruit 18–25 × 19–27 mm, subglobose, somewhat bilobed. Pyrene 17 × 13 ×
11 mm, almost semicircular in outline, gradually rounded at point of attachment,
strongly rugulose and grooved.

Zambia. N: Mpika, fl. 26.i.1955, *Fanshawe* 1858 (K). W: Mwinilunga Distr., Matonchi Farm,
corollas fallen 7.ii.1938, *Milne-Redhead* 4478 (K; PRE). **Malawi**. N: Mzimba Distr., South Viphya,
9 km from Chikangawa towards Mzuzu, young fl. & fr. 26.ii.1982, *Brummitt et al.* 16127 (K;
SRGH). C: Nkhota Kota Distr., Chinthembwe (Cintembwe) Mission, fr. 7.v.1961, *Chapman* 1294
(SRGH).
    Also in southern Tanzania and Zaire (Dem. Rep. Congo) (Shaba Prov.). In evergreen and
riverine forests, also *Brachystegia* woodland and mushitu; 1450–1825 m.

3. **Multidentia exserta** Bridson in Kew Bull. **42**: 647, fig. 3, map 1 (1987); in F.T.E.A., Rubiaceae:
    843 (1991). Type from Tanzania.

Shrub or small tree 1–7 m tall, glabrous; young stems pale brown, lenticellate.
Leaves restricted to apices of branches and young lateral shoots, mature or less often
immature at time of flowering; blades 4.5–14 × 2.5–9.5 cm, elliptic to broadly elliptic
or broadly ovate, acute to subacuminate at apex, obtuse or truncate at base, papery,
strongly discolorous when dry, glaucous beneath; lateral nerves in 5–6 main pairs;
tertiary nerves apparent, coarsely reticulate; domatia present as tufts of hair,
sometimes spread along the midrib; petioles 5–9 mm long; stipules 2–3 mm long,
sheath-like at the base and then somewhat triangular, terminating in a linear lobe
4–7 mm long, pubescent inside. Flowers 5-merous, borne in pedunculate cymes, up
to 30-flowered, arising from nodes from which the leaves have fallen; peduncles
16–25 mm long, sparsely puberulous (or puberulous); pedicels 2–6 mm long,
puberulous; bracts and bracteoles inconspicuous. Calyx tube 1–1.5 mm long; limb
c. 1 mm long, truncate, repand or shortly toothed. Corolla tube 3–5 mm long,
pubescent at throat and then with a ring of deflexed hairs inside; lobes 3.5–5.5 × 1–2
mm, oblong, thickened towards apex, but not apiculate, puberulous at margins.
Style long-exserted, up to twice as long as corolla tube; pollen presenter 1–1.25 mm
in diameter. Fruit up to 35 mm in diameter, globose, or broadly elliptic in 1-seeded
fruit, not lenticellate. Pyrene not known in detail, 25 × 15 mm.

Subsp. **exserta** —Bridson in Kew Bull. **42**: 648, fig. 3 (1987); in F.T.E.A., Rubiaceae: 843 (1991).
        Canthium stuhlmannii sensu Brenan, Check-list For. Trees Shrubs Tang. Terr.: 489
        (1949), pro parte, quoad *Gillman* 1476. —Vollesen in Opera Bot. **59**: 67 (1980), non
        Bullock.

Corolla tube 4.5–5 mm long; lobes 4.5–5.5 × c. 1 mm; style 9–10 mm long.

Zimbabwe. E: Chipinge Distr., 2 km west of Umzilizwe R., near road from Mt. Selinda to Save (Sabi) R., fl. 22.xii.1974, *Müller* 2229 (SRGH). **Mozambique**. N: Angoche (António Enes), Boila, near Namatil (Nametil), fr. 25.i.1968, *Torre & Correia* 17380 (LISC). MS: *Swynnerton* 1407a (BM).

Also in southern Tanzania. In *Brachystegia* or *Colophospermum mopane* woodlands and in *Manilkara sulcata* thickets, also in forest clearings, on sandy soil or in limestone valleys (Zimbabwe); 30–500(700) m.

Subsp. **robsonii** Bridson in Kew Bull. **42**: 649, map 1 (1987). Type: Zambia, Sasare to Petauke, mile 9, *Robson* 881 (K, holotype; BM; LISC; PRE).

    Canthium crassum sensu A.E. Gonçalves in Garcia de Orta, Sér. Bot. **5**: 187 (1982), pro parte, quoad *Macêdo* 5124, *Torre et al.* 13890, 18272 & 18953, non Hiern.

Corolla tube 3.5–4 mm long; lobes 3.25–4.5 × 1–1.75 mm; style 6–7 mm long.

Zambia. C: Lusaka Distr., Nachitete R., Chileba Village, corolla fallen 1.i.1993, *Bingham* 8764 (K; MO). E: Sasare to Petauke, mile 9, fl. 9.xii.1958, *Robson* 881 (K; BM; LISC; PRE). S: Mukulaikwa Agric. Station, 72 km west of Lusaka, sterile 22.iii.1963, *Angus* 3615 (FHO). **Malawi**. S: Liwonde National Park, Namelembo Thicket, fl. 31.xii.1983, *Dudley* 927 (K). **Mozambique**. T: edge of Zambezi R., near Cahora (Cabora) Bassa, margem esquerda do Zambeze, Monte Murumbuè, fr. 30.iii.1972, *Macêdo* 5124 (K; LISC; LMA; PRE).

Not known elsewhere. Thickets with *Combretum* and *Commiphora* or *Brachystegia*, sometimes in gullies, on sandy soils; (300)500–1200 m.

Subsp. uncertain
    Mozambique. N: Marrupa, fr. 21.ii.1982, *Jansen & Boane* 8006 (K; SRGH).

The above specimen, in immature fruit, was collected from an area intermediate between the known distributions of the two subspecies; the altitude, 850 m, is above that recorded for subsp. *exserta*. Additional gatherings from this area would be of interest.

4. **Multidentia crassa** (Hiern) Bridson & Verdc. in Kew Bull. **42**: 652, fig. 1F & M, fig. 2, A–C (1987); in F.T.E.A., Rubiaceae: 845, fig. 149 (1991). TAB. **47**/C1. Type from Sudan.

    Canthium crassum Hiern in F.T.A. **3**: 145 (1877), pro parte (*Schweinfurth* 1695 excluded). —Bullock in Bull. Misc. Inform., Kew **1932**: 379 (1932), pro parte, excluding *C. platyphyllum* Hiern in syn. —Brenan, Check-list For. Trees Shrubs Tang. Terr.: 488 (1949). —Dale & Greenway, Kenya Trees & Shrubs: 428 (1961). —F. White, F.F.N.R.: 404, fig. 68, J (1962), pro parte, excluding *C. randii* in syn. —A.E. Gonçalves in Garcia de Orta, Sér. Bot. **5**: 186 (1982). Type as above.

Shrub or small tree 0.6–6 m tall, the main trunk with almost black bark or rough grey bark, which peels to expose a reddish undersurface; shoots usually rather stout with thick often powdery very white-grey or straw-coloured wrinkled corky bark; sap copious, sometimes apparently milky. Leaves mostly restricted to apices of branches; blades 3–27.5 × 1.7–15.5 cm, oblong-elliptic to oblong or sometimes round, obtuse or acute at the apex or sometimes tapering-acute, cuneate to rounded at the base, rather fleshy or coriaceous or sometimes papery, often drying yellow-green and mostly markedly paler and glaucous beneath, glabrous or occasionally densely velvety tomentose on lower surface; lateral nerves in 5–7 main pairs; tertiary nerves finely reticulate, sometimes drying pale yellowish above and dark beneath; domatia present as tufts of hair; petioles 3–25 mm long; stipules with broad basal part 3.5–5 mm long and subulate apex 1.5–7.5 mm long, pubescent within. Flowers 5-merous, borne in axillary many-flowered divaricate cymes, all parts of inflorescence mostly pubescent and sometimes densely woolly tomentose; peduncle 10–50 mm long; pedicels 1–3 mm long; bracts 3–5 mm long, linear. Calyx tube 1.5–2 mm long, obconic, glabrous or sometimes pubescent; limb-tube short, 1–3 mm long, almost truncate or undulate or with short triangular teeth about 0.5–1 mm long. Corolla greenish-yellow, fleshy, thick, wrinkled when dry; tube 2.5–4.5 mm long, pilose at throat and with a ring of deflexed hairs one third of the way from the top; lobes 3 mm long, narrowly triangular. Pollen presenter distinctly exserted, green, 2 mm long, ellipsoid, ridged. Fruit green mottled brown, yellow or dull red or brown spotted white, fleshy and edible, 22–30 × 26–37 mm (probably at least 35–40 mm wide in life), depressed globose, lenticellate. Pyrenes 18–22 × 11–17 × 11–15 mm, ellipsoid; the walls ridged, exceedingly thick and woody, up to 5–7 mm thick, with a deep lateral semicircular groove on either side above the point of attachment, meeting on the apical edge.

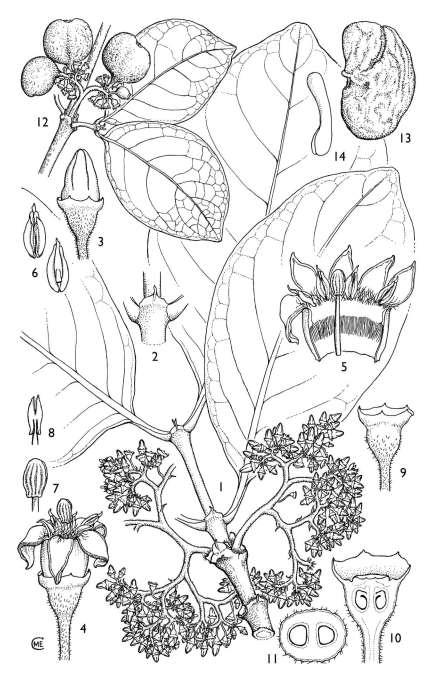

Tab. 61. MULTIDENTIA CRASSA var. CRASSA. 1, flowering branch (× ²/₃); 2, stipule (× 1); 3, flower bud (× 4); 4, flower (× 4); 5, corolla opened out, with style and pollen presenter (× 4); 6, anther, 2 views (× 6); 7, pollen presenter (× 6); 8, longitudinal section through pollen presenter (× 6); 9, calyx (× 4); 10, longitudinal section through ovary (× 6); 11, transverse section through ovary (× 10), 1–11 from *Richards* 25838; 12, portion of fruiting branch (× ²/₃), from *Burtt* 3591; 13, pyrene (× 2); 14, embryo (× 2), 13 & 14 from *Angus* 185. Drawn by M.E. Church. From F.T.E.A.

Var. **crassa** —Bridson in Kew Bull. **42**: 652 (1987); in F.T.E.A., Rubiaceae: 847 (1991). TAB. **61**.
  Craterispermum orientale K. Schum. in Engler, Pflanzenw. Ost-Afrikas **C**: 387 (1895). —
  Brenan, Check-list For. Trees Shrubs Tang. Terr.: 493 (1949). Type from Tanzania.
  Canthium opimum S. Moore in J. Linn. Soc., Bot. **37**: 308 (1906). —Krause in R.E. Fries, Wiss.
  Ergebn. Schwed. Rhod.-Kongo-Exped. (Erganzungsheft): 15 (1921). Type from Angola.
  Canthium dictyophlebum S. Moore in J. Bot. **57**: 87 (1919). Type from Zaire (Dem. Rep.
  Congo).
  Plectronia opima (S. Moore) Mildbr. in Notizbl. Bot. Gart. Berlin-Dahlem **9**: 204 (1924).

Leaf blades glabrous even when young, or with tomentose domatia beneath.

Zambia. N: Mbala Distr., Lucheche River, Muswilo Village, fl. bud 4.ii.1965, *Richards* 19594
(K). W: Ndola, fr. 1933, *Duff* 153/33 (FHO; K). C: Luangwa Valley, near Kapamba River, fl.
12.i.1966, *Astle* 4361 (K). E: Petauke Distr., Great East Road, Nyimba to Chipata (Fort
Jameson), mile 12, fr. 20.iv.1952, *White* 2425 (FHO; K). **Zimbabwe**. W: Victoria Falls, fr. *Rogers*
5984 (K). **Malawi**. N: Mzimba Distr., 11 km north west of Chikangawa, fl. 3.xii.1978, *E. Phillips*
4335 (K; LMU). C: Lilongwe, Nature Sanctuary Zone A, outside fence along Kenyatta Drive, fl.
9.xi.1984, *Seyani* 1678 (K). S: Zomba, fl. 1933, *Clements* 359 (FHO; K). **Mozambique**. N: 16 km
NNW of Mandimba, fl. 10.i.1942, *Hornby* 2516 (PRE). T: Macanga, between Furancungo and
Mualadze (Vila Gamito), fr. 10.vii.1949, *Barbosa & Carvalho* 3563 (LISC; PRE).
  Also in Sudan, Uganda, Burundi, Kenya, Tanzania, Zaire (Dem. Rep. Congo) (Shaba Prov.)
and Angola. In *Brachystegia* woodland, or less often in open bushland with scattered trees or at
forest edges, often on rocky outcrops; 600–1675 m.

Var. **ampla** (S. Moore) Bridson & Verdc. in Kew Bull. **42**: 653 (1987); in F.T.E.A., Rubiaceae: 847
  (1991). Type: Zambia, Chilanga, *Rogers* 8446 (BM, holotype; K).
  Canthium amplum S. Moore in J. Bot. **57**: 87 (1919).
  Plectronia buarica Mildbr. in Notizbl. Bot. Gart. Berlin-Dahlem **9**: 203 (1924). Type from
  Cameroon.

Leaf blades velvety beneath, with short matted hairs.

Zambia. N: Mbala Distr., Lake Chila, fl. 15.xi.1952, *Angus* 754 (FHO; K). W: Chingola, fl.
6.xi.1953, *Fanshawe* 484 (K). C: Lusaka Distr., Mt. Makulu Research Station, 3.3 km west of
Chilanga, fl. 21.xii.1958, *Robson & Angus* 979 (BM; K; LISC; PRE). S: Mazabuka, fr. 20.v.1961,
Fanshawe 6588 (K). **Malawi**. N: Chitipa Distr., 16 km south of Chisenga at border with Zambia,
fr. 23.iv.1972, *Pawek* 5162 (K). C: Dedza Distr., Chongoni, fl. 17.xi.1960, *Chapman* 1048 (K;
SRGH). **Mozambique**. N: Marrupa, corollas fallen 21.ii.1982, *Jansen* 8004 (K).
  Also in southern and eastern Tanzania. In *Brachystegia* woodland and *Brachystegia–Uapaca*
woodland, often on hill tops; 850–1250 m.
  Some specimens with a sparser indumentum on the leaf lower surface do occur, and one
from Zambia (*Silungwe* s.n.) has densely tomentose inflorescence axes and calyx tube.
  Bullock cited *Canthium platyphyllum* in synonymy for this species, noting "*C. platyphyllum* is a
further form with a lanate tomentum on the inflorescence and young branchlets; this falls away
and the plant develops a rusty-red coloured bark." It is in fact a synonym of *Vangueriopsis
lancifolia*. Sterile material of both *V. lanciflora* and *Ancylanthos rubiginosus* can resemble *M. crassa*
var. *ampla* but usually have a stronger indumentum, and rusty-red bark.

5. **Multidentia concrescens** (Bullock) Bridson & Verdc. in Kew Bull. **42**: 653, fig. 1E, 2D (1987);
  in F.T.E.A., Rubiaceae: 847 (1991). Type from Tanzania.
  Pygmaeothamnus concrescens Bullock in Bull. Misc. Inform., Kew **1933**: 471 (1933); in
  Hooker's Icon. Pl.: t. 3242 (1934). —Brenan, Check-list For. Trees Shrubs Tang. Terr.: 525
  (1949). —F. White, F.F.N.R.: 418 (1962). Type as above.
  Multidentia verticillata Gilli in Ann. Naturhist. Mus. Wien **77**: 21, taf. 1 (1973). Type from
  Tanzania.

Pyrophytic erect subshrub (suffrutex), (7.5)15–35 cm tall; stem simple or
slightly branched near the base, arising from a thick woody rootstock, glabrous,
dark, wrinkled, with elongate lenticels. Leaves opposite or in whorls of 3, drying
the bright yellow-green of an aluminium accumulator, mostly paler and often
glaucous beneath; blades 2.5–16(21.5) × 0.8–8.5(12.5) cm, oblong, elliptic,
oblong-elliptic or obovate to oblanceolate, very shortly acutely or obtusely
acuminate or less often merely acute or rounded at the apex, cuneate at the base,
sometimes quite coriaceous, glabrous to rather sparsely shortly pubescent on both
surfaces, lateral nerves in 5–7 main pairs, tertiary nerves apparent and coarsely
reticulate; domatia absent; petioles 3–10(20) mm long; stipules with a broad basal

(usually sheath-like) part 2–3 mm long, hairy inside and along the margin, with a glabrous subulate lobe 3–12 mm long. Inflorescences slightly to distinctly supra-axillary, divaricately cymose, usually 10–many-flowered, the axes mostly pubescent; peduncle 2–4.5 cm long, forming an acute angle with main stem; pedicels 1–11 mm long; bracts and bracteoles 3–8 mm long, linear-lanceolate. Flowers 5(6)-merous. Calyx tube 2–4 mm long, obconic, glabrous or pubescent; tubular part of limb 0.5–2.5 mm long, drying distinctly wrinkled; lobes (1)2.5–9 mm long, oblong to linear-lanceolate or narrowly triangular, acute or subacute at the apex. Corolla greenish or greenish-white, the lobes white inside, thickened and often drying wrinkled; tube 3–5 mm long, pilose at the throat and with a ring of deflexed hairs set one third down from the top; lobes reflexed, 4–5 × 2 mm, narrowly triangular, appearing to have lateral margins thickening towards the apex in the dry state. Style 6–8.5 mm long; pollen presenter green, 1.5–2 × 1–2 mm, oblong-ellipsoid, dividing at the apex. Fruit yellowish-green becoming red, 18–22 mm in diameter, globose; calyx limb persistent. Pyrenes 11–14 × 8–9 × 8 mm, ellipsoid, with deep irregular longitudinal ridges.

Zambia. N: Ishiba Ngandu (Shiwa Ngandu), fl. 5.ii.1955, *Fanshawe* 2000 (K). W: Solwezi, fr. 22.vii.1964, *Fanshawe* 8828 (K). **Malawi**. N: Chitipa Distr., Misuku Hills, near Chatu School, fr. 8.vii.1973, *Pawek* 7128 (K).

Also in southern Tanzania. In *Brachystegia* or miombo woodland, and in open woodlands with *Uapaca, Parinari* and *Harungana*, also in grassland and on kopjes, and reported from copper bearing soil on new mine sites; 1400–2000 m.

Apart from the type of *Multidentia verticillata* none of the fairly extensive material examined shows leaves in whorls of 3.

50. CANTHIUM Lam.

Canthium Lam., Encycl. Méth., Bot. **1**: 602 (1785). —Bridson in F.T.E.A., Rubiaceae: 861–885 (1991); in Kew Bull. **47**: 353–401 (1992).
Plectronia sensu auctt. div., non L.

Shrubs or small trees, sometimes scandent; spines absent or present. Heterophylly sometimes apparent; leaves either deciduous and restricted to brachyblasts (contracted lateral spurs) or apex of stem, or evergreen and spaced along the branches, opposite, paired, petiolate; domatia absent, pit-like or present as tufts of hair; stipules shortly sheathing, apiculate or aristate, glabrous or pubescent within. Flowers hermaphrodite or unisexual, 4–5 merous, pedicellate or not, borne in pedunculate to subsessile few–many-flowered cymes; inflorescence branches present, sometimes reduced; bracteoles inconspicuous. Calyx tube broadly ellipsoid to ovoid, ± obsolete or somewhat reduced, scarcely equalling the disk, bearing triangular or linear lobes. Corolla white or yellowish, glabrous or rarely pubescent outside; tube broadly cylindrical, ± equal to lobes or sometimes longer or shorter, with or without a ring of deflexed hairs inside, often pubescent at throat; lobes reflexed or less often erect, obtuse, acute or shortly apiculate. Stamens set at throat; anthers subsessile or borne on short filaments, with or without darkened connective tissue on dorsal face. Style slender, shortly exceeding the corolla tube, glabrous; pollen presenter ± spherical, point of attachment within a distinct basal recess, 2(3)-lobed; ovary 2(3)-locular. Disk glabrous (pubescent in some South African species). Fruit a 2(3)-seeded drupe or often 1-seeded by abortion, small to moderately large, strongly dorsiventrally flattened or not, strongly or scarcely indented at apex; pyrenes ellipsoid to ovoid or obovoid, often flattened on ventral face, thinly woody, slightly to distinctly triangular or truncate at point of attachment, with a shallow crest extending to apex, usually rugulose. Seeds with endosperm entire; testa finely reticulate; embryo ± straight to curved with radicle erect; cotyledons small, perpendicular to ventral face of seed.

A diverse and poorly defined genus of approximately 100 species, widely distributed throughout eastern and southern tropical Africa, South Africa, Seychelles, Madagascar, Asia and Malesia. The above description applies to the tropical African members only.

Multidentia, Pyrostria, Psydrax and *Keetia* were originally included in *Canthium* in Africa (see Bullock in Bull. Misc. Inform., Kew **1932**: 353–389 (1932)).

Key to subgenera of Canthium

1. Spines paired, always borne above brachyblasts (contracted lateral spurs), TAB. **46**/12 · ·
·· Subgen. 1. **Canthium** (p. 316)
 − Spines absent or, if present, paired or ternate but not strictly associated with brachyblasts · · 2
2. Flowers unisexual; inflorescence fasciculate or subumbellate, female flowers always fewer
than male flowers; corolla tube with a ring of deflexed hairs inside; stipules lacking a
conspicuous tuft of hair inside ···················· Subgen. 4. **Bullockia** (p. 333)
 − Flowers hermaphrodite (rarely gynodioecious); inflorescence cymose or sometimes
reduced and appearing subumbellate, occasionally solitary to few-flowered; corolla tube
lacking a ring of deflexed hairs inside; stipules often conspicuously hairy inside · · · · · 3
3. Spines always absent; stems often lenticellate; leaves always deciduous, flowers typically
borne at naked nodes, TAB. **46**/10; stipules sheathing when young, but usually quickly
becoming corky and triangular to ovate, sometimes deciduous; calyx limb reduced to a rim;
anthers with little or no darkened connective on dorsal face; pyrenes usually triangular or
beaked at point of attachment (Tabs. **63**/A10 & B6; **64**/10) ···················
··· Subgen. 2. **Afrocanthium** (p. 319)
 − Spines present or sometimes absent; stems not markedly lenticellate; leaves deciduous or
not strictly so, flowers occasionally borne at naked nodes; stipules sheathing, only
occasionally becoming corky; calyx limb lobed or shortly cupular; anthers with all but
marginal area covered with darkened connective on dorsal face; pyrenes ± truncate at point
of attachment (TAB. **65**/12) ···4
4. Leaves larger, discolorous when dry, dull green above, greyish-green beneath; calyx limb
lobed almost to base, TABS. **47**/A7; **65**/4; inflorescence 3–30-flowered; pedicels not
markedly long and slender ···················· Subgen. 3. **Lycioserissa** (p. 330)
 − Leaves small (0.5–4 × 0.5–2.6 cm), blackish when dry; calyx limb cupular, at least hiding the
disk, truncate or shortly dentate, TAB. **47**/A3; inflorescence 1–3-flowered; pedicels very
long and slender ······························ 16. *C. kuntzeanum* (p. 337)

Subgen. 1. CANTHIUM

Canthium subgen. **Canthium** —Bridson in F.T.E.A., Rubiaceae: 862 (1991); in Kew
Bull. **47**: 365 (1992).

Shrubs or small trees, sometimes scandent; spines usually present, situated above
brachyblasts (contracted lateral spurs), straight. Leaves deciduous, restricted to
brachyblasts and apex of stem, opposite, paired, petiolate, never large, papery to
chartaceous, often discolorous; tertiary nerves apparent; domatia present as small
tufts of hair; stipules triangular, apiculate or aristate, pubescent within. Flowers
hermaphrodite, 4–5-merous, pedicellate, borne in pedunculate to subsessile few- to
several-flowered cymes; inflorescence branches present or reduced; bracteoles
inconspicuous. Calyx tube broadly ellipsoid to ovoid; limb with tube rather reduced,
scarcely equalling the disk, dentate or bearing triangular-linear lobes. Corolla
whitish, glabrous, rarely pubescent outside; tube broadly cylindrical, equalling the
lobes or sometimes shorter, with a ring of deflexed hairs inside (apparently absent in
one Indian species) and pubescent at the throat; lobes reflexed, acute or shortly
apiculate. Stamens inserted at the throat; anthers with the dorsal face central area
covered by a darkened connective. Style slender, shortly exceeding the corolla tube,
glabrous; pollen presenter spherical, point of attachment within a basal recess about
one third to half of the receptacle in depth, 2-lobed; disk glabrous; ovary 2-locular.
Fruit a 2-seeded drupe, small to moderately large, strongly dorsiventrally flattened,
scarcely indented at the apex. Pyrenes ellipsoid with the ventral face flattened, thinly
woody, distinctly (or slightly in non-African species) triangular at the point of
attachment, with a shallow crest extending to the apex, rugulose. Seeds narrowly
obovate, flattened on ventral area, very briefly crested around the apex; endosperm
entire; testa finely reticulate; embryo ± straight with radicle erect; cotyledons small,
perpendicular to ventral face of seed.

   Canthium sensu stricto is considered to comprise 3 species only; the type *C. coromandelicum*
(Burm.f.) Alston and *C. travancoricum* Bedd., both from southern India and Sri Lanka, and the
African species *C. glaucum* occurring from Somalia to Zimbabwe. *Species No. 1.*

1. **Canthium glaucum** Hiern in F.T.A. **3**: 134 (1877), pro parte (excl. specim. *Kirk* s.n.). —
Bullock in Bull. Misc. Inform., Kew **1932**: 359 (1932). —Bridson in F.T.E.A., Rubiaceae: 862, fig.
153 (1991); in Kew Bull. **47**: 365 (1992). Type from Somalia.

Plectronia glauca (Hiern) K. Schum. in Engler, Pflanzenw. Ost-Afrikas **C**: 386 (1895). —
Chiovenda, Fl. Somalia **2**: 245 (1932). Type as above.

Shrub, sometimes scandent, 1–5 m tall; young branches glabrous or rarely
pubescent, often lenticellate, armed with spines up to 15(21) mm long, situated
above brachyblasts (very abbreviated lateral branches), apparently not developing
into lateral branches; bark light to dark grey or brown. Leaves restricted to lateral
spurs and apex of stem; blades 2.4–6 × 1.4–3 cm, narrowly to broadly elliptic, acute
to subacuminate or sometimes obtuse at apex, acute to cuneate at base, strongly
or moderately discolorous, glabrous, or glabrous above and sparsely pubescent at
least on nerves beneath, or rarely pubescent on both faces; lateral nerves usually
in 3 main pairs; tertiary nerves apparent; domatia present as small tufts of hair;
petiole 1–1.6 mm long, glabrous to sparsely pubescent, or rarely pubescent;
stipules 1–4 mm long, triangular, apiculate, densely pubescent within. Flowers
(4)5-merous, in 2–8-flowered cymes; cymes subsessile to pedunculate; peduncles
up to 10 mm long; pedicels 3–10 mm long, glabrous or sparsely pubescent;
bracteoles inconspicuous. Calyx tube 1 mm long, glabrous or rarely pubescent;
limb divided often almost to base into triangular or narrowly triangular lobes,
0.5–1 mm long. Corolla glabrous or rarely pubescent outside; tube 1–1.25 mm
long, with a ring of deflexed hairs at the throat and pubescent immediately above;
lobes 2–2.5 × 1.5 mm, triangular-lanceolate, scarcely apiculate to apiculate. Style
slightly exceeding the corolla tube; pollen presenter spherical, 0.75 mm across;
disk not prominent. Fruit pale red, edible, 9–12 mm long and wide, square in
outline, scarcely indented at apex, hardly tapered at base. Pyrene 7–12 × 4–6 mm,
ellipsoid with ventral face flattened, triangular at point of attachment, crested at
the apex, rugulose.

Subsp. **frangula** (S. Moore) Bridson in Kew Bull. **47**: 366 (1992). Type: Mozambique, Manica e
Sofala Distr., Madanda Forest, *Swynnerton* 1403 (BM, syntype) and 1894 (BM, syntype; K,
isosyntype).

Canthium glaucum sensu Hiern in F.T.A. **3**: 134 (1877), pro parte, quoad *Kirk* s.n.

Canthium frangula S. Moore in J. Linn. Soc., Bot. **40**: 89 (1911). —Miller in J. S. African
Bot. **18**: 82 (1952). —J.G. Garcia in Mem. Junta Invest. Ultramar **6** (sér. 2): 30 (1959)
[Contrib. Conhec. Fl. Moçamb. IV (1959)], pro parte. —F. White, F.F.N.R.: 402, fig. 68, H
(1962). —Drummond in Kirkia **10**: 276 (1975). —A.E. Gonçalves in Garcia de Orta, Sér.
Bot. **5**: 187 (1982). —K. Coates Palgrave, Trees Southern Africa, ed. 3, rev.: 880 (1988).

Var. **frangula** —Bridson in Kew Bull. **47**: 366, figs. 2B, 3C (1992). TAB. **62**.

Leaves always glabrous on upper surface, glabrous to sparsely pubescent on lower
surface; calyx and corolla glabrous.

Caprivi Strip. Mpilila Island, fr. 13.i.1959, *Killick & Leistner* 3353 (COI; PRE; SRGH).
Botswana. N: Chobe Distr., near Kasane, fl. *Henry* 32 (M; PRE; SRGH). **Zambia**. B: Sesheke
Distr., Simongoma Forest Reserve, fr. 26.i.1952, *White* 1963 (FHO; K). C: Lusaka Distr.,
Nachitete R., Chileba Village, 1.i.1993, *Bingham* 8766 (K; MO). S: Livingstone Distr., Victoria
Falls, 915 m, fl. 24.xi.1947, *Wild* 2198 (K; SRGH). **Zimbabwe**. N: Shamva Distr., Mazowe R.,
fl. 9.xii.1960, *Rutherford-Smith* 440 (K; LISC; PRE; SRGH). W: Hwange Hospital, fl.
12.xii.1974, *Raymond* 304 (K; SRGH). C: Kwe Kwe Distr., Sable Park, fr. 19.ii.1976, *Stephens*
394 (SRGH). E: Chipinge Distr., near Chivirira Falls, fr. 16.xii.1957, *Phelps* 217 (K; PRE;
SRGH). S: Bikita Distr., Lower Save (Sabi), west bank, Umkondo, 550 m, fl. 2.ii.1948, *Wild*
2487 (K; SRGH). **Malawi**. N: Chitipa Distr., Stevenson's Road, 1375 m, fl. 4.i.1974, *Pawek* 7753
(K). S: Mangochi (Fort Johnson), fl. 5.xii.1966, *Eccles* 135b (SRGH). **Mozambique**. N: Imala,
Mt. Melita, 7 km from Muecate, fr. 15.i.1964, *Torre & Paiva* 9983 (LISC). T: near Estima,
young fr. 16.v.1972, *Bond* E36 (LISC; SRGH). MS: Madanda Forest, sterile 5.xii.1906,
Swynnerton 1894 (BM; K, LISC).

Not known outside the Flora Zambesiaca area. In *Brachystegia* or *Colophospermum mopane*
woodlands, and mixed thickets and mutemwa on Kalahari Sands, also on clay soils, often on
termite mounds and sometimes on granite outcrops; 370–1380 m.

The specimen *White* 1963, from Zambia B:, and one other sterile gathering from the same
area are somewhat atypical but the material is insufficient to allow more detailed comment.

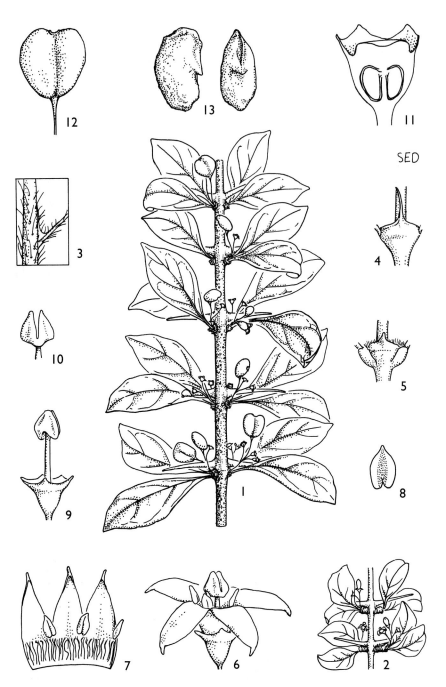

SED

Tab. 62. CANTHIUM GLAUCUM subsp. FRANGULA var. FRANGULA. 1, fruiting branch (×
²/₃), from *Pawek* 10811; 2, portion of flowering branch (× ²/₃), from *Pawek* 7753; 3,
domatium (× 8), from *Pawek* 10811; 4, juvenile stipule (× 4); 5, mature stipule (× 4), 4 &
5 from *Chase* 4718; 6, flower (× 8), from *Rutherford-Smith* 440; 7, portion of corolla
opened out (× 8); 8, dorsal view of anther (× 16); 9, calyx, style and pollen presenter (×
8); 10, pollen presenter (× 10); 11, longitudinal section through ovary (× 10), 7–11 from
Pawek 7753; 12, fruit (× 2); 13, pyrene, 2 views (× 3), 12 & 13 from *Goodier* 121. Drawn
by Sally Dawson.

Var. **pubescens** Bridson in Kew Bull. **47**: 366 (1992). Type: Zambia, Mazabuka Distr., north access road to Kariba Hills, *Goodier* 445 (K, holotype; BR; LISC; SRGH).

Leaves densely pubescent on upper and lower surfaces; calyx and corolla pubescent.

Zambia. S: Mazabuka Distr., north access road to Kariba Hills, fl. 4.xii.1959, *Goodier* 445 (BR; K; LISC; SRGH). **Zimbabwe**. N: Hurungwe Distr., Mhenza (Mensa) Pan area, young fr. 22.i.1957, *Phelps* 187 (K; SRGH). E: Mutare/Chimanimani Distr., Umvumvumu R., fl. 19.xii.1947, *Chase* 486 (K; SRGH).

Not known outside the Flora Zambesiaca area. In thickets or *Colophospermum mopane* woodland; c. 550 m.

Canthium glaucum subsp. *glaucum* is a comparatively poorly known taxon from the coastal regions of Kenya and southern Somalia. It differs from subsp. *frangula* in the leaves always glabrous and glaucous beneath, the light grey to light greyish-brown stems and the somewhat shorter pedicels.

Subgen. 2. AFROCANTHIUM

Canthium subgen. **Afrocanthium** Bridson in F.T.E.A., Rubiaceae: 864 (1991); in Kew Bull. **47**: 366 (1992).

Unarmed deciduous shrubs or trees; brachyblasts (contracted lateral spurs) sometimes present; lenticels frequently apparent on young stems. Leaves usually confined to new growth at apex of stems, paired, petiolate, papery to subcoriaceous; heterophylly observed in only one species (*C. ngonii*); tertiary nerves finely, or less often coarsely, reticulate or obscure; domatia present, pubescent, or sometimes absent; stipules sheathing on apical node and awned, but soon triangular to ovate, obtuse to apiculate and often becoming corky, sometimes caducous, with tufts of white or less often rust-coloured hair within. Flowers hermaphrodite, or gynodioecious (see note after *C. gilfillani*), 4–5-merous, pedicellate or subsessile, borne in pedunculate few–many-flowered cymes; inflorescence branches present or sometimes very reduced, often with the flowers borne along one side of the ultimate branch; bracteoles obscure or absent. Calyx tube broadly ellipsoid to ovoid; limb reduced to a rim, sometimes shortly dentate, scarcely equalling disk in length. Corolla white or greenish-yellow, usually rather small, glabrous outside; tube broadly cylindrical, ± as long as the lobes or a little longer, sparsely hairy to hairy within, often hairy at the throat; lobes reflexed, obtuse to acute or subacuminate. Stamens set at the corolla throat; filaments short; anthers ovate, dorsal face lacking a darkened connective or with only a central band. Style slender, shortly exceeding the corolla tube, or rarely somewhat longer, glabrous; pollen presenter ± spherical, 2-lobed; point of attachment within a basal recess about one third of the receptacle in depth; disk glabrous; ovary 2-locular. Fruit a 2-seeded drupe, small to moderately-sized, strongly dorsiventrally flattened, typically obcordate in lateral view and strongly or rarely scarcely indented at the apex. Pyrenes ellipsoid with the ventral face flattened, thinly woody, usually somewhat rugulose, sometimes triangular at point of attachment with a shallow but distinct crest extending from the point of attachment to the apex. Seeds ± obovoid with the ventral area flattened and the apex briefly crested; endosperm entire; testa finely reticulate; embryo scarcely curved, radicle erect; cotyledons perpendicular to ventral face of the seed.

A subgenus of 22 species, restricted to eastern and southern tropical Africa, extending to eastern Zaire (Dem. Rep. Congo) in the west and to South Africa in the south. *Species Nos.* 2–11.

1. Stipules with the marginal area thinner and shiny (obscured by inrolling at the apical node), soon caducous, seldom persisting beyond second node; fruit not or only slightly indented at the apex; corolla tube 2–3 mm long; domatia never conspicuous · · · · · · · · 2
  – Stipules not thin and shiny at the margins, often breaking down but bases at least persistent, sometimes becoming corky; fruit clearly bilobed and indented at the apex; corolla tube 1–2.25 mm long; domatia conspicuous, or less often inconspicuous · · · · · · · · · · · · · · · 3
2. Petioles 1–5 mm long; fruit not or scarcely bilobed or indented; leaves glabrous or sometimes pubescent, with (4)5–6(7) main pairs of lateral nerves · · · · · · · · · 10. *burttii*

– Petioles 5–13 mm long; fruit discernibly bilobed and indented; leaves pubescent, with 6–8 main pairs of lateral nerves ···························· 11. *pseudorandii*

3. Leaves large, 7–18 cm long, with 8–10 main pairs of lateral nerves; stems thick; inflorescence often much branched, 20–50-flowered ················· 8. *lactescens*

– Leaves smaller, sometimes up to 14.5 cm long, but usually much less, with up to 7 pairs of lateral nerves; stems not markedly thick; inflorescence with 2 or sometimes more branches or subumbellate by reduction, 1–30-flowered ···························· 4

4. Leaves with 3–4 main pairs of lateral nerves; lateral branches (some at least) reduced to cushion shoots (brachyblasts) or almost entirely suppressed so that each node appears to have four leaves; inflorescence seldom more than 6-flowered, obscurely branched; stipules with fawn to rust-coloured hairs inside; corolla tube 1 mm long ················ 5

– Leaves with (3)4–7 main pairs of lateral nerves; lateral branches always more developed; inflorescences 1–30-flowered, shortly to clearly branched; stipules usually with white hairs inside; corolla tube 1–2.25 mm long ······································· 6

5. Petioles 1–9 mm long; young stems not remarkably slender; leaves on young and older branches similar, glabrous or ciliate ····················· 2. *pseudoverticillatum*

– Petioles up to 2 mm long; young stems very slender; leaves on young stems strongly attenuated at apex and base in contrast to leaves on the reduced spurs, never ciliate ···· ··· 3. *ngonii*

6. Mature leaves small, 2–7.5 cm long, with (3)4–5 main pairs of lateral nerves ········· 7

– Mature leaves usually larger, 5–15 cm long, with (4)5–7(8) main pairs of lateral nerves ··· 8

7. Leaves glabrous, or sometimes sparsely pubescent to pubescent beneath, acute to cuneate, occasionally obtuse or rarely truncate at base; petioles 2–10 mm long; domatia conspicuous ·· 4. *mundianum*

– Leaves densely pubescent to tomentose beneath, rounded to truncate or rounded then acute at base; petioles 1–3 mm long; domatia inconspicuous (obscured by indumentum) ··· 5. *gilfillanii*

8. Leaf blades pubescent or glabrescent above, sparsely pubescent to tomentose beneath with midrib densely hairy even when mature ······························· 9

– Leaf blades glabrous above, occasionally pubescent on nerves beneath ············ 10

9. Fruit smaller, 8–10 × 8–10 mm; domatia visible as tufts of hair; bark red-brown ·· 7. *salubenii*

– Fruit larger, 12 × 11–13 mm; domatia inconspicuous; bark mid to dark grey ·········· ································· 9. *racemulosum* var. *nanguanum*

10. Tertiary nerves coarsely reticulate in mature leaves; apical stipule with lobe linear; corolla throat with conspicuous projecting hairs; fruit larger, 12 × 11–13 mm ·············· ······························· 9. *racemulosum* var. *racemulosum*

– Tertiary nerves finely reticulate; apical stipule with lobe filiform to subulate; corolla throat lacking projecting hairs; fruit smaller, 6–9 × 7–9 mm ·············· 6. *parasiebenlistii*

2. **Canthium pseudoverticillatum** S. Moore in J. Bot. **43**: 352 (1905). —Bullock in Bull. Misc. Inform., Kew **1932**: 377 (1932). —Dale & Greenway, Kenya Trees & Shrubs: 430 (1961). — Brenan, Check-list For. Trees Shrubs Tang. Terr.: 489 (1949). —Bridson in F.T.E.A., Rubiaceae: 866, fig. 154 (1991); in Kew Bull. **47**: 370 (1992). —Beentje, Kenya Trees, Shrubs Lianas: 506 (1994). Type from Kenya.

 Plectronia microterantha K. Schum. & K. Krause in Bot. Jahrb. Syst. **39**: 541 (1907). Type from Kenya.

 Canthium robynsianum Bullock in Bull. Misc. Inform., Kew **1932**: 377, fig. 2 (1932). — Dale & Greenway, Kenya Trees & Shrubs: 430 (1961). —Beentje, Kenya Trees, Shrubs Lianas: 506 (1994). Type from Kenya.

Shrub ?sometimes scandent, or small tree 2–6 m tall; young branches sparsely pubescent to pubescent; bark reddish or pale grey when older. Leaves paired at apex of main branches or brachyblasts, often giving impression of 4 per node due to extreme reduction of branches, mature or sometimes not fully mature at time of flowering; blades 2.5–7.5 × 1–5 cm, elliptic to broadly elliptic, acute to subacuminate at apex, acute to cuneate at base, papery or chartaceous when mature, glabrous or sometimes ciliate above, with glabrous to pubescent (rather prickle-like hairs) on nerves beneath; lateral nerves in 3–4 main pairs; tertiary nerves obscure or apparent, rather coarsely reticulate; domatia conspicuous; petioles 1–9 mm long, glabrous or sparsely pubescent to pubescent beneath; stipules with a filiform to subulate lobe present at apical node, later with triangular base 2–3 mm long which becomes corky and soon breaks down to reveal tufts of rust-coloured hair within. Flowers (4)5-merous, in small 1–6(10)-flowered shortly pedunculate cymes; peduncles 0.5–3 mm long; pedicels 1–5 mm long, glabrous or with a line of short hairs on either side; bracteoles inconspicuous. Calyx tube 1–2 mm long, glabrous or sparsely pubescent; limb reduced to a truncate or repand rim sometimes shortly dentate or with unequal linear lobes up to 1 mm long. Corolla yellow or creamy-green; tube 1 mm long, densely but finely pubescent at throat; lobes 1.5–2.25 × 1–1.75 mm, ovate or ovate-triangular at base, acute. Pollen presenter ± spherical, c. 1 mm in diameter, ribbed. Fruit ± oblong in lateral view, c. 10 × 9 mm, indented at apex; pedicel lengthening to 5–13 mm long. Pyrene 9 × 4 mm, narrowly oblong-ovoid, rugulose.

Subsp. **pseudoverticillatum**

Mozambique. Z: Pebane Distr., 15 km from River Ligonha, near Nabúri, fl. 17.i.1968, *Torre & Correia* 17222 (LISC).
Also in Kenya and Tanzania. On termite mounds.
Subsp. *somaliense* Bridson, is restricted to Somalia, and differs in having inflorescences up to 20-flowered, and leaves without domatia.

3. **Canthium ngonii** Bridson in Kew Bull. **47**: 371, fig. 4 (1992). TAB. **63**/A1–A10. Type: Zimbabwe, Chimanimani Distr., Ngorima Reserve, 3 km from Haroni/Rusitu confluence upstream of Rusitu (Lusitu) R., *Ngoni* 69 (K, holotype; LISC; SRGH; WAG).
    Canthium pseudoverticillatum sensu Drummond in Kirkia **10**: 276 (1975). —K. Coates Palgrave, Trees Southern Africa, ed. 3, rev.: 885 (1988), non S. Moore.

Shrub or small tree 1–2 m tall, with very short or almost suppressed brachyblasts which in growing season develop into very slender branchlets, often with appressed hairs; bark dark brown, sometimes with a greyish tinge. Leaves subsessile to shortly petiolate, restricted to the apex of short lateral spurs or well spaced along new branchlets but not along the main axis; blades of leaves on older growth, 2–5 × 0.5–2.8 cm, elliptic to broadly elliptic, perhaps not always fully mature at time of flowering; blades of leaves on young branchlets 2–6.5 × 0.4–1.7 cm, narrowly elliptic, strongly attenuate and sometimes apiculate at the apex, attenuate at the base, papery, glabrous; lateral nerves in 3 main pairs, tertiary nerves obscure; domatia present as tufts of hair; petiole up to 2 mm long; stipules 2.5–4 mm long when young, triangular at base bearing a linear to aristate lobe occasionally somewhat spathulate at apex, usually becoming corky and losing the lobe when mature, with pale or rusty-coloured hairs inside. Flowers 5-merous, in 1–5-flowered cymes restricted to the lateral spurs, inflorescences very reduced; peduncles 0.5–1.5 mm long, sparsely pubescent; pedicels 1–4 mm long, glabrous or with very few hairs; bracts inconspicuous. Calyx tube 0.75 mm long, glabrous; limb reduced to a rim. Corolla yellowish; tube 1 mm long with fine dense erect and deflexed hairs inside; lobes 1.5 × 1 mm, ovate, acute and slightly thickened at the apex. Anthers set at throat, subsessile, very shortly apiculate. Style slightly exceeding the corolla tube; pollen presenter 0.7 mm, long and wide. Fruit 9–10 × 8–9 mm, almost square in lateral view, slightly tapered towards the base, indented at the apex; pedicel lengthening to 8–10 mm long. Pyrene 8 × 4 mm, oblong-ellipsoid with a shallow crest at the apex, rugulose.

Zimbabwe. E: Chipinge Distr., 9 km NE of Umzilizwe (Musirizwe)–Bwazi R. confluence, fr. 23.i.1975, *Pope, Biegel & Russell* 1387 (K; SRGH). **Mozambique**. MS: southern foothills of Chimanimani Mts., Makurupini Falls, fl. 25.xi.1967, *Simon & Ngoni* 1305 (K; LISC; PRE; SRGH).
    Not known outside the Flora Zambesiaca area. In riverine or evergreen forest and high rainfall woodland, in *Terminalia gazensis* groves or *Brachystegia* woodland; 460–1130 m.
    P. van Wyk BSA 3338, from Harare National Botanic Garden, with immature leaves and closely spaced nodes most likely belongs here.

4. **Canthium mundianum** Cham. & Schltdl. in Linnaea **4**: 131 (1829), as "*mundianum*". —De Candolle, Prodr. **4**: 474 (1830). —Ecklon & Zeyher, Enum. Pl. Afr.: 301, 2295 (1837). —S. Moore in J. Linn. Soc., Bot. **40**: 90 (1911). —Palmer & Pitman, Trees Southern Africa **3**: 2097, figs. on 2098 & 2099 (1973). —Ross, Fl. Natal: 335 (1973). —van Wyk, Trees Kruger Nat. Park **2**: 557 (1974). —Drummond in Kirkia **10**: 276 (1975). —Compton, Fl. Swaziland [J. S. African Bot., Suppl. 11]: 581 (1976). —K. Coates Palgrave, Trees Southern Africa, ed. 3, rev.: 16j of appendix (1988); 884 (1988), pro parte excluding *C. gilfillanii*. —Moll, Trees

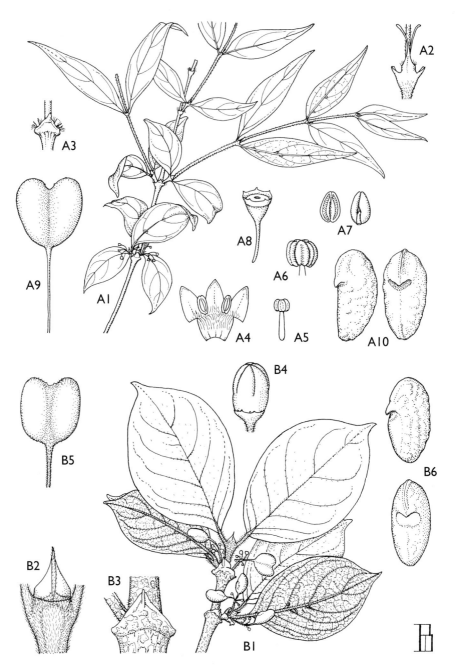

Tab. 63. A. —CANTHIUM NGONII. A1, flowering branch (× ²/₃); A2, juvenile stipule (× 3); A3, mature stipule (× 3), A1–A3 from *Goldsmith* (1/71); A4, portion of corolla opened out(× 6); A5, style and pollen presenter (× 6); A6, pollen presenter (× 10); A7, anther, 2 views (× 10); A8, calyx (× 6), A4–A8 from *Chase* 1498; A9, fruit (× 2); A10, pyrenes, 2 views (× 3), A9 & A10 from *Pope et al.* 1387. B. —CANTHIUM SALUBENII. B1, apical portion of fruiting shoot (× ²/₃); B2, juvenile stipule (× 3); B3, mature stipule (× 3), B1–B3 from *Salubeni* 5; B4, flower bud (× 8), from *Salubeni* 479; B5, fruit (× 2); B6, pyrene, 2 views (× 3), B5 & B6 from *Salubeni* 5. Drawn by Diane Bridson. From Kew Bull.

of Natal: 259 & 266 (1981). —von Breitenbach in J. Dendrol. **5**: 79, 82, figs. & map on p. 80 & 81 (1985). —Bridson in Kew Bull. **47**: 373 (1992). —Pooley, Trees of Natal, Zululand & Transkei: 476, figs. (1993). Type from South Africa.

Canthium mundianum var. *pubescens* Cham. & Schltdl. in Linnaea **4**: 132 (1829). —Sonder in F.C. **3**: 17 (1865). Type from South Africa.

Plectronia mundiana (Cham. & Schltdl.) Pappe, Silva Cap.: 19 (1854). —Sonder in F.C. **3**: 17 (1865).

Plectronia mundii Sim, For. Fl. Col. Cape Good Hope: 242, pl. 88 (1907).

Shrub or small tree 1–10(12) m tall; branches slender, often pubescent when young, covered with light grey bark. Leaves paired at apex of lateral and main branches, or with several pairs on current season's growth, not always fully mature at time of flowering; blades 2–7.5 × 1.2–4.5 cm, elliptic to round, obtuse to rounded or subacuminate at apex, acute to cuneate or occasionally obtuse at the base, rarely almost truncate, glabrous or less often sparsely pubescent on both faces; lateral nerves in 4–5 main pairs, tertiary nerves finely reticulate; domatia usually present as conspicuous tufts of hair; petioles 2–10 mm long, glabrous or sparsely pubescent; stipules 2–6 mm long, triangular at base, subulate above, gradually breaking down, lacking conspicuous hairs inside. Flowers 5-merous, in 3–20-flowered cymes; cymes shortly pedunculate, either borne along the length of mature stems or restricted to the apex; peduncles 2–8(12) mm long, sparsely pubescent to pubescent; pedicels 0.75–2(3) mm long; bracteoles inconspicuous. Calyx tube 1–2(2.75) mm long, glabrous or pubescent; limb reduced to a rim, with or without short filiform lobes (rarely with linear lobes up to 3 mm long). Corolla greenish-white or yellowish; tube 1–1.75(2) mm long, pubescent at throat; lobes 2–3 × 1.25–2(3) mm, triangular-ovate to ovate, acute and thickened at apex, often finely puberulous at the margin. Anthers set at throat, subsessile, sometimes apiculate. Style very slightly exceeding the calyx tube; pollen presenter 0.75–1 mm long and wide, ribbed. Fruit 7–10 × 9–11 mm, ± square in lateral view, slightly narrower or broader than long, often tapered to base, deeply indented at apex; pedicel lengthening. Pyrene 7–9 × 5 mm, oblong-ellipsoid with a shallow apical crest, slightly rugulose.

Zimbabwe. E: Nyanga Distr., Vukutu, 10 km west of Juliasdale, east of main road, fl. 3.i.1976, *Müller* 2430 (K; SRGH). S: Mberengwa Distr., Mt. Buhwa, fr. 10.xii.1953, *Wild* 4330 (K; PRE; SRGH). **Mozambique**. MS: Mt. Umtereni, fl. 1.x.1906, *Swynnerton* 644 (K; SRGH). M: Namaacha Distr., Namaacha, fr. 11.iii.1983, *Groenendijk & de Koning* 247 (LMU; SRGH).

Also in South Africa (North Prov., Gauteng, Mpumalanga, KwaZulu-Natal, Western Cape Prov. and Eastern Cape Prov.). On forest margins or in thickets, often near streams; 1060–1980 m.

The epithet for this species has often been spelt as *mundtianum*, but *mundianum* is accepted as the correct form (see Gunn and Codd, South African Collectors, Botanical Exploration of Southern Africa, 1981). See remarks after *C. gilfillanii*.

5. **Canthium gilfillanii** (N.E. Br.) O.B. Mill. in J. S. African Bot. **18**: 82 (1952). —Codd in Kirkia **1**: 109 (1961). —Jeppe, Trees & Shrubs Witwatersrand: 120 (1964). —Palmer & Pitman, Trees Southern Africa **3**: 2092, fig. (1973). —von Breitenbach in J. Dendrol. **5**: 79 & 82, fig. & map on p. 82 (1985). —K. Coates Palgrave, Trees Southern Africa, ed. 3, rev.: 16j of appendix (1988). —Bridson in Kew Bull. **47**: 374 (1992). Type from South Africa.

Plectronia gilfillanii N.E. Br. in Bull. Misc. Inform., Kew **1906**: 105 (1906).

Canthium mundianum sensu K. Coates Palgrave, Trees Southern Africa, ed. 3, rev.: 884 (1988), pro parte, non Cham. & Schltdl.

Shrub or small tree 2–4.5 m tall; young stems densely covered with whitish pubescence; bark grey or dark grey. Leaves restricted to apices of main and lateral branches, usually not fully mature at time of flowering; blades 2.5–5.5 × 1.5–4 cm, ovate or broadly elliptic to round, obtuse to rounded, or sometimes ± subacuminate at the apex, rounded to truncate or rounded then acute at the base, pubescent to densely pubescent above, densely pubescent to tomentose beneath; lateral nerves in 3–4 main pairs; tertiary nerves obscure; domatia not apparent; petioles 1–3 mm long, densely pubescent; stipules 3–7 mm long, triangular at base, subulate above, pubescent outside, lacking conspicuous hairs inside. Flowers 5-merous, in 3–16-flowered cymes; cymes shortly pedunculate, usually restricted to top node of previous season's growth; peduncles 1–5 mm long, sparsely pubescent to pubescent; pedicels 1–3 mm long, sparsely pubescent to pubescent; bracteoles inconspicuous. Calyx tube

c. 1 mm long, sparsely pubescent to pubescent; limb reduced to a shortly dentate rim.   Corolla tube 1–1.25 mm long, pubescent at throat; lobes 2 × 1.25 mm, triangular-ovate, acute, thickened and very shortly apiculate at the apex. Anthers set at throat, subsessile, often shortly apiculate.   Style slightly exceeding corolla tube; pollen presenter c. 0.75 × 0.75 mm, strongly ribbed. Fruit 7–11 × 10–11 mm, ± square in lateral view, slightly tapered towards base and deeply indented at apex; pedicel only slightly lengthening.   Pyrene 9 × 5 mm, oblong-ellipsoid with a distinct apical crest and point of attachment rather prominent, slightly rugulose.

Botswana. SE: Gaborone Distr., Muhoro R., Baratani Hill, fr. 5.xii.1947, *Miller* B560 (K; PRE). Also in South Africa (North-West Prov., Gauteng and eastern KwaZulu-Natal).   Riverine vegetation, in valley on stream bank.

This species is very close to *C. mundianum*; apart from differences given in the key, dissimilarities in habit and/or the times of leaf fall are indicated by Jeppe, Palmer & Pitman and von Breitenbach (all cited above).   *C. mundianum* is widely distributed while *C. gilfillanii* is relatively restricted, and although partly disjunct both species occur in Gauteng (Pretoria, Witwatersrand and Heidelberg), where they are sympatric (fide Tilney *pers. com.*).   Recent field studies in this region have demonstrated that under certain fire regimes plants form very extensive genets (colonies of interconnected individuals) which can give the false impression of numerous individuals and this should be borne in mind when interpreting local variation (K. Balkwill *pers. com.*).   Palmer and Pitman state that the distribution of *C. gilfillanii* extends eastwards to Weenan in Natal.   I have seen only one sheet from KwaZulu-Natal (*Codd* 2457 from Estcourt (near Weenen)) which appears indistinguishable from *C. gilfillanii*. It may well prove desirable either to sink *C. gilfillanii* in *C. mundianum* or to maintain it at infraspecific level. However, this decision is best deferred pending wider survey.   The two are not likely to be confused in the Flora Zambesiaca area.

Field observations have shown the plants of *C. mundianaum* (including *C. gilfillanii*) to be gynodioecious (hermaphrodite or female) but are functionally nearly unisexual (K. Balkwill, J.R. Sebola & E.R. Robinson in L.J.G. van der Masen et al. (eds.), The Biodiversity of African Plants, Proceedings XIVth AETFAT Congress: 650–655 (1996)).

6. **Canthium parasiebenlistii** Bridson in F.T.E.A., Rubiaceae: 870 (1991); in Kew Bull. **47**: 375, fig. 6 (1992).   Type: Zambia, Sunzu Hill, near Mbala, *Robson* 507 (BM; BR; K, holotype; LISC).

Canthium sp. 3 of F. White, F.F.N.R.: 404 (1962).

Shrub or small tree 1–7 m tall; lateral shoots well developed or short, with nodes close together; young stems glabrous, covered with purple-brown or red-brown lenticellate bark.   Leaves in 1(2) pairs at apex of main stem and lateral branches, usually immature and often sticky at time of flowering; blades 5–12(14.5) × 3–9 cm, broadly elliptic to round or sometimes oblong-elliptic, acuminate at the apex, acute to obtuse then somewhat attenuate or sometimes rounded or unequal at the base, glabrous or sometimes with a few hairs on the nerves above, glabrous or sparsely pubescent beneath, papery; lateral nerves in 5–7(8) main pairs; tertiary nerves finely reticulate; domatia present as conspicuous tufts of hair; petioles (3)5–10 mm long, glabrous or sparsely pubescent; stipules triangular, acuminate, the apical stipule with a filiform to subulate lobe, 3–12 mm long altogether, with a few white hairs inside, eventually caducous.   Flowers 4–5-merous, in (1)3–13-flowered cymes; cymes restricted to the uppermost node or sometimes the upper two nodes of mature stems; peduncles 2–15 mm long, glabrous or with lines of pubescence; pedicels 1–3 mm long, sparsely pubescent to pubescent; bracteoles inconspicuous. Calyx tube 1 mm long, glabrous or sparsely puberulous; limb not exceeding disk.   Corolla greenish-yellow; tube 2 mm long, pubescent at throat; lobes 1.75–2.5 mm long, triangular-ovate, acute and thickened at the apex. Style distinctly longer than corolla tube; pollen presenter c. 0.5 mm long and wide, ribbed. Fruit 6–9 × 7–9 mm, broadly oblong to obcordate in lateral view, indented at apex, tapering at base; pedicel lengthening to 5–7 mm. Pyrene 5–8(9) × 3.5–4(5) mm, ellipsoid or oblong-ellipsoid with a shallow crest around the apex, rugulose.

Zambia. N: Chinsali Distr., west of bridge over Mansha R. on Ishiba Ngandu (Shiwa Ngandu) to Kapisha School road, fr. 2.vi.1977, *Bingham* 2246 (K). W: Chingola, fl. 28.xi.1959, *Fanshawe* 5263 (K). **Malawi**. N: Mzimba Distr., Champhila (Champira) Forest, fr. 20.iv.1974, *Pawek* 8443 (K). C: Ntchisi Forest, fr. 2.ii.1983, *Dowsett-Lemaire* 619 (K). S: Kasupe Distr., Chikala Hills, fr. 17.ii.1975, *Brummitt et al.* 14363 (BR; K; WAG).

Also in Zaire (Dem. Rep. Congo) (Shaba Prov.) and Tanzania. In *Brachystegia* woodland, evergreen thickets or forest edges, very often on stony hillsides or rock outcrops; 1065–1620 m.
   A broad view of the taxon has been taken because of the lack of correlated collections at different stages from any one geographical area and habitat. Future revision may be necessary when better data are available, however, some variation noted from present observation is worth recording. The specimens from southern Malawi tend to have a longer stipule lobe, 5–12 mm long, and slightly bigger fruit than specimens from Zambia. The specimens seen from Zambia were all recorded as being collected in rocky places or near boulders; those from the north have reduced internodes while those from the south, and Shaba Prov. in Zaire (Dem. Rep. Congo), have longer internodes.

7. **Canthium salubenii** Bridson in Kew Bull. **47**: 377, fig. 7 (1992). TAB. **63**/B1–B6. Type: Malawi, Dedza Distr., Chongoni Mt., 4.i.1967, *Salubeni* 479 (K, holotype; LISC; MAL; SRGH).

Shrub or small tree 1.5–7 m tall; young stems densely pubescent; older stems often with numerous nodes close together, covered with flaking red-brown bark. Leaves restricted to apices of main and short lateral branches, perhaps not fully developed at time of flowering; blades 4–10.5 × 2.5–9.3 cm, broadly elliptic, subacuminate or sometimes acuminate at apex, acute or less often ± truncate at base, glabrescent above, sparsely pubescent to pubescent and more densely hairy on nerves beneath; lateral nerves in 6 main pairs; tertiary nerves finely to moderately finely reticulate, sometimes impressed above; domatia present; petioles 4–6 mm long, densely pubescent; stipules 4–5(10) mm long, triangular, finely acuminate to shortly aristate or with a filiform lobe, lacking conspicuous hairs inside. Flowers 5-merous, in up to 7-flowered cymes; cymes restricted to the top, or sometimes top three, nodes of mature stem; peduncles 1–12 mm long, pubescent; inflorescence branches reduced; pedicels 1–7 mm long. Calyx tube c. 0.75 mm long, pubescent; limb reduced to a rim. Only one detached corolla seen; tube 2 mm long, pubescent at throat; lobes 2 mm long, 1 mm wide at base, triangular, acute. Anthers apparently reflexed; filaments short. Style apparently exserted; pollen presenter c. 0.5 × 0.5 mm. Fruit 8–10 × 8–10 mm, ± square in lateral view, very slightly tapered towards base, indented at apex, yellowish-green; pedicel lengthening to 5–6 mm. Pyrene 8 × 4 mm, oblong-ellipsoid, with a shallow crest at apex, slightly rugulose; point of attachment relatively blunt.

Malawi. C: Dedza Distr., Lengwe Hill, on east slopes of Dedza Mt., T.A. Kasumbu, corolla fallen 13.i.1985, *Patel & Kaunda* 1957 (K; MAL; MO).
   Not known elsewhere. In forest patches and *Brachystegia* woodland below forest; c. 1525–1600 m.

8. **Canthium lactescens** Hiern, Cat. Afr. Pl. Welw. **1**: 511 (1898). —Bullock in Bull. Misc. Inform., Kew **1932**: 379 (1932). —Brenan, Check-list For. Trees Shrubs Tang. Terr.: 489 (1949). —F. White, F.F.N.R.: 404 (1962). —Drummond in Kirkia **10**: 276 (1975). —Bridson & Troupin in Fl. Pl. Lign. Rwanda: 548, fig. 184 (1982); in Fl. Rwanda **3**: 148, fig. 43.1 (1985). —K. Coates Palgrave, Trees Southern Africa, ed. 3, rev.: 882 (1988). —Bridson in F.T.E.A., Rubiaceae: 871, figs. 131/1, 131/13 & 155 (1991); in Kew Bull. **47**: 379 (1992). —Beentje, Kenya Trees, Shrubs Lianas: 505 (1994). TAB. **64**. Type from Angola.
   Canthium crassum sensu Hiern in F.T.A. **3**: 145 (1877), pro parte quoad *Schweinfurth* 1695, non Hiern sensu stricto.
   Canthium umbrosum Hiern, Cat. Afr. Pl. Welw. **1**: 479 (1898). Type from Angola.
   Plectronia lactescens (Hiern) K. Schum. in Just's Bot. Jahresber., 26th year, part 1: 393 (1900).
   Plectronia umbrosa (Hiern) K. Schum. in Just's Bot. Jahresber., 26th year, part 1: 393 (1900).
   Plectronia psychotrioides K. Schum. ex De Wild., Études Fl. Katanga [Ann. Mus. Congo, Sér. IV, Bot.] **1**: 228 (1903). Types from Zaire (Dem. Rep. Congo).
   Canthium lactescens var. *grandifolium* S. Moore in J. Linn. Soc., Bot. **37**: 161 (1905). Type from Uganda.
   Canthium randii S. Moore in J. Bot. **49**: 152 (1911). Type: Zimbabwe, near Harare, *Rand* 1393 (BM, holotype).
   Plectronia randii (S. Moore) Eyles in Trans. Roy. Soc. South Africa **5**: 493 (1916). Type as above.

Tree 2.5–9 m tall; stems thick, often with short internodes, covered with dark grey to reddish bark, glabrous or occasionally pubescent, lenticels usually apparent. Leaves restricted to new growth at apex of main shoots and short lateral branches,

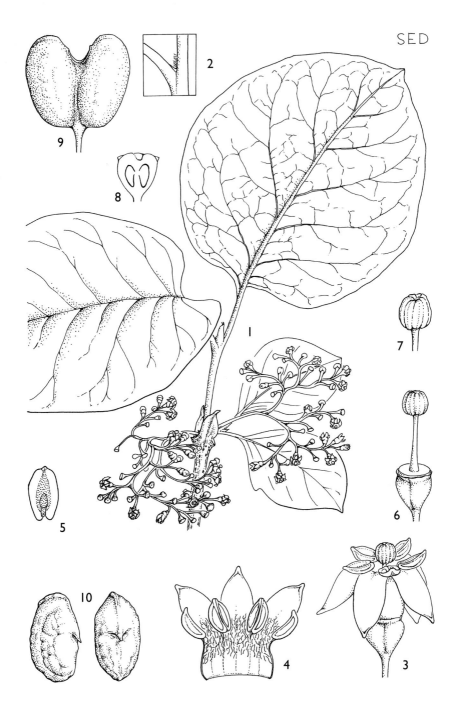

SED

Tab. 64. CANTHIUM LACTESCENS. 1, apical portion of flowering branch (× ²/₃); 2, portion of leaf showing domatium (× 3); 3, flower (× 6); 4, part of corolla opened out (× 6); 5, dorsal view of anther (× 8); 6, calyx with style and pollen presenter (× 6); 7, pollen presenter (× 8); 8, section through ovary (× 6), 1–8 from *Gereau & Lovett* 2784; 9, fruit (× 3); 10, pyrene, 2 views (× 3), 9 & 10 from *Tanner* 5900. Drawn by Sally Dawson. From F.T.E.A.

usually only one pair on each branch; blades 7–18 × 5–13 cm, broadly elliptic to round, acute to subacuminate at apex, acute to obtuse or truncate at base, glabrous or occasionally pubescent beneath and sparsely pubescent above; lateral nerves in 8–10 main pairs, acutely angled, often turning black when dry; tertiary nerves apparent but not conspicuous; domatia usually present, but not markedly conspicuous; petioles 5–20 mm long; stipules 6–12 mm long, broadly triangular to ovate, often long-acuminate when young, resembling bark when mature, not conspicuously hairy within. Flowers (4)5-merous, 20–50 in pedunculate cymes, restricted to leafless nodes of previous season's growth; cymes branched 3–5 times, but ultimate segments with flowers arranged on one side; peduncle 1–35 mm long, pubescent on either side; pedicels 0.5–3 mm long; bracteoles inconspicuous. Calyx tube 1.5 mm long, glabrous or sometimes sparsely pubescent; limb reduced to a rim. Corolla cream to yellowish; tube 2–2.5 mm long, finely pubescent at throat, rather wider than calyx tube; lobes 1.5–3 mm long, 1–2 mm wide at base, triangular-ovate, acute, thickened towards the apex and sometimes appendaged inside, spreading or reflexed. Style shortly exceeding corolla tube; pollen presenter hemispherical, strongly ribbed below, c. 1 mm wide. Fruit yellowish, edible, 7–12 × 8–12 mm, ± square in lateral view, straight-sided or tapering at base, strongly bilobed and indented at apex. Pyrene 8–10 mm long, ellipsoid to ovoid, slightly rugulose.

Zambia. N: Chambeshi R. terrace, corolla fallen 9.i.1962, *Astle* 1216 (K). W: Kitwe, fr. 3.v.1955, *Fanshawe* 2265 (BR; K). C: Lusanga River, sterile 8.ix.1930, *D. Stevenson* 87/30 (FHO; K). S: Bombwe For. Res., fr. iv.1933, *Martin* 649 (FHO; K). **Zimbabwe**. N: Makonde Distr., Mhangura (Mangula) area, Gudubu Farm, young fr. 31.i.1965, *W. Jacobsen* 2655 (K; PRE; SRGH). W: Matobo Distr., Hope Fountain Mission, on top of hill with church bell, fl. 3.xiii.1973, *Norrgrann* 435 (K; PRE; SRGH). C: Makoni Distr., near Maidstone Village, 30.xii.1930, *Fries, Norlindh & Weimarck* 4042 (K; PRE; SRGH). E: Mutare, east of Darlington, fl. 17.i.1950, *Chase* 1922 (K; LISC; SRGH). S: Masvingo, Mushandike National Park, fl. 1.ix.1983, *Mahlangu* 789 (SRGH).

Also in Ethiopia, Sudan, Uganda, Rwanda, Burundi, Kenya, Tanzania, Zaire (Dem. Rep. Congo) and Angola. On rocky hillsides and granite outcrops, and often on termite mounds in miombo woodlands, also in thickets on Kalahari Sand (mutemwa); 1000–1800 m.

9. **Canthium racemulosum** S. Moore in J. Linn. Soc., Bot. **40**: 87 (1911). —Bridson in F.T.E.A., Rubiaceae: 875 (1991); in Kew Bull. **47**: 383 (1992). Type: Mozambique, Madanda Forest, *Swynnerton* 541 (BM, holotype; K).

Shrub or small tree, 1–7 m tall; green twigs glabrous or densely pubescent; bark flaking in small scales, mid to dark grey. Leaves paired at apex of main stem and short lateral branches, immature at time of flowering; blades 3.5–10 × 1–11.5 cm, elliptic to broadly elliptic or less often round, obtuse to acute or sometimes subacuminate at apex, acute to cuneate or sometimes obtuse at base, chartaceous, entirely glabrous or pubescent above and tomentose beneath; lateral nerves in 6–7 main pairs; tertiary nerves moderately coarsely reticulate, often discolorous; domatia inconspicuous; petioles 2–5 mm long, glabrous to tomentose; stipules at apical nodes of green shoots 8–12 mm long, with well developed linear lobes, stipules at woody nodes 5–11 mm long, ovate-triangular, glabrous or pubescent when young becoming corky and resembling bark when mature, with tufts of white hair inside, eventually caducous. Flowers 5-merous, usually restricted to leafless terminal nodes of the previous season, borne in 4–30-flowered pedunculate 2-branched cymes, with flowers arranged on one side; peduncle 4–15 mm long, with lines of pubescence, or densely pubescent; pedicels 0.5–2 mm long, glabrous or pubescent; bracteoles not apparent. Calyx tube 1–2 mm long, glabrous or densely covered with spreading hairs; limb reduced to a repand rim, equalling or exceeding the disk. Corolla pale green or yellow, glabrous or sometimes pubescent outside; tube 2–2.5 mm long, densely pubescent at throat; lobes 2–2.75 × 1.5–1.75 mm wide at base, triangular-ovate, acute. Style distinctly longer than corolla tube; pollen presenter 0.75 × 0.5 mm, grooved. Fruit 12 × 11–13 mm, ± obcordate in lateral view; pedicel lengthening to 4–5 mm long. Pyrenes 11–12 × 5–6 mm, narrowly oblong-obovoid, rugulose.

Var. **racemulosum** —Bridson in F.T.E.A., Rubiaceae: 875 (1991); in Kew Bull. **47**: 383 (1992). *Canthium siebenlistii* sensu Vollesen in Opera Bot. **59**: 67 (1980), non (K. Krause) Bullock.

Leaf blades glabrous to glabrescent above, glabrous to sparsely pubescent beneath, midrib pubescent beneath; calyx tube glabrous or sometimes pubescent; corolla always glabrous outside.

Zimbabwe. E: Chipinge Distr., Nyamagamba (Nyangambe) River valley, fr. ii.1962, *Goldsmith* 112/62 (K; LISC; M; PRE; SRGH). S: Ndanga Distr., c. 3 km north of Chipinda Pools, fl. 17.xii.1959, *Goodier* 727 (K; LISC; SRGH). **Malawi**. S: Liwonde National Park, Chiloli Hill, fl. 17.xii.1983, *Dudley & Lingren* 872 (K). **Mozambique**. N: c. 40 km from Malema (Entre Rios), to Ribáuè, Serra Murripa, young fr. 16.xii.1967, *Torre & Correia* 16581 (LISC). MS: Chemba, andados 29 km de Nhacolo (Tambara), para Planalto de Serra Lupata (Montes Nhamalongo), fr. 14.v.1971, *Torre & Correia* 18399 (LMU; PRE; SRGH). GI: Inhambane, Govuro andados 27 km da Povoação Banamana, para Maxaila (Machaila), fr. 26.iii.1974, *Correia & Marques* 4203 (LMU). Also in southern Tanzania (Lindi District). Mixed woodland and thicket in hot low altitude river valley vegetation, often on termitaria; 137–1200 m.

Var. **nanguanum** (Tennant) Bridson in F.T.E.A., Rubiaceae: 875 (1991); in Kew Bull. **47**: 383 (1992). Type from Tanzania.
    Canthium burttii var. *nanguanum* Tennant in Kew Bull. **22**: 440 (1968).
    Canthium burttii sensu Vollesen in Opera Bot. **59**: 67 (1980), non Bullock.

Leaf blades pubescent to densely pubescent above, densely pubescent to tomentose beneath; calyx tube tomentose; corolla glabrous or hairy outside.

Mozambique. N: Cabo Delgado Prov., Mt. Ancuabe, fr. 7.ii.1984, *de Koning & Groenendijk* 9497 (LMU). Z: 32 km from Vila de Mocuba, fr. 5.iii.1966, *Torre & Correia* 15028 (LISC). Also in southern Tanzania (Lindi District). On basalt rock outcrops; 200–550 m.

10. **Canthium burttii** Bullock in Bull. Misc. Inform., Kew **1933**: 146 (1933). —Brenan, Checklist For. Trees Shrubs Tang. Terr.: 488 (1949). —Bridson in F.T.E.A., Rubiaceae: 876 (1991); in Kew Bull. **47**: 383 (1992). Type from Tanzania.

Small tree or shrub 1.5–6.5 m tall; green twigs glabrous or densely covered with spreading hairs; older branches covered with reddish to dark grey, smooth bark; lenticels apparent but not conspicuous. Leaves paired at apex of main stem and short lateral branches, not fully mature at time of flowering; blades 5–12 × 2.8–10.5 cm, broadly elliptic to round, acute to shortly acuminate at apex, cuneate to obtuse or occasionally truncate at base, papery, pubescent to velutinous on both faces or sometimes glabrous; lateral nerves in (4)5–6(7) main pairs, acutely angled; tertiary nerves finely reticulate; domatia obscure, or present as lines of pubescence in nerve angles; petioles short, 1–5 mm long; stipules 5–11 mm long, ovate, acuminate, central area pubescent or sometimes glabrous, membranous and shiny at margin, occasionally hyaline at very edge, readily caducous. Flowers 5-merous, borne in 3–11(15)-flowered pedunculate cymes; cymes restricted to leafless terminal nodes of the previous season, either 2-branched with flowers borne on one side or ± subumbellate; peduncle 4–20 mm long, pubescent or with 2 pubescent lines; pedicels 1–5 mm long; bracteoles not apparent. Calyx tube 1–1.25 mm long, glabrous or densely pubescent; limb reduced to a rim. Corolla greenish-yellow, glabrous or with very few hairs outside; tube 2–3 mm long, pubescent inside; lobes 1.75–2.25 × 1.25 mm, oblong-ovate, acute. Style equalling or shortly exceeding corolla tube in length; pollen presenter c. 1 mm long and wide, ± match-head-like, grooved. Fruit yellow or sometimes orange, 6–8 mm long and wide, broadly oblong-ovate in lateral view, scarcely bilobed, not indented at apex, glabrous to glabrescent. Pyrene 7 × 4 mm, obovate with ventral face flattened, rugulose.

Subsp. **burttii** —Bridson in F.T.E.A., Rubiaceae: 876 (1991); in Kew Bull. **47**: 384 (1992).

Plants pubescent.

Zambia. N: Mbala Distr., in gorge at Kalambo Falls, sterile 15.v.1936, *B.D. Burtt* 6406 (K). Also in Tanzania. Among *Burttia prunoides*, *Haplocoelum foliolosum* etc.; 1370 m. This sterile specimen is not quite so densely pubescent as the typical forms from Tanzania.

Subsp. **glabrum** Bridson in F.T.E.A., Rubiaceae: 876 (1991); in Kew Bull. **47**: 385 (1992). Type: Zambia, S: Pemba, *Trapnell* 1304 (K, holotype).
   Canthium burttii sensu F. White, F.F.N.R.: 404 (1962), pro parte, quoad *Fanshawe* 1815, non Bullock.

Plants glabrous, or rarely sparsely hairy.

Zambia. B: Mongu Distr., Kataba Mine, young fr. xi.1959, *Armitage* 203/59 (SRGH). N: Musesha (Museshia), fl. 8.x.1958, *Fanshawe* 4900 (BR; K). C: Kapiri Mposhi, fr. 22.i.1955, *Fanshawe* 1815 (K). S: Mazabuka Distr., Pemba to Choma, (lower road) mile 19.7, fr. 4.iii.1960, *White* 7615 (K).
   Also in Zaire (Dem. Rep. Congo) (Shaba Prov.) and Tanzania (Ufipa District). *Brachystegia* woodland and in thickets, often on granite kopjes, also in forest understorey on Kalahari Sand; 780–1000 m.

11. **Canthium pseudorandii** Bridson in Kew Bull. **47**: 385, fig. 12 (1992). Type: Zimbabwe, Hwange, near cricket ground, *Raymond* 229 (K, holotype; PRE; SRGH).
   Canthium randii sensu Miller in J. S. African Bot. **18**: 83 (1952), non S. Moore.
   Canthium burttii sensu F. White, F.F.N.R.: 404 (1962), pro parte, quoad *Fanshawe* 4115 & *Morze* 101 (not seen). —sensu Drummond in Kirkia **10**: 276 (1975), sensu. K. Coates Palgrave, Trees Southern Africa, ed. 3, rev.: 879 (1988), non Bullock.

Small tree 3–8.5 m tall; green stems pubescent; older stems with dark red-brown or purplish bark, sometimes lenticellate. Leaves paired at apex of main and lateral branches, sometimes immature at time of flowering; blades 4.5–13 × 2.5–12.5 cm, elliptic to broadly elliptic when young, broadly to very broadly ovate when mature, distinctly acuminate at apex, acute (when young) or rounded to subcordate at base; papery, sparsely pubescent to pubescent above, pubescent beneath; lateral nerves in 6–8 main pairs, acutely angled; tertiary nerves very finely reticulate; domatia inconspicuous; petioles 5–13 mm long, pubescent; stipules 8–11 mm long, triangular-ovate, acuminate with marginal area rather membranous and shiny, scarcely hairy within, soon caducous, seldom persisting beyond second node. Flowers 5-merous, in 10–30-flowered cymes; cymes restricted to apical node of mature stem, pedunculate; peduncles 5–25 mm long, pubescent; pedicels 0–2 mm long, pubescent; bracteoles small. Calyx tube c. 1 mm long, densely pubescent; calyx limb not exceeding disk. Corolla bright yellow or greenish; tube 2–3 mm long, pubescent at throat; lobes 2 × 1.25 mm, triangular or oblong-lanceolate, acute and thickened at apex. Style slightly exserted; pollen presenter 1 × 0.75 mm, ribbed. Fruit 7–9 × 8–9 mm, ± square in lateral view, slightly tapered towards the base and slightly indented at apex, sparsely pubescent; pedicel lengthening to 2–4 mm long. Pyrene 7–8 × 4–5 mm, oblong-ovoid, with a well or poorly defined apical crest, rugulose.

Botswana. N: Chobe, 13 km west of Serondela, fl. xi.1950, *Miller* 1122 (K). **Zambia**. N: North Luangwa National Park, 11°47', 32°21', fl. 19.xi.1993, *P.P. Smith* 1440 (K). C: Kapiri Mposhi, corolla fallen 5.xii.1957, *Fanshawe* 4115 (K; LISC). E: Chipata Distr., Machinje Hills, corolla fallen 15.v.1965, *B.L. Mitchell* 2964 (K; SRGH). S: Livingstone Distr., Katambora, fl. 2.xii.1955, *Gilges* 494 (K; PRE; SRGH). **Zimbabwe**. N: Murewa Distr., Pfungwe Tribal Trust Land, 5 km north of Nyadire Bridge, fl. 5.xii.1968, *Müller & Burrows* 961 (K; SRGH). W: Victoria Falls Village, fr. 6.iii.1974, *Gonde* 48/74 (K; SRGH). S: Bikita Distr., Save (Sabi) Valley, Umkondo Hills, fr. 9.ii.1966, *Wild* 7526 (K; LISC; PRE; SRGH). **Malawi**. S: Mangoche Distr., Namaso Bay, 12 km SE of Monkey Bay, corolla fallen 11.xii.1986, *Brummitt* 18301 (BR; K; MAL; SRGH).
   Not known outside the Flora Zambesiaca area. Mixed deciduous woodland on rocky hills and outcrops, and on Kalahari Sands and sandy soils, also in thickets (jesse bush and pemba thickets, floodplain thickets and Kalahari Sand mutemwa); 450–1000 m.
   This species has often been confused with pubescent forms of *C. lactescens* (as represented by *C. randii*) and with *C. burttii*. It may be distinguished from *C. lactescens* by the more slender branches, the readily caducous stipules with a membranous margin, the finer reticulum of tertiary nerves and the fruit which is only slightly though discernibly indented at the apex. From *C. burttii* it may be distinguished by its longer petioles and fruit which is discernibly bilobed and indented at the apex; furthermore the distribution of the pubescent subspecies, *C. burttii* subsp. *burttii*, is quite distinct from that of *C. pseudorandii*.

Subgen. 3. LYCIOSERISSA

Canthium subgen. **Lycioserissa** (Roem. & Schult.) Bridson in F.T.E.A., Rubiaceae:
876 (1991); in Kew Bull. **47**: 387 (1992).
Lycioserissa Roem. & Schult., Syst. Veg. **4**: 353 (1819).
Plectronia sensu Sond. in F.C. **3**: 17 (1865) et auctt. div., non L.

Shrubs or trees, sometimes with spines present on the trunks, young coppice
shoots and saplings; spines paired or often ternate, supra-axillary but not associated
with brachyblasts (contracted lateral spurs); lenticels not markedly apparent on
young stems. Leaves tending to be confined towards the apex of branches but not
strictly so, occasionally subtending inflorescences, mature or immature at time of
flowering, paired but often ternate on young shoots, petiolate, papery to
subcoriaceous, glabrous, somewhat discolorous; heterophylly sometimes present, the
leaves on young and coppice shoots smaller and broader than normal leaves, those
on saplings very much narrower than leaves on mature plants. Tertiary nerves
obscure, or sometimes apparent near the margin; domatia pit-like, ciliate or densely
pubescent; stipules sheathing at the base, bearing a linear lobe, occasionally
becoming corky with age, pubescent within. Flowers hermaphrodite, 5-merous,
pedicellate, borne in pedunculate or sessile (1)3–40-flowered cymes; inflorescence
branches present or reduced; bracteoles inconspicuous. Calyx glabrous; tube
broadly ellipsoid to ovoid; limb with tube ± obsolete, lobed almost to the base; lobes
small, triangular to linear, often rather unequal. Corolla white or yellow-green,
glabrous outside; tube broadly cylindrical, subequal to lobes or a little longer,
glabrous to sparsely hairy inside; lobes reflexed, often shortly apiculate, often
puberulous at the margins. Stamens set at corolla throat, erect; filaments short;
anthers oblong-ovate, attached shortly above the base, the dorsal face covered with a
darkened connective, except for marginal band. Style slender, shortly exceeding the
corolla tube, glabrous; pollen presenter ± spherical, 2–3 lobed, point of attachment
within a basal recess about one third of the receptacle in depth; disk glabrous; ovary
2–3-locular. Fruit a 2–3-seeded drupe, of moderate size, bilobed in 2-seeded species
often dorsiventrally flattened and indented at apex, 3-lobed in 3-seeded species.
Pyrenes ellipsoid with ventral face flattened, thinly woody, truncate at point of
attachment, with a slight or well developed crest extending from the point of
attachment to the apex, usually rugulose. Seeds narrowly obovoid with the ventral
area flattened and very shortly crested at the apex; endosperm entire; testa finely
reticulate; embryo curved with radicle erect; cotyledons small, perpendicular to
ventral face of seed.

A subgenus of three species restricted to eastern and tropical Africa and South Africa,
extending as far west as eastern Zaire (Dem. Rep. Congo). *Species* 12 and 13.

Fruit 13–24 × 14–20 mm, distinctly bilobed and indented at the apex. Pedicels 0.5–4(7) mm
long. Leaves ± uniformly spaced, not restricted to apices of stems; tertiary nerves obscure
· 12. *oligocarpum*
Fruit 12–15 × 9–10 mm, narrowly oblong-ovoid, scarcely bilobed or indented at the apex.
Pedicels 3–7 mm long. Leaves restricted to new growth at apex of stems; tertiary nerves
often apparent, at least near the margins · 13. *inerme*

12. **Canthium oligocarpum** Hiern in F.T.A. **3**: 138 (1877). —Bridson & Troupin in Fl. Pl. Lign.
Rwanda: 550, fig. 183.2 (1982); in Fl. Rwanda **3**: 148, fig. 43.2 (1985). —Bridson in F.T.E.A.,
Rubiaceae: 877, figs. 131/8, 131/14, 132/20 & 156 (1991); in Kew Bull. **47**: 388 (1992).
TAB. **47**/B2. Type from Ethiopia.
Canthium ruwenzoriense Bullock in Bull. Misc. Inform., Kew **1932**: 376 (1932). Type from
Uganda/Zaire (Dem. Rep. Congo).
Canthium captum sensu Robyns, Fl. Sperm. Parc Nat. Alb. **2**: 352 (1947), non Bullock.
Canthium sidamense Cufod. in Senckenberg. Biol. **46**: 100, taf. 10, fig. 6 (1965); in Bull.
Jard. Bot. État **35**, Suppl. [Enum. Pl. Aethiop. Sperm.]: 1011 (1965). Type from Ethiopia.

Shrubs or trees 1.5–20 m tall, glabrous; trunks, coppice shoots and young branches
usually armed with ternate or occasionally paired spines; bark greyish. Leaves not
restricted to stem apex; blades 3–14.5 × 1.5–6 cm, elliptic, oblong-elliptic,

occasionally broadly elliptic or narrowly elliptic, acuminate or less often obtuse to acute at apex, rounded or obtuse to cuneate at the base, leaves on coppice shoots or young shoots smaller and rounded, leaves on saplings very much narrower than mature foliage, papery to subcoriaceous, strongly discolorous, tertiary nerves obscure in dry specimens; domatia pit-like, ciliate to densely hairy; petiole 2–15 mm long; stipules sheathing at the base, 2–3 mm long, bearing a linear lobe 1–3 mm long, sometimes rather corky when old, pubescent inside. Flowers 5-merous, borne in pedunculate 3–25(40)-flowered cymes; peduncles 4–15 mm long, glabrous or sparsely and unequally pubescent, inflorescence branches sometimes reduced; pedicels 0.5–4(7) mm long, glabrous or sparsely pubescent; limb-tube ± obsolete; lobes triangular or linear, unequal, up to 1.5 mm long. Corolla white or yellow-green, acute in bud; tube 2–4.25 mm long, 1–2 mm wide, glabrous or sparsely hairy inside; lobes 1.5–4 × 1.5–2 mm wide at base, narrowly triangular to triangular or oblong-ovate, shortly apiculate, puberulous at margins outside. Style shortly exceeding the tube; pollen presenter as broad as long, c. 1 mm across. Fruit slightly longer than wide, subequal or wider than long, 13–25 × 14–20 mm, bilobed and indented at the apex. Pyrenes 12–14 × 7–8 mm, ellipsoid with ventral face flattened, scarcely or distinctly curved, truncate at point of attachment with crest running from point of attachment to apex, sometimes not well developed, somewhat rugulose.

Subsp. **captum** (Bullock) Bridson in F.T.E.A., Rubiaceae: 878 (1991); in Kew Bull. **47**: 389 (1992). Type from Tanzania.
    Canthium captum Bullock in Bull. Misc. Inform., Kew **1932**: 376 (1932). —Brenan, Check-list For. Trees Shrubs Tang. Terr.: 488 (1949). —Brenan in Mem. N.Y. Bot. Gard. **8**: 452 (1954).

Leaf blades 4–11 × 2–4.5 cm, mostly elliptic or sometimes broadly elliptic, thinly chartaceous to subcoriaceous and moderately shiny above; lateral nerves in 6–7 main pairs, often impressed above.

Malawi. N: Chitipa Distr., Misuku Hills, Wilindi Forest Reserve, fl. 16.ix.1970, *Müller* 1676 (K; SRGH). S: Mulanje Mt., Likhubula-Thuchila Divide (Likubula-Tuchila), fr. 9.vii.1946, *Brass* 16771 (K; PRE). **Mozambique.** Z: Gurué, fr. 18.iii.1966, *Torre & Correia* 14942 (LISC).
Also in Tanzania. In mixed evergreen rainforest; 1300–2100 m.

Subsp. **angustifolium** Bridson in Kew Bull. **47**: 389 (1992). Type: Zimbabwe, Chimanimani Distr., Muchira River, 1.5 km west of Glencoe Forest House, *Goldsmith* 76/69 (K, holotype; PRE; SRGH).
    Canthium captum sensu Drummond in Kirkia **10**: 276 (1975). —sensu K. Coates Palgrave, Trees Southern Africa, ed. 3, rev.: 879 (1988), non Bullock.

Leaf blades 3.5–7.5 × 1.3–2.5 cm, narrowly elliptic or narrowly oblong-elliptic, papery to chartaceous, usually dull above; lateral nerves in (4)5(6) main pairs, not impressed above.

Zimbabwe. E: Nyanga Distr., from Henkel's Nek across the Nyumkombe to iNyamariro Forest Patch, fr. 1934, *Gilliland* Q763 (K). **Mozambique.** MS: Chimanimani Mts., along path between Skeleton Pass and Namadima, young fr. 30.ii.1959, *Goodier & Phipps* 359 (K; PRE; SRGH).
Not known outside the Flora Zambesiaca area. In mixed evergreen rainforest, often in ravines; 1200–2000 m.
Three additional subspecies have been described: subsp. *friesiorum* Bridson, restricted to central Kenya, characterised by acute to obtuse leaf apicies; subsp. *intermedium* Bridson restricted to the Kenya/Tanzania border, characterised by pale greyish bark; and subsp. *oligocarpum* extending from Ethiopia, Sudan, Uganda, Kenya and eastern Zaire (Dem. Rep. Congo) south to Tanzania, resembling subsp. *captum* but differing in having thinner leaves with fewer lateral nerves, and inflorescences with more flowers.

13. **Canthium inerme** (L.f.) Kuntze, Revis. Gen. Pl. **3**: 545 (1898). —Drummond in Kirkia **10**: 276 (1975). —Gibson, Wild Fls. Natal: 8, pl. 102 (1975). —Moll, Trees of Natal: 262, 263, 266, 267 & 279 (1981). —von Breitenbach in J. Dendrol. **5**: figs. & map on p. 84 & 85 (1985). —K. Coates Palgrave, Trees Southern Africa, ed. 3, rev.: 882, pl. 304 (1988). —Bridson in Kew Bull. **47**: 392 (1992). —Pooley, Trees of Natal, Zululand & Transkei: 474, figs. (1993). TAB. **65**. Type from South Africa.
    Lycium inerme L.f., Suppl. Pl.: 150 (1782). Type as above.

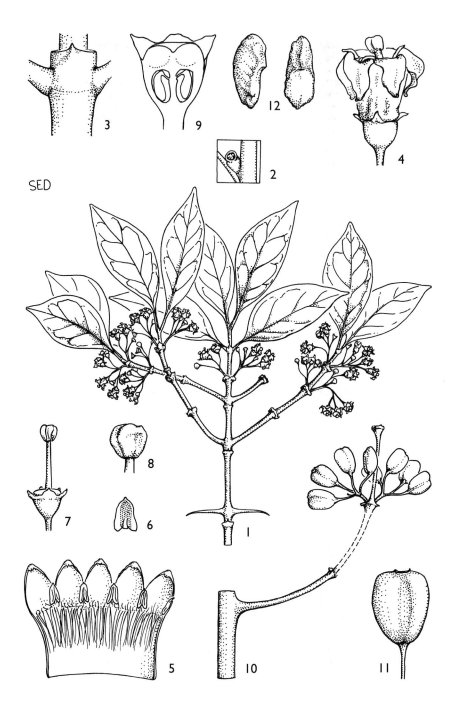

SED

Tab. 65. CANTHIUM INERME. 1, apical portion of flowering branch (× ²/₃), from *Müller* 477; 2, portion of leaf showing domatium (× 6), from *Mogg* 28368; 3, stipule (× 6), from *Lemos et al.* 274; 4, flower (× 6); 5, corolla opened out (× 6); 6, dorsal view of anther (× 8); 7, calyx with style and pollen presenter (× 4); 8, pollen presenter (× 8); 9, longitudinal section through ovary (× 8), 4–9 from *Maera* 40; 10, part of fruiting branch (× ²/₃); 11, fruit (× 2); 12, pyrenes, 2 views (× 2), 10–12 from *de Koning & Zunguze* 8065. Drawn by Sally Dawson.

Lycium barbatum Murray, Syst. Veg., ed. 14: 228 (1784). —Thunberg, Prodr. Pl. Cap., part 1: 37 (1794). —Willdenow, Sp. Pl. 1, part 2: 1060 (1798), nom. superfl. for *L. inerme* L.
Serissa capensis Thunb., Fl. Cap. 2: 66 (1818); Fl. Cap., ed. 2: 193 (1823), pro parte, excl. *Serissa foetida* Willd. in syn., nom. superfl. for *Lycium inerme*.
Lycioserissa capensis (Thunb.) Roem. & Schult., Syst. Veg. 4: xxiv & 353 (1819) based on *Lycium barbatum*, *L. inerme* & *Serissa capensis*, comb. invalid.
Canthium thunbergianum Cham. & Schltdl. in Linnaea 4: 130 (1829), nom. superfl. for *Serissa capensis* Thunb. and *Lycium barbatum* Thunb., excluding *Serissa foetida* Willd. in syn., based on *Bergius, Mund & Maire* s.n., not seen.
Plectronia ventosa sensu Sond. in F.C. 3: 17 (1865). —sensu Sim, For. Fl. Col. Cape Good Hope: 240, pl. 89 (1907). —sensu Eyles in Trans. Roy. Soc. South Africa 5: 494 (1916), non *Plectronia ventosa* L., the lectotype of *Plectronia* L. is *Olinia ventosa* (L.) Cufod.
Canthium ventosum sensu S. Moore in J. Linn. Soc., Bot. 40: 91 (1911), pro parte, excl. type. —sensu Palmer & Pitman, Trees Southern Africa 3: 2099, figs. on 2098–9 (1973). —sensu Ross, Fl. Natal: 335 (1973). —sensu van Wyk, Trees Kruger Nat. Park 2: 557 (1974), non *Plectronia ventosa* L.
Canthium swynnertonii S. Moore in J. Linn. Soc., Bot. 40: 88 (1911). Type: Zimbabwe, Chirinda Forest, *Swynnerton* 546 (BM, holotype; K).
Plectronia swynnertonii (S. Moore) Eyles in Trans. Roy. Soc. South Africa 5: 493 (1916). Type as above.

Shrub or tree 1–10(14) m tall, occasionally said to be scandent, often armed with paired supra-axillary spines, glabrous; bark light grey to fawn-coloured. Leaves restricted to new growth at apex of stems, not usually fully mature at time of flowering; blades 3–10 × 1.3–4.5 cm, narrowly to broadly elliptic, obtuse to acute or subacuminate at apex, usually cuneate at base, papery, often discolorous, drying brownish above and glaucous green beneath; lateral nerves in 4 main pairs; tertiary nerves apparent, at least near the margin; domatia glabrous to pubescent; petioles 5–13 mm long; stipules 2–4 mm long, shortly sheathing then broadly triangular, usually apiculate, with dense white hairs inside. Flowers 5-merous, borne in shortly pedunculate 4–40-flowered cymes with rather reduced branches; peduncles 0.5–5(7) mm long, glabrous; pedicels 3–7 mm long, glabrous. Calyx tube 1–1.25 mm long, glabrous; limb a dentate rim scarcely equalling the disk. Corolla cream to yellowish; tube 2–3 mm long, pubescent at throat; lobes 2–3 × 1–2 mm, triangular to ovate, acute. Style shortly but distinctly exserted; pollen presenter 0.75–1 × 0.75–1 mm. Fruit 12–15 × 9–10 mm, narrowly oblong-ovoid, scarcely bilobed or indented. Pyrene 12–13 × 4–5 mm, ellipsoid with ventral face flattened or oblong-ellipsoid, with a crest along the ventral face above the truncate point of attachment, rugulose.

Zimbabwe. E: Chimanimani Distr., Mutzarara Farm, fl. & fr. 19.x.1950, *A.O. Crook* M202 (K; SRGH). **Mozambique**. MS: Chimanimani Mts., along path between Skeleton Pass and Nhamadima, corolla fallen 30.xii.1959, *Goodier & Phipps* 354 (K; SRGH). GI: Xai Xai (Vila de João Belo), Chipenhe, Régulo Chiconela, floresta de Chirindzeni, fr. 1.iv.1959, *Barbosa & Lemos* 8433 (K; LISC; PRE; SRGH). M: Marracuene, Bobole, Reserva Botânica, fl. 3.x.1957, *Barbosa & Lemos* 7948 (COI; K; LISC).
Also in South Africa (North Prov., Mpumalanga, KwaZulu-Natal, Eastern Cape Prov. and Western Cape Prov.) and Swaziland. Evergreen rain forest patches and forest margins, also in woodland, often near streams, and on rocky outcrops; 0–2130 m.

Subgen. 4. BULLOCKIA

Canthium subgen. **Bullockia** Bridson in Kew Bull. **42**: 630–637 (1987); in F.T.E.A., Rubiaceae: 880 (1991).

Shrubs, sometimes scandent, or small trees. Leaves not deciduous, petiolate or subsessile; blades drying brown, bronze, greyish or green, chartaceous to stiffly chartaceous, usually dull above, glabrous or pubescent; tertiary nerves obscure; domatia absent, or present but rather small; stipules narrowly ovate to ovate, or triangular at the base, sometimes produced into a lobe, caducous or somewhat persistent, with a few silky hairs and colleters inside. Flowers unisexual (4)5–6-merous; male inflorescence 1–20-flowered, fasciculate, or occasionally umbellate with short peduncles; female flowers 1(2); small free bracteoles present on pedunculate inflorescences. Calyx glabrous or pubescent; tube ± ovoid in female flowers and very reduced in male flowers; limb lobed almost to the base; lobes usually irregular. Corolla small, rapidly caducous; tube with a ring of deflexed hairs inside; lobes subequal to tube, erect, thickened towards the apex, not apiculate. Stamens set at throat, subsessile; anthers erect, ovate, the dorsal face with a central band of darkened connective tissue, not apiculate, usually smaller in female flowers and producing little or no pollen. Style moderately slender, slightly longer than the corolla tube, slightly tapered at the apex; pollen presenter ± as wide as long, 2-lobed, hollow with style attached well inside; disk annular, glabrous; ovary 2-locular, each locule containing one ± pendulous ovule. Fruit yellow to dark red, fleshy, obovoid to broadly obovoid, laterally flattened, somewhat bilobed, not or scarcely indented at the apex; pyrenes thinly woody, narrowly obovoid, flat on ventral face, crested around the apex, eventually dehiscing from point of attachment back along the crest, slightly rugulose. Seeds narrowly obovoid with the ventral area flattened and the apex shortly crested; endosperm entire; testa finely reticulate; embryo slightly curved with radicle erect; cotyledons small, perpendicular to ventral face of seed.

A subgenus of five species occurring in eastern tropical Africa and South Africa, possibly with additional species in the Seychelles and in Madagascar. *Species* 14–15.

Young stems, pedicels and calyx tube glabrous, or occasionally sparsely covered with fine hairs; leaves glabrous above, drying brown, bronze or greyish; stipules ovate to narrowly ovate or sometimes lanceolate · 14. *mombazense*
Young stems, pedicels and calyx tube covered with straw- to rust-coloured rather stiff hairs; leaves glabrescent to pubescent or rarely glabrous above, usually drying greenish; stipules triangular at the base with a somewhat decurrent lobe · · · · · · · · · · · · · · · · 15. *setiflorum*

14. **Canthium mombazense** Baill. in Adansonia **12**: 188 (1878). —Dale & Greenway, Kenya Trees & Shrubs: 430 (1961). —Bridson in Kew Bull. **42**: 631, fig. 1B (1987); in F.T.E.A., Rubiaceae: 881 (1991). —Beentje, Kenya Trees, Shrubs Lianas: 505 (1994). Type from Kenya.
  Plectronia diplodiscus K. Schum. in Engler, Pflanzenw. Ost-Afrikas **C**: 385 (1895). Type from Tanzania.
  Plectronia pallida K. Schum. in Bot. Jahrb. Syst. **28**: 77 (1899). Type from Zanzibar Island.
  Canthium pallidum (K. Schum.) Bullock in Bull. Misc. Inform., Kew **1932**: 387, fig. 4 (1932). —Brenan, Check-list For. Trees Shrubs Tang. Terr.: 484 (1949). —Dale & Greenway, Kenya Trees & Shrubs: 430 (1961).
  Canthium diplodiscus (K. Schum.) Bullock in Bull. Misc. Inform., Kew **1932**: 387 (1932). —Brenan, Check-list For. Trees Shrubs Tang. Terr.: 484 (1949).
  Canthium greenwayi Bullock in Bull. Misc. Inform., Kew **1932**: 387 (1932). —Brenan, Check-list For. Trees Shrubs Tang. Terr.: 484 (1949). Type from Tanzania.
  Canthium inopinatum Bullock in Bull. Misc. Inform., Kew **1932**: 389 (1932). —Dale & Greenway, Kenya Trees & Shrubs: 429 (1961). Type from Kenya.

Compact, occasionally straggling, shrub or small tree 2–7 m tall (rarely a liana); bark pale grey, glabrous, or occasionally pubescent. Leaf blades pale greyish to bronze above when dry, 2.2–13 × 1–7.5 cm, broadly elliptic to round or broadly oblong-elliptic, obtuse to rounded at apex, obtuse to rounded or sometimes truncate at base, often unequal, glabrous or sometimes pubescent beneath; lateral nerves in 4 main pairs; tertiary nerves always obscure; domatia absent or inconspicuous; petiole 2–10 mm long; stipules light brown when dry, 4–16 × 2–10 mm, ovate or sometimes narrowly ovate, acuminate. Flowers 5–6-merous: functionally male flowers borne in 3–20-flowered, sessile or rarely shortly pedunculate (peduncle up to 2 mm long) umbels; functionally female flowers solitary; pedicels glabrous or sometimes pubescent, 3–7 mm long in functionally male flowers and 4–10 mm long in functionally female flowers. Calyx tube reduced in functionally male flowers, 1.75–2.5 mm long in functionally female flowers; calyx limb lobed almost to the base; lobes in functionally male flowers somewhat unequal, 0.75–2 mm long, linear-triangular, occasionally slightly spathulate; lobes in functionally female flowers usually broader, 2–3 mm long, more strongly spathulate and often with a few hairs at the apex. Corolla whitish or yellowish-green, readily caducous; tube c. 1.5 mm long with a ring of deflexed hairs at the throat inside; lobes 1.25–1.5 × 1–1.5 mm wide at base, triangular-ovate, acute, often with a few hairs at the apex outside, erect, apex generally more thickened in functionally female flowers than in functionally male flowers. Anthers present in both flower forms, exserted but almost sessile, c. 0.75 mm long in functionally male flowers, c. 0.5 mm long in functionally female flowers. Style shortly exceeding corolla tube; pollen presenter 0.25 mm long and wide, persistent in functionally male flowers and caducous in functionally female flowers. Fruit dark maroon when mature, 7–12 × 7–9 mm, flattened obovoid, not indented at the apex; calyx lobes persistent; pedicel lengthening to 20 mm. Pyrene 7 × 3.5 mm, obovoid, with ventral face flattened, crustaceous, surface ± smooth except for a shallow crest around the apex.

Mozambique. N: Nampula, Mossuril, Serra de Mesa, young fr. 19.ii.1984, *de Koning et al.* 9760 (K; LMU).
Also in Tanzania, Kenya and Somalia. On calcareous soil; 250–300 m.

15. **Canthium setiflorum** Hiern in F.T.A. **3**: 134 (1877). —J.G. Garcia in Mem. Junta Invest. Ultramar **6** (sér. 2): 29 (1959) [Contrib. Conhec. Fl. Moçamb. IV (1959)]. —Palmer & Pitman, Trees Southern Africa **3**: 2105, fig. on 2106 (1973). —Ross, Fl. Natal: 335 (1973). —Moll. Trees of Natal: 266, 268 & 280 (1981). —A.E. Gonçalves in Garcia de Orta, Sér. Bot. **5**: 188 (1982). —Bridson in Kew Bull. **42**: 634 (1987). —K. Coates Palgrave, Trees Southern Africa, ed. 3, rev.: 886 (1988). —Bridson in F.T.E.A., Rubiaceae: 883 (1991). —Pooley, Trees of Natal, Zululand & Transkei: 472, figs. (1993). Type: Mozambique near Tete, *Kirk* s.n. (K, holotype).

Canthium microdon S. Moore in J. Linn. Soc., Bot. **40**: 87 (1911). —Bullock in Bull. Misc. Inform., Kew **1932**: 366 (1932), pro parte. —J.G. Garcia in Mem. Junta Invest. Ultramar **6** (sér. 2): 29 (1959) [Contrib. Conhec. Fl. Moçamb. IV (1959)]. Type: Mozambique, Manica e Sofala Distr., Madanda Forest, *Swynnerton* 552 (BM, holotype; K).

Scandent shrub 1–4 m tall; young branches covered with dark greyish bark, not conspicuously marked with lenticels, densely covered with upwardly directed straw-coloured, golden or rust-coloured hairs. Leaf blades greenish when dry, 1.2–6.5 × 0.6–3.4 cm, elliptic, oblong-elliptic or sometimes broadly elliptic, acute to obtuse and apiculate at apex, obtuse to rounded at base, glabrescent to pubescent beneath, with hairs more dense on nerves beneath, upper surface sparsely pubescent to glabrescent or rarely glabrous save for midrib; lateral nerves in (3)4(5) main pairs; tertiary nerves obscure; domatia absent or inconspicuous; petiole 1–2 mm long, pubescent; stipules 2–5 mm long, triangular at the base with a somewhat decurrent lobe, pubescent outside. Flowers (4)5(6)-merous: functionally male flowers borne in subsessile to pedunculate 2–12-flowered subumbellate cymes, functionally female flowers solitary or occasionally paired; peduncles 1–5 mm long, pubescent; pedicels 1–4 mm long in functionally male flowers and 1.5–8 mm long in functionally female flowers, pubescent. Calyx tube 1.5–2 mm long and densely pubescent in functionally female flowers, reduced and rather sparsely pubescent in functionally male flowers; limb lobed almost to the base; lobes linear to spathulate, 0.5–3 mm long in functionally male flowers and 1.5–3 mm long in functionally female flowers. Corolla yellowish-green to cream; tube 1.5–2 mm long in functionally male flowers and 1–1.25 mm long in functionally female flowers, with a ring of deflexed hairs at the throat; lobes 1 mm long, 0.75–1 mm wide at base, triangular-ovate, sparsely pubescent toward the apex outside, acute and thickened at the apex, erect or spreading. Anthers exserted, but almost sessile, somewhat larger in functionally male flowers. Style ± equalling corolla tube; pollen presenter larger in functionally female flowers, up to 0.75 mm long and wide. Fruit yellow, or turning black when mature, 7–12 × 7–12 mm, broadly obovoid, flattened, scarcely indented at the apex, pubescent; calyx lobes persistent; pedicel lengthening to 6–12 mm. Pyrene 8 × 4 mm, obovoid with ventral face flattened, slightly rugulose; apical crest slightly defined.

Subsp. **setiflorum** —Bridson in Kew Bull. **42**: 634 (1987). TAB. **66**.

Calyx lobes 0.5–3 mm long in functionally male flowers and 1.5–2 mm long in functionally female flowers. Male inflorescences often pedunculate; peduncle up to 5 mm long. Fruit 7.5–12 × 7.5–12 mm.

Zimbabwe. N: Gokwe Distr., near Tare River, male corolla fallen 19.iv.1963, *Bingham* 632 (K; SRGH). E: Chimanimani Distr., Umvumvumvu R. Gorge, near old drift at East Bridge, male fl.

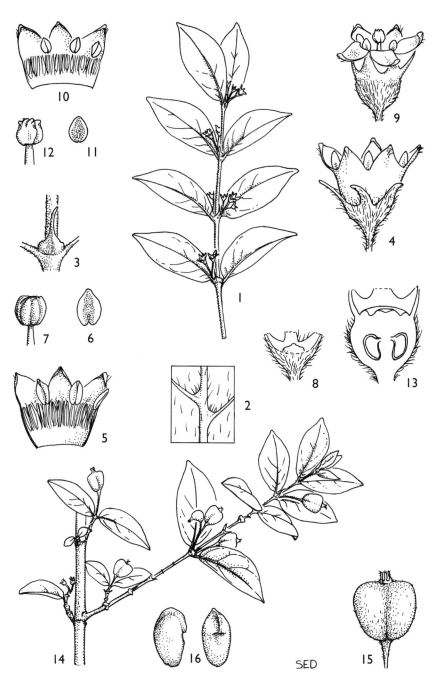

Tab. 66. CANTHIUM SETIFLORUM subsp. SETIFLORUM. 1, twig with male flowers (× ²/₃), from *Groenendijk & Dungo* 1336; 2, domatia (× 6), from *Barbosa* 8342; 3, stipule (× 4); 4, male flower (× 6); 5, part of male flower corolla (× 6); 6, dorsal view of anther from male flower (× 10); 7, pollen presenter from male flower (× 14); 8, calyx of male flower with lobes cut to show disk (× 6), 3–8 from *Groenendijk & Dungo* 1336; 9, female flower (× 8); 10, part of female flower corolla (× 8); 11, anther from female flower (× 14); 12, pollen presenter from female flower (× 14); 13, longitudinal section through ovary from female flower (× 10), 9–13 from *Davies* 1737; 14, fruiting branch (× ²/₃); 15, fruit (× 2), 14 & 15 from *Lemos & Balsinhas* 7; 16, pyrene, 2 views (× 2), from *Davies* 1675. Drawn by Sally Dawson.

20.i.1957, *Chase* 6300 (K; PRE; SRGH). S: Mwenezi Distr., Runde (Lundi) R., Rhino Hotel, fl. & fr. xii.1955, *Davies* 1675 (K; LISC; PRE; SRGH). **Malawi**. S: Chikwawa Distr., Kapichira Falls (Livingstone Falls), west bank of Shire River, male fl. 21.iv.1970, *Brummitt* 9999 (K; SRGH). **Mozambique**. N: Nampula, Monapo, male fl. 11.ii.1984, *de Koning et al.* 9548 (K; LMU). Z: Pebane, female fl. & fr. 9.iii.1966, *Torre & Correia* 15117 (LISC). T: Massanângua, male fl. xii.1908, *Dawe* 373a (K). MS: Sofala Prov., Gorongosa National Park, Sangarassa Forest, male fl. i.1972, *Tinley* 2335 (K; SRGH). GI: Macia, Nhangone (S. Martinho do Bilene), male fl. 8.x.1958, *Barbosa* 8342 (K). M: Costa do Sol, Polana, fr. 20.i.1960, *Lemos & Balsinhas* 7 (K; LISC; PRE; SRGH).

Also in South Africa (KwaZulu-Natal and North Prov.) and Swaziland. In *Brachystegia* or mixed deciduous woodland on rocky hillsides or rock outcrops, in riverine forest and coastal dune forest, frequently on sandy soils; 0–1000 m.

Subsp. *telidosma* (K. Schum.) Bridson occurs in Tanzania and southern Kenya. It differs from the typical subspecies chiefly in having shorter calyx lobes and slightly smaller fruit.

Canthium subgen. Uncertain

16. **Canthium kuntzeanum** Bridson in Kew Bull. **47**: 393 (1992). —Pooley, Trees of Natal, Zululand & Transkei: 474, figs. (1993). Type from South Africa.

Canthium pauciflorum (Klotzsch ex Eckl. & Zeyh.) Kuntze, Revis. Gen. Pl. **3**: 545 (1898). —Palmer & Pitman, Trees Southern Africa **3**: 2103, fig. on 2104 (1973). —Ross, Fl. Natal: 335 (1973). —Moll, Trees of Natal: 256 (1981). —von Breitenbach in J. Dendrol. **5**: fig. on p. 83 (1985) —K. Coates Palgrave, Trees Southern Africa, ed. 3, rev.: 885 (1988). non Blanco 1837, nec Baillon (1878).

Plectronia pauciflora Klotzsch ex Eckl. & Zeyh., Enum. Pl. Afr.: 363 (1837). —Sonder in F.C. **3**: 18 (1865). —Sim, For. Fl. Col. Cape Good Hope: 242 (1907). Type as above.

Canthium sp. no. 1 of Drummond in Kirkia **10**: 276 (1975), quoad *Chase* 3576.

Shrub or small tree, or sometimes a liana, 1.7–10 m tall, often armed with supra-axillary spines; young stems slender, glabrous or occasionally pubescent; bark grey. Leaves well spaced along older branches, or less often restricted to new growth; blades 0.5–4 × 0.5–2.6 cm, broadly elliptic or sometimes ovate, obtuse to acute at the apex, cuneate or sometimes obtuse to rounded at the base, glabrous, papery, turning blackish when dry; lateral nerves, in 3–4 main pairs, inconspicuous; tertiary nerves obscure; domatia present as ciliate pits, sometimes incompletely formed or absent; petioles 3–6 mm long; stipules 2–3 mm long, sheathing at base then triangular, apiculate, with a few hairs inside. Flowers 5-merous, solitary, or with 2 or 3 borne on a common peduncle; peduncle 10–14 mm long; pedicels 8–15 mm long, or 16–28 mm long in solitary flowers. Calyx tube c. 1 mm long; limb 0.5–1 mm long, repand and very shortly dentate. Corolla greenish-white; tube 2.5–4 mm long, inside with a ring of deflexed hairs near the top and with some hairs projecting from throat; lobes 1.75–2.25 × 1.25–1.5 mm, oblong-triangular, acute. Style scarcely exserted; pollen presenter c. 0.75 × 0.75 mm; disk glabrous. Anthers set at throat, apiculate with a central band of darkened connective tissue on the dorsal side. Fruit (11)13–15 × 9–15 mm, either narrowly obovate or obcordate in lateral view, scarcely or distinctly indented at the apex. Pyrene c. 10 × 5 mm, narrowly oblong-ellipsoid, point of attachment ± truncate with shallow crest running to apex, rugulose.

Zimbabwe. E: Mutare Distr., Himalaya Mts., Engwa Farm, fl. 27.xi.1966, *Müller* 481 (BR; K; SRGH). **Mozambique**. MS: Manica, Serra Zuira, Tsetserra, 6 km from cowshed on road to Chimoio (Vila Pery), 1800 m, 2.iv.1966, *Torre & Correia* 15613 (LISC).

Also in South Africa (Mpumalanga, KwaZulu-Natal, Western Cape Prov. and ?Eastern Cape Prov.). In mixed evergreen rain forest, *Diospyros* forest, bush clumps and evergreen scrub; 1500–2200 m.

51. PLECTRONIELLA Robyns

Plectroniella Robyns in Bull. Jard. Bot. État **11**: 243 (1928). —Bremekamp in Ann. Transvaal Mus. **15**: 259 (1933).

Shrubs or small trees, armed with spines set above brachyblasts (reduced lateral branches). Leaves restricted to lateral branches, paired but congested, petiolate;

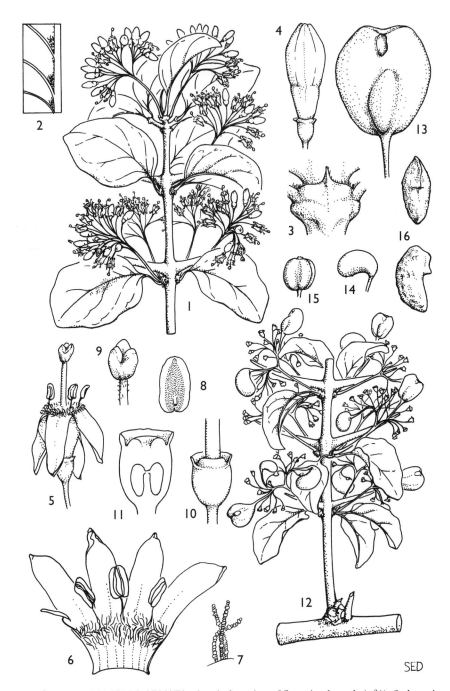

Tab. 67. PLECTRONIELLA ARMATA. 1, apical portion of flowering branch ($\times \frac{2}{3}$); 2, domatia ($\times 4$), 1 & 2 from *Leach & Bayliss* 10625; 3, stipule ($\times 4$), from *Schlieben* 10664; 4, bud ($\times 4$); 5, flower ($\times 4$); 6, corolla opened out ($\times 6$); 7, hairs from corolla throat ($\times 16$); 8, dorsal view of anther ($\times 14$); 9, pollen presenter ($\times 10$); 10, calyx with base of style ($\times 8$); 11, longitudinal section through ovary ($\times 10$), 4–11 from *Leach & Bayliss* 10625; 12, fruiting branch ($\times \frac{2}{3}$); 13, fruit ($\times 3$); 14, 1-seeded fruit ($\times 1$), 12–14 from *Coetzee* 1337; 15, 4-seeded fruit ($\times 1$), from *Balsinhas* 3172; 16, pyrene, 2 views ($\times 2$), from *Coetzee* 1337. Drawn by Sally Dawson.

domatia present; stipules small, connate at the base, with silky white hairs inside, soon caducous. Flowers 4-merous, borne in pedunculate corymbose cymes; bracts and bracteoles inconspicuous. Calyx tube ± ovate; limb short, repand or shortly toothed. Corolla creamy-white; tube cylindrical, somewhat shorter than the lobes, densely choked with moniliform hairs at the throat; lobes reflexed, thickened but not apiculate at the apex. Anthers distinctly exserted, but not reflexed. Style narrowly cylindrical, slightly more than twice the length of the corolla tube. Pollen presenter very slightly longer than broad, hollow at the base, 3-lobed at the apex. Disk not markedly prominent, glabrous. Ovary 3(4)-locular; ovule 1 per loculus, attached near apex to the septum. Fruit a 1–3(4)-seeded, fleshy drupe. Pyrenes flattened ellipsoid, clearly crested around the apex. Seeds flattened ellipsoid, ± winged at the apex, very finely reticulate; embryo slightly curved with an erect radicle, and small cotyledons positioned perpendicular to ventral face of seed.

A small, southern African genus, with 2 species described so far.

Plectroniella armata (K. Schum.) Robyns in Bull. Jard. Bot. État **11**: 243, figs. 23 & 24 (1928). —J.G. Garcia in Mem. Junta Invest. Ultramar **6** (sér. 2): 28 (1959) [Contrib. Conhec. Fl. Moçamb. IV (1959)]. —Palmer & Pitman, Trees Southern Africa **3**: 2109, photogr. (1973). —Ross, Fl. Natal: 335 (1973). —van Wyk, Trees Kruger Nat. Park **2**: 558 (1974). —Moll, Trees of Natal: 255 & 262 (1981). —K. Coates Palgrave, Trees Southern Africa, ed. 3, rev.: 888 (1988). —Herman, Fl. Pl. Africa **51**: pl. 2029 (1991). —Pooley, Trees of Natal, Zululand & Transkei: 480, figs. (1993). TAB. **67**. Type: Mozambique, Delagoa Bay, near Masinga *Schlechter* 12135 (B†, holotype; T; Z).
  Vangueria armata K. Schum. in Bot. Jahrb. Syst. **28**: 69 (1899). —Schinz in Mém. Herb. Boissier, No. 10: 68 (1900). —De Wildeman in Bull. Jard. Bot. État **8**: 43 (1922). Type as above.
  Plectronia sp. of Schinz in Mém. Herb. Boissier, No. 10: 68 (1900), quoad *Junod* 521.
  Plectronia ovata Burtt Davy in Bull. Misc. Inform., Kew **1921**: 191 (1921). Type from South Africa.
  Canthium ovatum (Burtt Davy) Burtt Davy in Bull. Misc. Inform., Kew **1935**: 568 (1935).

Shrub or small tree 1–5(8) m tall; young branches glabrous or sometimes pubescent, often with a few lenticels, armed with spines 1–4.5 cm long, situated above brachyblasts (abbreviated lateral branches) perpendicular or upwardly angled; bark rusty-brown when young, grey when older. Leaves mostly restricted to brachyblasts, blades 2.8–5 × 1.4–4.5(5) cm, elliptic to ovate-elliptic (or round to broadly cordate when on main branches), obtuse to rounded at apex, shortly attenuated at base, often somewhat discolorous but not glaucous beneath, thickly chartaceous, usually glabrous or occasionally pubescent beneath; lateral nerves in 3 main pairs; tertiary nerves finely reticulate, apparent beneath; domatia present as small tufts of hair; petiole (3)5–9 mm long, slender, often pubescent above; stipules c. 1.5 mm long, triangular, apiculate, with silky white hairs inside, soon caducous. Flowers 4-merous, borne in pedunculate many-flowered, corymbose cymes arising from brachyblasts; peduncles 8–35 mm long, glabrous to pubescent; pedicels 3–12 mm long, glabrous to pubescent; bracts and bracteoles inconspicuous. Calyx tube 1–1.25 mm long, glabrous; limb 0.5–0.75 mm long, repand or shortly toothed. Corolla creamy-white; tube 2–2.5(3) mm long, densely choked with finely moniliform hairs at throat, those above erect and those beneath deflexed; lobes (2)3.5–3.75(4) × 1.5–1.75 mm, narrowly oblong, thickened and obtuse at apex, reflexed. Anthers exserted but not reflexed, dorsal face covered with darkened connective tissue except for margin. Style exceeding corolla tube by a little more than twice; pollen presenter 1–1.25 mm across, cylindrical, 3–4-lobed at apex; ovary 3(4)-locular. Fruit orange, 1–3(4)-seeded, 10–15 mm long, varying in shape according to number of seeds developing. Pyrene 8 × 5 × 4 mm, flattened ellipsoid, clearly crested around the apex, rugulose. Seeds 6.75 mm long, flattened ellipsoid, ± winged at the apex.

Mozambique. M: between Boane and Ribeira do Movene, fr. 7.ii.1961, *Carvalho* 464 (K); Inhaca Island, fr. 4.ii.1960, *Mogg* 29914 (PRE).
  Also in South Africa (North Prov., Mpumalanga and KwaZulu-Natal) and Swaziland. In thicket and coastal forest, usually on sand, also in deciduous woodland on rocky slopes, associated with *Sclerocarya caffra, Lannea kirkii, Bolusanthus speciosus* etc.

52. PYROSTRIA Comm. ex Juss.

Pyrostria Comm. ex Juss., Gen. Pl.: 206 (1789). —Lamarck, Tabl. Encycl. 1, 1: t. 68 (1791); 1, 2: 289 (1792). —De Candolle, Prodr. 4: 464 (1830), pro parte majore. — A. Richard, Mém. Fam. Rubiac.: 136 (1830); in Mém. Soc. Hist. Nat. Paris 5: 216 (1834). —Bentham & Hooker f., Gen. Pl. 2, 1: 111 (1873). —Drake in Bull. Mens. Soc. Linn. Paris, nouv. sér. 6: 41 (1898). —K. Schumann in Engler & Prantl, Nat. Pflanzenfam. IV, 4: 94, fig. 33 H (1891). —Verdcourt in Kew Bull. 37: 563 (1983). — Bridson in Kew Bull. 42: 611–630 (1987); in F.T.E.A., Rubiaceae: 885–892 (1991). *Psydrax* sensu DC., Prodr. 4: 476 (Sept. 1830). —sensu A. Richard, Mém. Fam. Rubiac.: 110 (Dec. 1830); in Mém. Soc. Hist. Nat. Paris 5: 190 (1834), pro parte, excluding type species, non Gaertn. *Canthium* sect. *Psydracium* Baill. in Adansonia 12: 199 (1878); Hist. Pl. 7: 425 (1880). *Dinocanthium* Bremek. in Ann. Transvaal Mus. 15: 259 (1933). *Dinocanthium* sect. *Hypocrateriformes* Robyns in Bull. Jard. Bot. État 17: 94 (1943). *Dinocanthium* sect. *Rotatae* Robyns in Bull. Jard. Bot. État 17: 94 (1943). *Pseudopeponidium* Arènes, Notul. Syst. (Paris) 16: 19 (1960). *Pyrostria* sect. *Involucratae* Cavaco in Adansonia, sér. 2, 11: 393 (1971).

Small shrubs to medium-sized trees. Leaves petiolate, less often subsessile or occasionally sessile; blades drying blackish-brown, slate-grey or less often dull green, typically subcoriaceous but occasionally chartaceous or coriaceous, usually glabrous, very infrequently pubescent; tertiary nerves obscure (or rarely apparent); domatia present as glabrous to pubescent cavities, or sometimes absent; stipules triangular at their base, sometimes produced into a lobe, or linear to ovate, persistent or readily caducous, with a few silky hairs and colleters inside. Flowers unisexual or hermaphrodite, 4- or 5-merous borne in pedunculate umbels, sometimes with both peduncle and pedicels somewhat reduced, in dioecious species the female inflorescence containing fewer flowers (often solitary) than the male; bracts paired, connate and entirely surrounding the inflorescences in bud, persistent, with silky hairs and colleters towards base inside. Calyx tube ± ovate in female and hermaphrodite flowers, and very reduced in male flowers; limb reduced to a rim, sometimes shortly dentate or bearing unequal lobes, or sometimes (in Mauritian species) cupular. Corolla rather fleshy, usually drying dark with the lobes pale above; tube shorter than or longer than the lobes, throat densely congested with crisped or less often straight moniliform hairs, lacking a well-defined ring of deflexed hairs inside; lobes spreading, reflexed or sometimes erect, thickened towards the the apex, sometimes shortly apiculate, upper surface of lobes usually graniculate or very finely colliculate. Stamens set at throat; filaments short, attached near the base of the anther; anthers erect, less often spreading or rarely reflexed, ovate to oblong-ovate, the dorsal face, except for the margin, covered with darkened connective tissue, sometimes extending into a short apiculum. Style slender, slightly longer than corolla tube, somewhat tapered at apex; pollen presenter as wide as long, 2–several-lobed, but lobes only separating to reveal stigmatic surface in female and hermaphrodite flowers, solid, attached to style at base or sometimes slightly recessed; disk annular or hemispherical; the region between the ovary and disk sometimes elongating (in Madagascan species only); ovary 2–10-locular, each loculus containing one pendulous ovule. Fruit yellow to red, fleshy, when 2-locular subspherical to obcordate in lateral view, or less often didymous, bilaterally flattened, often elaborated with extra lobes or flanges, when multi-locular lobes correspond to number of locules. Pyrenes thinly woody to woody, ellipsoid or obovoid with ventral face flattened, crested around the apex, eventually dehiscing from the point of attachment back along the crest. Seeds narrowly obovoid with ventral area flattened and apex somewhat crested; endosperm entire; testa finely reticulate; embryo straight or slightly curved with radicle erect and cotyledons small and set perpendicular to ventral face of the seed.

A genus of c. 45 species distributed in Africa, Madagascar and the Mascarenes. Taxa from Malesia and Indochina should possibly also be included in *Pyrostria*. A total of 15 species are known from Africa with 6 from the Flora Zambesiaca area.

1. Leaf blades small, 8–23 mm long, elliptic or narrowly obovate, obtuse to rounded at apex; lateral branches usually perpendicular to main stem, bearing reduced and often spine-like side shoots; inflorescences 1–3-flowered · 6. *hystrix*
   – Leaf blades larger, at least 25 mm long, but usually more; lateral branches not perpendicular to main stem; inflorescences 1–30-flowered · · · · · · · · · · · · · · · · · · · 2
2. Calyx lobes c. 1.75 mm long; peduncle c. 2 mm long; pedicels suppressed at least in bud; small deciduous tree of coastal dunes; leaf blades c. 35 × 10 mm, obtuse at apex, with three main pairs of lateral nerves, and with domatia · 5. *sp. B*
   – Calyx lobes up to 1 mm long, sometimes unequal; peduncle 2–7 mm long; pedicels 1.5–5 mm long; evergreen or deciduous trees or shrubs; leaf blades with characters not combined as above · 3
3. Leaves drying greyish-green, papery to stiffly papery; lateral nerves obscure, or only discernible in largest leaves; lateral branches short with numerous closely spaced nodes; inflorescences 1–3-flowered · 4. *sp. A*
   – Leaves drying blackish or brown, stiffly papery or more often subcoriaceous to coriaceous; lateral nerves clearly apparent; lateral branches usually with nodes well spaced; inflorescences (3)4–30-flowered · 4
4. Leaf blades oblong-elliptic or narrowly obovate, rounded to obtuse at apex, shiny above; domatia conspicuous; inflorescences mostly borne at nodes from which leaves have fallen; fruit distinctly tapered at base · 3. *chapmanii*
   – Leaf blades elliptic to broadly elliptic, or occasionally oblong-elliptic, obtuse to acute at apex; inflorescences mostly subtended by leaves; fruit scarcely tapered at base · · · · · · · 5
5. Lateral nerves of leaves in 4–5 main pairs; stipules triangular-acuminate or triangular at base, lobed above; fruit lacking supplementary lobes · · · · · · · · · · · · · · · · 1. *bibracteata*
   – Lateral nerves of leaves in (5)6 main pairs; stipules triangular to triangular-ovate; fruit with a small supplementary lobe on either side of each main lobe (difficult to observe in dry state) · 2. *lobulata*

1. **Pyrostria bibracteata** (Baker) Cavaco in Bull. Mus. Hist. Nat. (Paris) sér. 2, **39**: 1015 (1968). —Bridson in Kew Bull. **42**: 625 (1987). —K. Coates Palgrave, Trees Southern Africa, ed. 3, rev.: 16k of appendix (1988). —Bridson in F.T.E.A., Rubiaceae: 887 (1991). —Beentje, Kenya Trees, Shrubs Lianas: 540 (1994). Type from Seychelles.
   Plectronia bibracteata Baker, Fl. Mauritius & Seychelles: 146 (1877). Type as above.
   Canthium bibracteatum (Baker) Hiern in F.T.A. **3**: 145 (1877). —Bullock in Bull. Misc. Inform., Kew **1932**: 375 (1932). —Brenan, Check-list For. Trees Shrubs Tang. Terr.: 487 (1949). —J.G. Garcia in Mem. Junta Invest. Ultramar **6** (sér. 2): 31 (1959) [Contrib. Conhec. Fl. Moçamb. IV (1959)]. —Dale & Greenway, Kenya Trees & Shrubs: 427 (1961). Type as above.

Shrub or small tree 2–10 m tall, glabrous; young stems covered with pale grey bark. Leaves turning black-brown or brown when dry; blades 2.5–13.5 × 1–7 cm, elliptic to broadly elliptic or sometimes oblong-elliptic, acute to obtuse at apex, acute to cuneate or occasionally rounded at base, stiffly papery to subcoriaceous, dull above; lateral nerves in 4–5 main pairs; tertiary nerves obscure; domatia sometimes present as tufts of hair; petiole 2–7 mm long; stipules 4–14 mm long, triangular at their base with a lobe above, caducous. Flowers hermaphrodite (or mostly so), 4-merous, borne in 4–30-flowered umbels; peduncles 2–4 mm long; pedicels 1.5–5 mm long, glabrous; bracts 2–7 mm long, acuminate. Calyx tube ± globose, c. 1 mm in diameter; limb reduced to a rim, or unequally lobed, up to 1 mm long. Corolla yellowish-cream; tube 2–3 mm long, densely congested with hairs at throat; lobes 2–2.5 × 1.25 mm, ovate-triangular, shortly apiculate. Pollen presenter 0.75–1 mm across. Fruit yellow, edible, 5–8 mm long, almost globose or somewhat laterally compressed, scarcely indented at the apex. Pyrene 5.5–6 × 2–3 mm, obovoid with ventral face flattened, with a shallow crest extending around the apex, slightly rugulose.

Zimbabwe. E: Chimanimani Distr., near Chisengu R., Hayfield B, bud 25.viii.1977, *Müller* 3433 (K; SRGH). **Mozambique**. N: Angoche (António Enes), Aúbe, ao 6 km estrada para Boila, fr. 22.i.1968, *Torre & Correia* 17304 (LISC). Z: Pebane, praia de Pebane, juncto ao farol, fr. 12.i.1968, *Torre & Correia* 17095 (LISC). MS: Gorongosa National Park, Cheringoma Plateau, 5 km along Urema road from junction with Inhaminga–Beira road, fr. i.1972, *Tinley* 2336 (K; LISC; M; PRE; SRGH). GI: Nhachengue (Inhachengo), fr. 26.ii.1955, *Exell, Mendonça & Wild* 615 (BM; LISC).

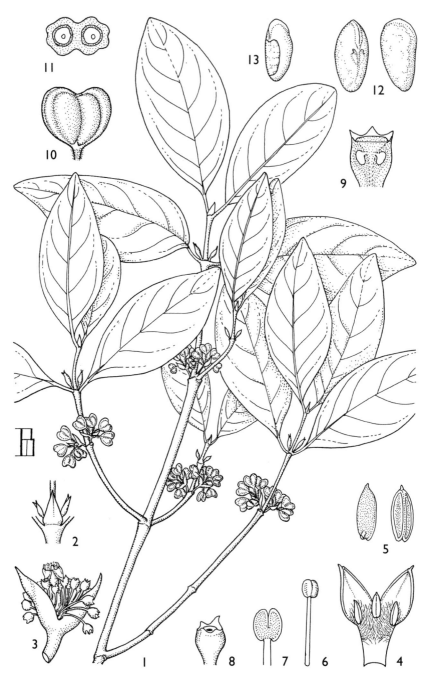

Tab. 68. PYROSTRIA LOBULATA. 1, fruiting branch ($\times \frac{2}{3}$), from *Bridson* 281; 2, stipule ($\times 1$); 3, inflorescence ($\times 2$); 4, portion of corolla opened out ($\times 6$); 5, anther, 2 views ($\times 10$); 6, style and pollen presenter ($\times 6$); 7, longitudinal section through pollen presenter ($\times 10$); 8, calyx ($\times 6$); 9, longitudinal section through ovary ($\times 9$), 2–9 from *Ford* 729; 10, fruit ($\times 2$); 11, transverse section through fruit ($\times 2$); 12, pyrene, 2 views ($\times 3$); 13, seed ($\times 3$), 10–13 from *Bridson* 281. Drawn by Diane Bridson. From F.T.E.A.

Also in Kenya, Tanzania, Madagascar, the Seychelles and ?Aldabra. Wooded grasslands and *Brachystegia* woodland and thickets, usually on high water-table sandy soils, also on margins of evergreen rain forest; 0–700 m.

2. **Pyrostria lobulata** Bridson in Kew Bull. **42**: 627, fig. 3 (1987); in F.T.E.A., Rubiaceae: 887, fig. 158 (1991). TABS. **47**/B1 & **68**. Type from Rwanda.

   Canthium bibracteatum sensu Bridson & Troupin in Fl. Pl. Lign. Rwanda: 546, fig. 182.2 (1982); in Fl. Rwanda **3**: 148, fig. 42.2 (1985), non (Baker) Hiern.

Shrub or small tree 3–8 m tall, glabrous; young stems square, very slightly winged, covered with grey or fawn-coloured bark. Leaves black-brown when dry; blades 2.5–13.5 × 1–6 cm, acute to obtuse at apex, acute or sometimes rounded and often slightly unequal at base, subcoriaceous, dull above; lateral nerves in (5)6 main pairs; tertiary nerves obscure; domatia present as inconspicuous tufts of hair in the nerve axils, or sometimes absent; petiole 2–10 mm long; stipules 5–10 mm long, triangular to triangular-ovate, caducous. Flowers unisexual, or ?sometimes hermaphrodite, 4-merous, borne in pedunculate umbels; functionally male flowers 15–30, functionally female flowers perhaps fewer; peduncles 4–7 mm long; pedicels 3–4 mm long, glabrescent to pubescent; bracts 5–7 mm long, ovate, acuminate, pubescent near the base inside. Calyx tube in functionally male or hermaphrodite flowers, ± globose, up to 1 mm in diameter, possibly ribbed in functionally female flowers judging from very immature fruit; limb a reduced rim, truncate or bearing unequal lobes up to 1 mm long. Corolla cream or greenish-white; tube 2 mm long, congested with crisped hairs at throat; lobes 2–2.5 × 1–1.25 mm, ovate-triangular, shortly apiculate. Style very slightly longer than corolla tube; pollen presenter ± globose, c. 0.5 mm in diameter. Fruits usually borne in umbels of 3–10, 7–8 × 7–13 mm, distinctly bilobed, with a small lobe on either side of each lobe (tending to disappear on drying), distinctly indented at the apex. Pyrene 6 × 3.25 mm, narrowly obovoid, with a shallow wing around the apex, scarcely rugulose.

Zambia. W: Kitwe Distr., Baluba R., fr. 14.i.1959, *Fanshawe* 5075 (K).
Also in Rwanda and Tanzania. Termite mounds in plateau woodland.

3. **Pyrostria chapmanii** Bridson, sp. nov.* Type: Malawi, Mt. Mulanje, along the Nessa Path, Litchenya Plateau, 7.xi.1986, *J.D. & E.G. Chapman* 8198 (K, holotype; FHO; MAL; MO, not all seen).

Subshrub or shrub 1–6 m tall, or tree 15–20 m tall; stems glabrous, covered with grey bark. Leaves tending to be restricted to apex of shoots, brown or black-brown when dry; blades 4–8.5 × 1.6–3.8 cm, oblong-elliptic to narrowly obovate or less often elliptic, obtuse to rounded at apex, acute at base, coriaceous, shiny above; lateral nerves in 3–5 main pairs, tertiary nerves obscure; domatia present as conspicuous tufts of hair; petioles 3–7 mm long; stipules 5–7 mm long, often triangular, acuminate, present at the top 1–2 nodes only. Flowers unisexual, 4-merous, borne in 3–8-flowered pedunculate umbels; peduncles 3–7 mm long; pedicels 2.5–4 mm long; bracts 4–5 mm long. Calyx tube 2 mm long in female flowers (corolla just fallen), tending to be ridged or reduced in male flowers; limb 0.5 mm long, dentate. Corolla cream-coloured; tube 1.25–2 mm long, congested with hairs at throat; lobes 1.5 × 1 mm, ovate, acuminate. Pollen presenter 1 mm across, very deeply cleft in two when mature. Fruit 8–9 × 8–9 mm, obcordate in outline, indented at apex and distinctly tapered at base, probably with a small lobe on either side of each lobe but these lobes are indistinct when dry. Pyrenes 9.75 × 4 mm, narrowly obovoid with a well defined crest around the apex, rugulose.

Subsp. **chapmanii**

Forest trees (6)10–20 m tall, with the boles smooth or distinctly spiralled and fluted and the branches widely spreading or horizontally inclined.

---

* A *Pyrostria bibracteata* foliis oblongo-ellipticis vel anguste obovatis, apice obtusis vel rotundatis, domatiis satis conspicuis vel conspicuis, floribus unisexualibus, fructibus versus basem angustatis differt.

Malawi. S: Mt. Mulanje, Chisongeli Forest, Muluzi Valley, male fl. 29.ix.1988, *Chapman* 9347 (K; FHO; MO).

Not known elsewhere. In tall *Newtonia buchananii* forest, riverine or moist montane evergreen forests; 1150–1850 m.

Subsp. A

Pyrostria sp. A of Bridson in Kew Bull. **42**: 627 (1987).

Subshrubs, shrubs or small trees, 1–6 m tall.

Mozambique. N: Ribáuè, Serra de Ribáuè, Mepáluè, fr. 25.i.1964, *Torre & Paiva* 10251 (LISC). Z: Gurué, Serra do Gurué, oeste dos Picos Namuli, prox. da nascente do rio Malema, male corolla fallen 8.xi.1967, *Torre & Correia* 16002 (LISC).

Not known elsewhere. In submontane mist-zone grassland to evergreen rainforest ecotone vegetation, and moist afromontane evergreen rainforest margins, also in submontane riverine forests; 1350–1800 m.

Apart from the differing habit, no other significant differences have been noted between the Malawi and Mozambique specimens. No mature flowers are yet known from Mozambique and additional material and field observations are needed to clarify the taxonomy.

4. Pyrostria sp. A

Shrub to c. 6 m tall; branches sarmentose; lateral twigs rather short, with nodes crowded together, glabrous, covered with greyish bark. Leaves tending to be crowded near apex of twigs, greyish-green when dry; blades 3.2–7.5 × 1.3–2.5 cm, elliptic or sometimes oblong-elliptic, acute at apex and base, stiffly papery, dull above; lateral nerves in c. 4 main pairs, only faintly apparent in all but the largest leaves; tertiary nerves obscure; domatia present as tufts of fawn-coloured hair. Petiole 2–3 mm long. Flowers not known. Fruiting inflorescences 1-fruited; peduncles up to 2 mm long; pedicels up to 3.5 mm long; bracts 1.5 mm long, ovate acute. Fruit 6 × 7 mm, broadly oblong in lateral view, distinctly bilobed, indented at the apex, scarcely tapered at the base; persistent calyx lobes c. 0.75 mm long, triangular. Pyrenes 5.5 × 3.5 mm, obovoid with ventral face flattened and with an apical crest discernible, slightly rugulose.

Mozambique. Z: Maganja da Costa, Floresta de Gobene, prox. praia Raraga, ao 35 km de Maganja da Costa (Vila de Maganja), fr. 10.i.1968, *Torre & Correia* 17041 (LISC).

Only known from the above specimen. Deciduous coastal forest on sandy soil, with *Brachystegia, Cynometra, Albizia, Mimusops, Trachylobium*, etc.; 20 m.

5. Pyrostria sp. B

Small deciduous tree 2–4 m tall; lateral stems glabrous, covered with greyish bark which rubs off to reveal a red-brown under-layer; twigs rather short, with nodes closely spaced. Leaves crowded near apex of twigs, light brown when dry; blades c. 3.5 × 1 cm, oblong-elliptic, obtuse at apex, acute at base; lateral nerves in c. 3 main pairs; tertiary nerves obscure; domatia present; petiole up to 3 mm long. Inflorescences mostly borne at naked nodes; peduncle c. 2 mm long; pedicels ± suppressed; bracts c. 2.25 mm long. Flowers apparently unisexual, only young male buds seen, 2 per inflorescence. Calyx lobes c. 1.75 mm long, lanceolate.

Mozambique. N: Angoche (António Enes), S13°14', E39°54', male bud 18.x.1965, *Mogg* 32331 (LISC).

Only known from the above specimen. *Mimusops–Eugenia* coastal scrub, on littoral dunes; 2 m.

This taxon seems most closely related to *Pyrostria sp. D* of F.T.E.A. from the Lindi District of Tanzania. The latter, only known in male flower, is similar in its compressed inflorescence and well developed calyx lobes. However, the habitat (undergrowth in dry forest at 750 m) is rather different.

6. Pyrostria hystrix (Bremek.) Bridson in Kew Bull. **42**: 629 (1987). —von Breitenbach in J. Dendrol. **5**: 92 (1985). —K. Coates Palgrave, Trees Southern Africa, ed. 3, rev.: 16k of appendix (1988). —Pooley, Trees of Natal, Zululand & Transkei: 480, figs. (1993). Type from South Africa.

Dinocanthium hystrix Bremek. in Ann. Transvaal Mus. **15**: 259 (1933). —Palmer & Pitman, Trees Southern Africa **3**: 2109 (1973). —K. Coates Palgrave, Trees Southern Africa, ed. 3, rev.: 888 (1988). Type as above.

Shrub or small tree 2–4 m tall; lateral branches perpendicular to the main stem, supporting reduced and sometimes spine-like side shoots, glabrous, covered with light to dark grey bark. Leaves grey-green when dry; blades 0.8–2.2 × 0.4–1 cm, narrowly obovate or elliptic, rounded or obtuse at apex, acute to cuneate at base, glabrous, papery, dull above; lateral and tertiary nerves indistinct; domatia absent; petiole very reduced, up to 1 mm long; stipules up to 2 mm long, triangular, apiculate. Flowers unisexual, 4-merous, borne in 1–3-flowered shortly pedunculate umbels; peduncles 1–2 mm long; pedicels very reduced, up to 1 mm long; bracts 1–2 mm long, ovate to broadly ovate, obtuse. Calyx tube reduced in male flowers, not known in female flowers; limb c. 1 mm long with tubular part reduced and four equal triangular or lanceolate to ovate lobes. Corolla cream-coloured; tube 1.5 mm long, sparsely hairy within; lobes 1.5 × 1–1.25 mm ovate, shortly apiculate. Style distinctly longer than the corolla tube; pollen presenter c. 0.3 mm across, broadly ovoid. Fruit 5–7 × 6–8 mm, broadly oblong in outline, bilobed, indented at the apex and somewhat narrowed at the base; pedicel up to 0.7(1.5) mm long. Pyrene c. 5 × 3.5 mm, ± obovoid, ventral face slightly flattened, with an apical crest scarcely discernible, slightly rugulose.

Mozambique. GI: 30 km from Mandlakazi (Manjacaze), fr. 3.xii.1942, *Mendonça* 1581 (LISC). M: Moamba, fr. 1.xii.1942, *Mendonça* 1525 (LISC).
Also in South Africa (North Prov., Mpumalanga and KwaZulu-Natal) and Swaziland. Low altitude *Acacia* woodland.

53. PSYDRAX Gaertn.

Psydrax* Gaertn., Fruct. Sem. Pl. **1**: 125, t. 26, fig. 2 (1778). —De Candolle, Prodr. **4**: 476 (1830), pro parte. —A. Richard, Mém. Fam. Rubiac.: 110 (1830); in Mém. Soc. Hist. Nat. Paris **5**: 190 (1834), pro parte. —Bridson in Kew Bull. **40**: 687–725 (1985); in F.T.E.A., Rubiaceae: 892 (1991).
Mitrastigma Harv. in London J. Bot. **1**: 20 (1842). —E. Phillips, Gen. S. African Fl. Pl. [Mem. Bot. Survey S. Afr. No. 10]: 587 (1926).
Mesoptera Hook.f. in Gen. Pl. **2**(1): 130 (1873). —K. Schumann in Engler & Prantl, Nat. Pflanzenfam. IV, **4**: 92 (1891).
Canthium sensu Sond. in F.C. **3**: 16–17 (1865), pro parte. —Hiern in F.T.A. **3**: 132–146 (1877), pro parte. —Bullock in Bull. Misc. Inform., Kew **1932**: 358–373 (1932). —Hepper in F.W.T.A. ed. 2, **2**: 181–185 (1963), pro parte, non Lam.

Trees, shrubs or scandent shrubs, sometimes climbers. Leaves mostly evergreen, sometimes deciduous and then restricted to apex of branches, paired or rarely ternate, petiolate or sometimes subsessile; blade typically coriaceous, less often chartaceous, glabrous or sometimes pubescent; domatia glabrous or pubescent, or sometimes absent; stipules usually with a truncate to broadly triangular base bearing a decurrent, slightly to very strongly keeled, sometimes foliaceous appendage, or less often lanceolate to ovate and soon caducous, never with silky white hairs inside. Flowers 4–5-merous, borne in sessile to pedunculate umbellate cymes, or in pedunculate clearly branched cymes, or flowers very rarely solitary; bracts and bracteoles typically inconspicuous, or occasionally conspicuous, or sometimes (not in Africa) entirely enclosing the developing inflorescence and eventually rupturing into 2 or more segments. Calyx tube broadly ellipsoid to ± semi-spherical; limb a truncate to dentate rim, only occasionally equalling the tube in length and often much

---

* The gender of the name *Psydrax* has been queried recently when some specific epithets have been given as masculine. In the original publication Gaertner included only one species, *P. dicoccos*. This epithet is a noun in apposition and does not dictate the gender of *Psydrax*. The first author to attribute gender to an epithet in *Psydrax* was A. Richard in 1830 who made it feminine. All the authors mentioned above have followed this and treated it as feminine.

shorter. Corolla white to yellow; tube broadly cylindrical mostly equal to the lobes in length, or somewhat longer or shorter, inside usually with a ring of deflexed hairs or sometimes with hairs not restricted to a well defined ring or absent, often pubescent at throat; lobes reflexed, often thickened towards the apex and obtuse to acute, rarely acuminate-apiculate, very rarely apiculate. Stamens set at corolla throat; filaments well developed; anthers lanceolate to narrowly ovate, attached shortly above the base, reflexed. Style long slender, glabrous, sometimes narrowing at apex; pollen presenter cylindrical, always longer than wide, occasionally flared or somewhat narrowed at base, hollow to the middle, bifid or rarely deeply cleft at apex when mature; disk glabrous or pubescent; ovary 2-locular, with 1 ovule per loculus, attached to the upper third of the septum. Fruit a 2-seeded drupe, small or sometimes large (in India), ellipsoid to bi-spherical; pyrenes cartilaginous to woody, ± plano-ellipsoid to laterally flattened spherical, usually with a very shallow crest from the point of attachment around the apex, eventually splitting and with grooves from point of attachment extending to lateral faces, scarcely bullate, rugulose or deeply furrowed; seeds with endosperm entire; testa very finely reticulate; embryo with radicle erect, almost straight or curved to a C-shape (according to the pyrene shape), cotyledons small, set parallel to ventral face of the seed.

A large genus of at least 100 species, occurring throughout the Old World tropics. Combinations for the majority of the non-African *Psydrax* species have not yet been made; 34 species occur in Africa of which 16 are represented in the Flora Zambesiaca area.

1 Climbers, or at least scandent shrubs; corolla lobes acuminate-apiculate, corolla tube inside with a ring of deflexed hairs below mid-point; pollen presenter widened at the base; fruit didymous, flattened [subgen. *Phallaria*] · 16. *kraussioides*
 – Trees or shrubs, occasionally scandent bushes; corolla lobes obtuse to acute, corolla tube inside with a ring of deflexed hairs above mid-point, or sometimes the ring not well defined; pollen presenter not markedly widened at the base; fruit ovoid to didymous, often flattened, or rarely ellipsoid [subgen. *Psydrax*] · 2
2. Leaves deciduous, usually restricted to stem apex or to apices of short lateral branches, chartaceous to subcoriaceous, dull or shiny on upper surface, pubescent or glabrous · · 3
 – Leaves not deciduous (or rarely deciduous), not restricted to apex of stem, typically subcoriaceous to coriaceous but occasionally chartaceous, shiny or less often dull on upper surface, glabrous or infrequently pubescent · 6
3. Inflorescences sessile; lateral branches patent or almost so; leaves sometimes clustered (if strongly coriaceous see couplet 16) ·7. *martinii*
 – Inflorescences clearly pedunculate; habit not as above · 4
4. Leaves drying discolorous (brown above, glaucous beneath), lateral nerves in 6–7 main pairs; stipules triangular, acute with a short decurrent apiculum; disk pubescent; fruit indented at apex · 4. *sp. A*
 – Leaves not drying as above, lateral nerves in 3–5 main pairs; stipules triangular acuminate; fruit scarcely indented at apex · 5
5. Leaf blades 1.5–13 cm long, obtuse to rounded or sometimes acute at base; lateral nerves not impressed above; corolla lobes 1.25–2 mm long ·5. *livida*
 – Leaf blades 1.3–3 cm long, rounded or sometimes truncate or subcordate at base; lateral nerves impressed above; corolla lobes 2.5–3 mm long · · · · · · · · · · · · · · · · · 6. *whitei*
6. Inflorescences clearly pedunculate, with branches compact to lax · · · · · · · · · · · · · · 7
 – Inflorescences sessile to subsessile or shortly pedunculate, with inflorescence branches absent or rudimentary · 13
7. Tall trees with horizontal branches (sometimes said to be palm-like), myrmecophilous (inhabited by ants); leaves chartaceous to subcoriaceous, often large with 7–10 main pairs of lateral nerves; stipules narrowly ovate to ovate, caducous; bracteoles apparent · · · · · · ·
 · 1. *subcordata*
 – Shrubs or trees, but not as above, myrmecophily not recorded; leaves subcoriaceous or coriaceous, small to large, with 2–8 main pairs of lateral nerves; stipules triangular at base, apiculate or terminating in a decurrent lobe; bracteoles inconspicuous · · · · · · · · · · · 8
8. Disk pubescent; fruit distinctly broader than long, often almost bi-circular in outline; inflorescence branches lax · 9
 – Disk glabrous; fruit somewhat broader than long, or longer than broad; inflorescence branches compact to moderately lax · 10

9. Leaves 5.5–15 cm long, acuminate at apex; lateral nerves of leaves in 4–8 main pairs; fruit ± flattened, bi-circular in lateral view · **2.** *parviflora*
– Leaves 2.2–4 cm long, obtuse to rounded at apex; lateral nerves of leaves in 3–4 main pairs; fruit broadly oblong in lateral view, scarcely indented at apex · · · · · · · · · · · **3.** *richardsiae*
10. Pedicels pubescent; inflorescence branches compact · 11
– Pedicels glabrous; inflorescence branches moderately lax · 12
11. Corolla tube 3–3.5 mm long, with a ring of deflexed hairs in throat; leaves acuminate at apex, lateral nerves in 5(6) main pairs · **11.** *mutimushii*
– Corolla tube 1.5–2.5 mm long, with deflexed hairs not restricted to a definite ring; leaves obtuse at apex, lateral nerves in 2–3 main pairs · **8.** *locuples*
12. Flowers larger, pollen presenter 1.5–2 mm long, corolla tube 2–3.25 mm long, lobes 2.75 × 4.5 mm long; leaves larger, 2–4.7 cm long · **9.** *obovata*
– Flowers smaller, pollen presenter 0.75–1.25 mm long, corolla tube 1.5–2.5 mm long, lobes 2.5–3 mm long; leaves smaller, 1.3–3.3 cm long · · · · · · · · · · · · · · · · **10.** *fragrantissima*
13. Leaves distinctly acuminate at apex, tertiary nerves obscure; short peduncles and rudimentary inflorescence branches mostly present; pedicels pubescent · · **11.** *mutimushii*
– Leaves obtuse to acute, or sometimes gradually acuminate or subacuminate at apex, tertiary nerves apparent to conspicuous; peduncles and rudimentary inflorescence branches usually absent; pedicels glabrous or pubescent · 14
14. Pedicels puberulous to pubescent; lateral nerves of leaves in 4(5) main pairs, tertiary nerves ± apparent, coarsely reticulate · **12.** *schimperiana*
– Pedicels glabrous; lateral nerves of leaves in 3–4 main pairs, tertiary nerves conspicuous, finely or rather coarsely reticulate · 15
15. Leaves often ternate; blades elliptic or less often narrowly elliptic; tertiary nerves rather finely reticulate · **15.** *micans*
– Ternate leaves not recorded; blades elliptic to ± round; tertiary nerves rather coarsely reticulate · 16
16. Petioles 2–4 mm long; pedicels rather stout; flowers larger (corolla tube 4–5 mm long, corolla lobes 6 mm long, pollen presenter 2–2.5 mm long); domatia present as erupted blisters; deciduous condition not recorded · **13.** *kaessneri*
– Petioles 4–7 mm long; pedicels slender; flowers smaller (corolla tube 3–3.5 mm long, corolla lobes 3–4 mm long, pollen presenter 1.5 mm long); domatia pocket-like; leaves sometimes said to be deciduous · **14.** *moggii*

Subgen. PSYDRAX

Trees or shrubs, occasionally scandent bushes; corolla lobes obtuse to acute, corolla tube inside with a ring of deflexed hairs above mid-point, or sometimes the ring not well defined; pollen presenter not markedly widened at the base; fruit ovoid to didymous, often laterally flattened, or rarely ellipsoid. *Species* 1–15.

1. **Psydrax subcordata** (DC.) Bridson in Kew Bull. **40**: 698 (1985). —Bridson & Troupin in Fl. Rwanda **3**: notes nomenclaturales, corrigenda for page 150 (1985). —Bridson in F.T.E.A., Rubiaceae: 894, fig. 159 (1991). Type from the Gambia.
    Canthium subcordatum DC., Prodr. **4**: 473 (1830). —Hiern in F.T.A. **3**: 141 (1877). — Hepper in F.W.T.A. ed. 2, **2**: 184, fig. 239 (1963). —Bridson & Troupin in Fl. Pl. Lign. Rwanda: 550, fig. 182.3 (1982); in Fl. Rwanda **3**: 150, fig. 44.2 (1985). Type as above.

Tree 5–15 m tall, with a palm-like habit, associated with ants, the branches often hollow and swollen with access pores present; young branches distinctly square in cross section, often slightly winged on the angles, glabrous to pubescent. Leaf blades 9.5–22(30) × 4.5–16.5 cm, oblong to ovate or sometimes broadly ovate, shortly acuminate at apex, obtuse to rounded or truncate to subcordate at base, glabrous or sometimes sparsely pubescent on upper surface, glabrous or pubescent on nerves or sometimes sparsely pubescent beneath, papery to subcoriaceous; lateral nerves in 7–10 main pairs; tertiary nerves moderately conspicuous beneath; domatia present as hair-lined cavities in the axils of lateral and sometimes tertiary nerves; petioles 10–15 mm long, glabrous or pubescent; stipules 4–10 mm long (or 14–18 mm long on saplings or coppice growth), narrowly ovate to ovate-caducous. Flowers with a very unpleasant smell, 5-merous, borne in 80–120-flowered, dichotomous corymbose pedunculate cymes; true peduncle rather short, the portion above the bract being

Tab. 69. PSYDRAX SUBCORDATA var. SUBCORDATA. 1, portion of flowering branch (× ²/₃);
2, flower (× 6); 3, calyx (× 10); 4, longitudinal section through ovary (× 16); 5, portion of
fruiting branch (× ²/₃), 1–5 collector unknown; 6, fruit (× 2); 7, pyrene, 2 views (× 3), 6 &
7 from *Dschang staff* in *CNAD* 1779. Drawn by Stella Ross-Craig, with 6 & 7 by Sally Dawson.
From F.W.T.A.

longer, 0.8–2(4) cm long altogether, pubescent or glabrous; pedicels 1.5–2.5 mm long, sparsely puberulous to pubescent; bracts 1–4 mm long, in connate pairs or sometimes free; bracteoles free, inconspicuous, or conspicuous and up to 3 mm long. Calyx tube c. 1 mm long, glabrous or sometimes pubescent towards the base; limb 0.5–1 mm long, truncate to dentate, ciliate, puberulous inside. Corolla white or creamy-yellow; tube 2–3.5 mm long, with a ring of deflexed hairs set above mid-point inside; lobes 2.5 × 1.25–1.5 mm, oblong ovate, acute. Anthers not persistent in mature flowers. Style 5–7 mm long; pollen presenter 1.5–2 mm long, ribbed; disk pubescent or sometimes glabrous. Fruit 6–8 × 9–12(14) mm, usually rather prominent at apex, black or greyish-black. Pyrenes 6–8 × 4.5–6 × 3–4 mm, strongly curved, with a groove from point of attachment to the centre of the lateral face, with a shallow crest at the apex, cartilaginous, scarcely rugulose.

Var. **subcordata** —Bridson in F.T.E.A., Rubiaceae: 894, fig. 159 (1991). TAB. **69**.

    Canthium glabriflorum Hiern in F.T.A. **3**: 140 (1877). Syntypes from São Tomé and Nigeria.
    Canthium polycarpum Schweinf. ex Hiern in F.T.A. **3**: 139 (1877). Type from Sudan.
    Plectronia glabriflora (Hiern) K. Schum. in Engler, Pflanzenw. Ost-Afrikas **C**: 386 (1895).
    Plectronia subcordata (DC.) K. Schum. in Engler, Pflanzenw. Ost-Afrikas **C**: 386 (1895), pro parte.
    Canthium welwitschii Hiern, Cat. Afr. Pl. Welw. **1**: 475 (1898). Type from Angola.
    Plectronia welwitschii (Hiern) K. Schum. in Just's Bot. Jahresber., 26th year, part 1: 393 (1900).
    Plectronia formicarum K. Krause in Bot. Jahrb. Syst. **54**: 351 (1917). Type from Cameroon.
    Plectronia laurentii var. *katangensis* De Wild., Notes Fl. Katanga **7**: 61 (1921). Type from Zaire (Dem. Rep. Congo) (Shaba Prov.).
    Canthium sp. 1 of F. White, F.F.N.R.: 403 (1962).

Bracts 1–2 mm long; bracteoles up to 1(2) mm long, inconspicuous. Calyx limb 0.5(1) mm long.

Zambia. N: Ishiba Ngandu (Shiwa Ngandu), young fr. 26.vii.1938, *Greenway* 5525 (EA; K). W: Solwezi Distr., 11 km north of Solwezi Boma, fr. 14.ix.1952, *Holmes* 864 (BM; FHO; K; PRE; WAG).
Also throughout west tropical Africa, Cameroon, Central African Republic, Zaire (Dem. Rep. Congo), Sudan, Rwanda, Uganda and Angola. In swamp forest (mushitu); 1525–1650 m.
Var. *connata* (De Wild. & T. Durand) Bridson occurs in the Central African Republic and Zaire (Dem. Rep. Congo), and may be distinguished by its larger bracts, bracteoles and calyx limb.

2. **Psydrax parviflora** (Afzel.) Bridson in Kew Bull. **40**: 700 (1985). —Bridson & Troupin in Fl. Rwanda **3**: notes nomenclaturales, corrigenda for page 152 (1985). —Bridson in F.T.E.A., Rubiaceae: 896 (1991). Type from Sierra Leone.

    Pavetta parviflora Afzel., Rem. Guin.: 47 (1815). —De Candolle, Prodr. **4**: 492 (1830).

Shrub or tall timber tree 2–27 m tall, sometimes with a fluted bole; young stems glabrous, or rarely puberulous, square or sometimes rounded. Leaf blades 5.5–15.5 × 2–8 cm, narrowly elliptic to elliptic, ovate or oblong-elliptic, acuminate to finely acuminate at apex, acute to obtuse or rounded at base, subcoriaceous to coriaceous, shiny on upper surface, glabrous; mid-rib whitish or red; lateral nerves in 4–8 main pairs; tertiary nerves apparent, sometimes raised above; domatia present as rather prominent blisters in the axils of the lateral nerves, with or without an eruption, glabrous, ciliate or pubescent; petioles 3–10 mm long; stipules 2–7 mm long, triangular at base, terminating in a usually abrupt acumen. Flowers 4-merous, borne in pedunculate, 20–100-flowered corymbs, 2–6 cm across; peduncles 3–20 mm long, glabrescent to pubescent; pedicels 2–9 mm long, pubescent; bracts and bracteoles inconspicuous. Calyx tube 0.75–1.25 mm long, glabrous to pubescent; limb 0.25–1.25 mm long, shorter or longer and often wider than the tube, truncate to repand, glabrous to pubescent outside and inside. Corolla whitish; tube 2–3 mm long, with deflexed hairs above mid-point inside and pubescent at throat; lobes 1.5–2.5 × 1.25–1.5 mm, oblong ovate, acute to obtuse. Anthers reflexed. Style up to 8 mm long; pollen presenter (0.5)0.75–1 mm long; disk pubescent. Fruit 5–8 × 8–14 mm, almost bispherical, black when mature. Pyrene 5–8 mm across, 3/4 circular to almost circular in outline with a groove extending from point of attachment on either side to centre of the plane face, cartilaginous, somewhat bullate.

Key to the subspecies

1. Fruit 8–11 mm wide; leaf midrib pale; corymbs larger, 30–100-flowered; calyx limb shorter
   or longer than the tube · subsp. *parviflora*
 – Fruit 10–14 mm wide; leaf midrib orange to red; corymbs somewhat smaller, 20–40(80)-
   flowered; calyx limb always shorter than the tube · 2
2. Leaves often drying brownish; leaf blades ovate or elliptic to broadly elliptic, 3.3–6.8 cm
   wide; petioles 6–10 mm long; calyx glabrous to glabrescent · · · · · · · · · subsp. *rubrocostata*
 – Leaves drying light green; leaf blades narrowly elliptic to elliptic, 2–4.5 cm wide; petioles
   3–6 mm long; calyx sparsely pubescent to pubescent · · · · · · · · · · · · · · subsp. *chapmanii*

Subsp. **parviflora** —Bridson in Kew Bull. **40**: 701 (1985). —Bridson & Troupin in Fl. Rwanda
  3: notes nomenclaturales, corrigenda for page 152 (1985). —Bridson in F.T.E.A.,
  Rubiaceae: 897 (1991). —Beentje, Kenya Trees, Shrubs Lianas: 539 (1994).
  Canthium afzelianum Hiern in F.T.A. **3**: 142 (1877), pro parte, nom. nov. for *P. parviflora*.
  Type as above.
  Plectronia vulgaris K. Schum. in Engler, Pflanzenw. Ost-Afrikas **C**: 386 (1895). —K. Krause
  in Mildbraed, Wiss. Ergebn. Deutsch. Zentr.-Afrika Exped., Bot., part 4: 326 (1911).
  Neotype from Tanzania.
  Canthium golungensis Hiern, Cat. Afr. Pl. Welw. **1**: 478 (1898). Type from Angola.
  Plectronia golungensis (Hiern) K. Schum. in Just's Bot. Jahresber., 26th year, part 1: 393
  (1900).
  Canthium golungensis var. *parviflorum* S. Moore in J. Linn. Soc., Bot. **37**: 161 (1905). Type
  from Uganda.
  Plectronia brieyi De Wild. in Repert. Spec. Nov. Regni Veg. **13**: 380 (1914). Type from
  Zaire (Dem. Rep. Congo).
  Canthium vulgare (K. Schum.) Bullock in Bull. Misc. Inform., Kew **1932**: 374 (1932). —
  Brenan, Check-list For. Trees Shrubs Tang. Terr.: 488 (1949). —F. White, F.F.N.R.: 403
  (1962). —Hepper in F.W.T.A. ed. 2, **2**: 184 (1963). —Bridson & Troupin in Fl. Pl. Lign.
  Rwanda: 552, fig. 183.4 (1982).
  Canthium vulgare subsp. *vulgare* —Bridson & Troupin in Fl. Rwanda **3**: 152 (1985).
  Canthium giordanii Chiov. in Atti Reale Accad. Italia, Mem. Cl. Sci. **11**: 35 (1941). —
  Cufodontis in Bull. Jard. Bot. État **35**, Suppl. [Enum. Pl. Aethiop. Sperm.]: 1009 (1965).
  Syntypes from Ethiopia.
  Canthium mutiflorum sensu Hepper in F.W.T.A. ed. 2, **2**: 182 (1963), pro parte, quoad *C.
  afzelianum* in syn.

Shrub or tree 2–20 m tall. Leaves drying green; blades 2–8 cm wide, narrowly
elliptic, elliptic, oblong-elliptic or ovate, acuminate at apex, obtuse to rounded or
occasionally acute at base; midrib pale; domatia glabrous; petioles 4–8 mm long.
Corymbs 30–100-flowered. Calyx pubescent; limb 0.25–1.25 mm long, shorter or
longer than the tube. Corolla tube 2–2.75 mm long. Fruit 8–11 mm wide.

Zambia. W: Solwezi Distr., Mlulungu Stream near Mutanda Bridge, fr. 28.vi.1930, *Milne-
Redhead* 618 (K).
Also in West Africa, Sudan, Ethiopia, Uganda, Kenya, Tanzania, Rwanda, Zaire (Dem. Rep.
Congo) and Angola. In evergreen woodland or mushitu, sometimes on termite-mounds.
This species is very variable and to a certain extent the variation can be grouped into
geographical units. However, the character disjunctions between them do not seem sufficiently
clear-cut for the recognition of additional subspecies. The specimens from Zambia, together
with an odd specimen from western Tanzania (Ufipa District) are characterised by thickly
coriaceous leaves which dry yellow-green. These may perhaps represent an ecotype associated
with certain metallic deposits.

Subsp. **rubrocostata** (Robyns) Bridson in Kew Bull. **40**: 702 (1985); in F.T.E.A., Rubiaceae: 898
  (1991). —Beentje, Kenya Trees, Shrubs Lianas: 539 (1994). Type from Kenya.
  Canthium rubrocostatum Robyns in Notizbl. Bot. Gart. Berlin-Dahlem **10**: 616 (1929). —
  Bullock in Bull. Misc. Inform., Kew **1932**: 373 (1932). —Brenan, Check-list For. Trees
  Shrubs Tang. Terr.: 488 (1949). —Dale & Greenway, Kenya Trees & Shrubs: 432 (1961).
  Canthium melanophengos sensu Bullock in Bull. Misc. Inform., Kew **1932**: 375 (1932),
  pro parte.

Shrub or tree, 6–18(24) m tall. Leaves often drying brownish; blades 3.3–6.8 cm
wide, oblong-elliptic or sometimes broadly elliptic or ovate, acuminate at apex,
obtuse or rounded at base, with midrib conspicuously red; domatia ciliate; petioles

6–10 mm long. Corymbs 20–40-flowered, 2–4 cm across. Calyx glabrous to glabrescent; limb 0.5–0.75 mm long, always shorter than calyx tube. Corolla tube 2–3 mm long. Fruit 10–14 mm wide.

Malawi. N: Misuku Hills, Mugesse Forest, sterile 1.iii.1983, *Dowsett-Lemaire* 668 (K).
Also in Uganda, Kenya, Tanzania and Sudan. In Afromontane evergreen rain forest; 1800–2000 m.

Subsp. **chapmanii** Bridson in Kew Bull. **40**: 702 (1985). —K. Coates Palgrave, Trees Southern Africa, ed. 3, rev.: 16k of appendix (1988). Type: Malawi, Zomba Plateau, east of Chingwe's Hole, *Salubeni & Tawakali* 3312 (BR; K, holotype; MAL; SRGH).
Canthium vulgare sensu Drummond in Kirkia **10**: 276 (1975).
Canthium rubrocostatum sensu A.E. Gonçalves in Garcia de Orta, Sér. Bot. **5**: 188 (1982) non Robyns.

Tree 6–27 m tall. Leaves drying green; blades 2–4.5 cm wide, narrowly elliptic to elliptic, finely acuminate at apex, acute or sometimes obtuse to rounded at base; midrib sometimes orange; domatia glabrous or pubescent; petioles 3–6 mm long. Corymbs 20–30(80)-flowered, 20–35 mm across. Calyx sparsely pubescent to pubescent; limb 0.5–0.75 mm long, always shorter than tube. Corolla tube 2 mm long. Fruit 10–13 mm wide.

Zimbabwe. E: Chipinge Distr., Chirinda Forest, fl. iii.1967, *Goldsmith* 37/67 (K; LISC; SRGH).
Malawi. S: Zomba Distr., Kasonga Village, along Domasi R., fr. 27.vii.1982, *Patel* 971 (K).
Mozambique. N: Serra de Ribáuè, Mepáluè, fl. 28.i.1964, *Torre & Paiva* 10283 (K; LISC). Z: Gurué, no cimo da serra do Gurué, prox. da nascente do rio Malema, fl. 5.i.1968, *Torre & Correia* 16947 (K; LISC). MS: Tsetserra, SE slopes, beneath villa of Carvalho, sterile 7.vi.1971, *Müller & Gordon* 1833 (K; SRGH).
Not known outside the Flora Zambesiaca area. In montane mixed evergreen rainforest; 1200–1800(2000) m.
This taxon would be expected to occur also in Malawi C:. The sterile ecological voucher *Dowsett-Lemaire* ecological series 315 (K) [Kirk Range, Chirobwe Forest Reserve, sterile, 28.xi.1983] may belong to this subspecies.
Subsp. *melanophengos* (Bullock) Bridson, from Uganda and Rwanda, is distinguished by its leaves which become blue-black when dry, and have (6)7–8 pairs of main lateral nerves.

3. **Psydrax richardsiae** Bridson in Kew Bull. **40**: 703, fig. 3 A–H (1985). Type: Zambia, Mbala Distr., Old Sumbawanga Rd., Kawimbe, *Richards* 15226 (K, holotype; PRE).

Tree 3–10 m tall; young branches square in cross section, pubescent; bark dark greyish with fine vertical fissures. Leaf blades 2.2–4 × 1.4–2.6 cm, broadly elliptic to round, rounded or obtuse at apex, obtuse to rounded or sometimes truncate at base, coriaceous, moderately shiny on upper surface, usually recurved at margins, glabrous; lateral nerves in 3–4 main pairs; tertiary nerves reticulate or almost obscure; domatia present as erupted or unerupted blisters, glabrous; petioles 1.5–2.5 mm long; stipules 1.5–2 mm long, triangular with an acuminate decurrent lobe, puberulous outside. Flowers 4-merous, borne in pedunculate 15–30-flowered corymbs; peduncles 5–12 mm long puberulous; pedicels 2–4(8) mm long, puberulous; bracteoles inconspicuous. Calyx tube 1–1.25 mm long, sparsely puberulous; limb 0.5 mm long, repand, ciliate. Corolla creamy-white; tube 3–3.5 mm long, pubescent and with a ring of deflexed hairs at throat; lobes 2.75–3 × 1.5–1.75 mm, ovate, acute and thickened towards the apex. Anthers reflexed. Style 7.5–8.5 mm long; pollen presenter 1 mm long; disk pubescent. Fruit 7 × 10 mm, broadly oblong in outline, bilobed but only slightly indented at the apex. Pyrene 6 × 4 × 4 mm, almost semicircular in outline with a groove extending from point of attachment to the lateral faces, and a small irregular crest from point of attachment to the apex, strongly sculptured.

Zambia. N: Mbala Distr., Kawimbe Rocks, fl. & young fr. 18.vi.1970, *Sanane* 1219 (K). **Malawi**. N: Chisenga, foothills of Mafinga Hills (Mts.), fr. 9.xi.1958, *Robson & Fanshawe* 527 (BM; BR; K; LISC).
Not known outside the Flora Zambesiaca area. In *Brachystegia* woodland or miombo, often on rocks; 1675–1900 m.

4. **Psydrax sp. A** —Bridson in Kew Bull. **40**: 703, fig. 3 J–L (1985).

Tree up to 14 m tall, glabrous; young stems covered with dark grey-black bark; lenticels not conspicuous. Leaves borne towards apex of stems, perhaps deciduous; blades 7–12 × 3.7–6.3 cm, broadly elliptic, oblong-elliptic or sometimes obovate, acute to subacuminate at apex, acute to rounded or sometimes tending to be cordate at base, stiffly papery, drying brown and shiny on upper surface and glaucous beneath; lateral nerves in 6–7 main pairs; tertiary nerves very coarsely reticulate, apparent but not conspicuous; domatia absent, or present as inconspicuous ciliate or glabrous pits; petioles 10–23 mm long, strongly grooved above; stipules 4–5 mm long, triangular, acute or with a short decurrent apiculum, glabrous inside. Inflorescence a pedunculate cyme, borne at leafless nodes; peduncles 20–40 mm long, sparsely puberulous; pedicels up to 8 mm long (in fruiting stage), puberulous; bracts and bracteoles c. 1 mm long, narrowly triangular, puberulous. Calyx limb c. 0.5 mm long, reduced to a 5-dentate rim; corolla not known. Disk pubescent. Fruit 9 × 13–14 mm, strongly bilobed and indented at the apex. Pyrene 8 × 6 × 5 mm, not quite semi-circular in outline, very strongly ridged and rugulose but with a fine shallow apical crest apparent.

Zambia. N: Kawambwa, fr. 5.viii.1958, *Fanshawe* 4660 (K).
Also in Zaire (Dem. Rep. Congo) (Shaba Prov.). In chipya miombo woodland.

5. **Psydrax livida** (Hiern) Bridson in Kew Bull. **40**: 705 (1985). —von Breitenbach in J. Dendrol. **5**: 88, figs. (1985). —K. Coates Palgrave, Trees Southern Africa, ed. 3, rev.: 16j of appendix (1988). —Bridson in F.T.E.A., Rubiaceae: 898 (1991). —Beentje, Kenya Trees, Shrubs Lianas: 538 (1994). TAB. **47**/C2. Type: Mozambique, Morrumbala (Moramballa), *Kirk* s.n. (K, holotype).
  Canthium lividum Hiern in F.T.A. **3**: 144 (1877).
  Plectronia livida (Hiern) K. Schum. in Engler, Pflanzenw. Ost-Afrikas **C**: 386 (1895). —Eyles in Trans. Roy. Soc. South Africa **5**: 493 (1916).
  Plectronia syringodora K. Schum. in Engler, Pflanzenw. Ost-Afrikas **C**: 386 (1895). Types from Tanzania.
  Canthium huillense Hiern, Cat. Afr. Pl. Welw. **1**: 476 (1898). —Warburg, Kunene-Samb.-Exped. Baum: 388 (1903). —Bullock in Bull. Misc. Inform., Kew **1932**: 370 (1932). —Brenan, Check-list For. Trees Shrubs Tang. Terr.: 487 (1949). —Miller in J. S. African Bot. **18**: 83 (1952). —Dale & Greenway, Kenya Trees & Shrubs: 429 (1961). —F. White, F.F.N.R.: 404 (1962). —Palmer & Pitman, Trees Southern Africa **3**: 2093 (1973). —van Wyk, Trees Kruger Nat. Park **2**: 557 (1974). —Drummond in Kirkia **10**: 276 (1975). —A.E. Gonçalves in Garcia de Orta, Sér. Bot. **5**: 187 (1982). —K. Coates Palgrave, Trees Southern Africa, ed. 3, rev.: 881, pl. 303 (1988). Types from Angola.
  Plectronia huillensis (Hiern) K. Schum. in Just's Bot. Jahresber., 26th year, part 1: 393 (1900). Type as above.
  Plectronia heliotropiodora K. Schum. & K. Krause in Bot. Jahrb. Syst. **39**: 540 (1907). Type from Tanzania.
  Plectronia sp. of Eyles in Trans. Roy. Soc. South Africa **5**: 494 (1916), based on *Allen* 250 (Zimbabwe, Gwaai Forest).
  Plectronia junodii Burtt Davy in Bull. Misc. Inform., Kew **1921**: 192 (1921). Type from South Africa.
  Canthium clityophilum Bullock in Bull. Misc. Inform., Kew **1932**: 382 (1932). —Brenan, Check-list For. Trees Shrubs Tang. Terr.: 485 (1949). Type from Tanzania.
  Canthium syringodorum (K. Schum.) Bullock in Bull. Misc. Inform., Kew **1932**: 373 (1932). —Brenan, Check-list For. Trees Shrubs Tang. Terr.: 488 (1949).
  Canthium junodii (Burtt Davy) Burtt Davy in Bull. Misc. Inform., Kew **1935**: 568 (1935).
  Plectronia wildii Suess. in Trans. Rhodesia Sci. Assoc. **43**: 59 (1951). Type: Zimbabwe, Marondera (Marandellas), *Dehn* 590 (M; SRGH, holotype).
  Canthium sp. of Miller in J. S. African Bot. **18**: 83 (1952), based on *Miller* B543 & B577 (not seen).
  Canthium gymnosporioides Launert in Mitt. Bot. Staatssamml. München **16**: 314 (1957). Type: Caprivi Strip, Okavango, Pupa (Popa) Falls, *Volk* s.n. (M, holotype).
  Canthium wildii (Suess.) Codd in Kirkia **1**: 109 (1961). —A.E. Gonçalves in Garcia de Orta, Sér. Bot. **5**: 188 (1982). Type as above.

Shrub or small tree 0.8–8 m tall; green twigs glabrescent to densely pubescent; older stems at first covered with pale tissue-like bark which flakes off to reveal a darker underlayer. Leaves restricted to young stems, or occasionally on older stems, not always fully mature at time of flowering; blades 1.5–13 × 1–7 cm, narrowly ovate to ovate or occasionally broadly ovate, acute to subacuminate or sometimes acuminate or obtuse to rounded at apex, obtuse to rounded or sometimes acute at the base, herbaceous, dull to slightly shiny on upper surface, glabrous to densely pubescent on both surfaces; lateral nerves in 4–5 main pairs; tertiary nerves apparent; domatia present as ciliate membranous pockets; petioles 2–5(10) mm long; stipules 4–7 mm long, triangular-acuminate with lobe eventually decurrent. Flowers 4-merous, borne on 6–70-flowered pedunculate compact corymbs; peduncle 3–15 mm long, pubescent; inflorescence branches proportionately short, not spreading; pedicels 3–10(15) mm long, pubescent; bracts and bracteoles inconspicuous. Calyx tube 1 mm long, broader than long, pubescent or glabrous save for a few hairs towards the base; limb 0.5 mm long, irregularly dentate, ciliate. Corolla greenish-white (or yellow); tube 1.25–3 mm long, pubescent at throat but without a well defined ring of deflexed hairs; lobes 1.25–2 × 1–1.5 mm long, oblong, obtuse to rounded. Anthers reflexed. Style up to 6 mm long; pollen presenter 1 × 0.5 mm long, obscurely ribbed; disk glabrous. Fruit 5–6 × 7.5– 8 mm, broadly oblong in outline, not or scarcely indented at the apex, black when mature. Pyrene 5–6 × 3.5 mm, ± ellipsoid, rugulose.

Botswana. SE: Aedume Park at Gaborone Dam, fl. 22.x.1978, *O.J. Hansen* 3504 (K). **Zambia**. B: Kalabo Distr., Kalabo–Sihole, mile 5, fr. 16.ii.1952, *White* 2087 (FHO; K). N: Kasama Distr., Misamfu Experimental Station, fr. 14.ii.1961, *I. Coxe* 191 (K; LISC; SRGH). W: Ndola, fl. 18.xii.1954, *Fanshawe* 1729 (K). C: 13 km east of Lusaka on Leopard's Hill Road, fl. 10.xii.1959, *Angus* 2089 (FHO; K). E: Chadiza, fl. 28.xi.1958, *Robson* 763 (BM; K; LISC). S: Namwala Distr., Kafue National Park, Ngoma, young fr. 16.i.1963, *B.L. Mitchell* 17/17 (K; NDO). **Zimbabwe**. N: Hurungwe (Urungwe) Reserve, fr. 27.i.1953, *Lovemore* 359 (K; LISC; SRGH). W: Gwanda Distr., Runde (Lundi) River, near Hotel, fl. 13.xii.1956, *Davies* 2378 (K; SRGH). C: Charter, fl. 27.xii.1926, *Eyles* 4611 (K). E: Mutare (Umtali), 10 miles south, fl. 21.xi.1948, *Chase* 974 (K; SRGH). S: Buhera, in grounds of club near swimming pool, fl. 19.x.1962, *Strang* in *CAH* 11012 (K; SRGH). **Malawi**. N: Nkhata Bay Distr., 4 km west of Bandawe Point, south of Chintheche, fr. 6.iii.1982, *Brummitt, Polhill & Banda* 16334 (K). S: Kasupe Distr., Chikala Hills, west side, fr. 17.ii.1975, *Brummitt, Banda, Seyani & Patel* 14357 (K). **Mozambique**. N: Marrupa, picada para Mucuaiaia andados c. 15 km proximo de Momopsus, fr. 23.ii.1981, *Nuvunga* 691 (K; LMU). Z: Morrumbala (Moramballa), fl. 18.i.1863, *Kirk* s.n. (K). T: Serra da Estima para a Meroeira, fl. 3.xii.1973, *Macêdo* 5404 (LISC; LMU). MS: Gorongosa National Park, west side of Rift Valley on Gorongosa (Vila Paiva) road, fr. xi.1971, *Tinley* 2228 (K; SRGH). GI: Inhambane Prov., Govuro andados 27 km da Povoação Banamana, para Maxaila (Machaila), fr. 26.iii.1974, *Correia & Marques* 4202 (LMU).

Also in Burundi, Kenya, Tanzania, Zaire (Dem. Rep. Congo), Angola, Namibia and South Africa (North Prov., North West Prov., Gauteng and Mpumalanga). In miombo woodlands and mixed deciduous woodlands, often below escarpments, on hillsides and rocky outcrops, also in thickets in miombo woodland on sandy soil, and in mixed evergreen forest and riverine vegetation; 300–1600 m.

6. **Psydrax whitei** Bridson in Kew Bull. **40**: 706, fig. 4 (1985); in F.T.E.A., Rubiaceae: 888, fig. 160 (1991). TAB. **70**. Type: Malawi, Nyika Plateau, *Richards* 22527 (K, holotype).
   Canthium sp. 2 of F. White, F.F.N.R.: 404 (1962).
   Canthium whitei (Bridson) F. White in Bull. Jard. Bot. Belg. **60**: 109 (1990).

Shrub or small tree 1.5–9 m tall (also described as a dwarf shrub 0.3 m tall); very young stems pubescent; bark greyish. Leaves restricted to green stems, or sometimes on mature stems; blades 1.3–3 × 0.9–3.5 cm, ovate or sometimes elliptic to round, obtuse or occasionally rounded at apex, rounded to truncate or occasionally subcordate at base, herbaceous to slightly coriaceous, shiny and glabrous on upper surface, sparsely pubescent to pubescent on midrib beneath; lateral nerves in 3–4 main pairs, impressed above; tertiary nerves apparent; domatia present as ciliate membranous pockets; petioles not exceeding 2 mm long; stipules 3–4 mm long, triangular-acuminate, keeled when mature. Flowers 4-merous, borne on (5)12–70-flowered, compact, pedunculate corymbs; peduncles 3–15 mm long, pubescent; inflorescence branches proportionately short, not spreading; pedicels 2–13 mm long, pubescent; bracteoles inconspicuous. Calyx tube 1 mm long, broader than long, glabrous or with a few hairs towards the base; limb 0.5–0.75 mm long, irregularly dentate. Corolla greenish-cream; tube 1.5–2 mm long, pubescent at throat but lacking a well defined ring of deflexed hairs; lobes 2.5–3 × 1.5 mm, oblong, obtuse. Anthers reflexed. Style 6–8 mm long; pollen presenter 1 × 0.5 mm

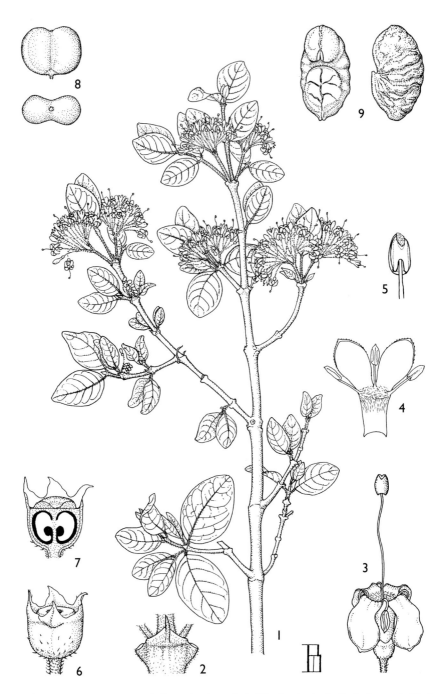

Tab. 70. PSYDRAX WHITEI. 1, flowering branch (× ⅔); 2, stipule (× 4); 3, flower (× 6); 4, section through corolla (× 6); 5, longitudinal section through pollen presenter (× 9); 6, calyx (× 9); 7, longitudinal section through ovary (× 9), 1–7 from *Richards* 22527; 8, fruit, 2 views (× 1 ½); 9, pyrene, 2 views (× 3), 8 & 9 from *Pawek* 13763. Drawn by Diane Bridson. From F.T.E.A.

long, obscurely ribbed; disk glabrous. Fruit 7 × 7 × 4 mm, broadly oblong in outline, not indented at the apex, black when mature. Pyrenes 7 × 4 × 3 mm, ± ellipsoid with ventral face flattened, rugulose.

Zambia. N: Isoka Distr., Mafinga Hills (Mts.), near Chisenga, corolla fallen 21.xi.1952, *Angus* 818 (FHO; K). E: Nyika, fr. 29.xii.1962, *Fanshawe* 7299 (K). **Malawi**. N: Rumphi Distr., Nyika Plateau, Chilinda, fl. xi.1959, *Adlard* 300 (K; SRGH).

Also in Tanzania. In evergreen rain forest and forest margins, also on rocky outcrops in submontane grassland; 2100–2300 m.

7. **Psydrax martinii** (Dunkley) Bridson in Kew Bull. **40**: 707 (1985). —K. Coates Palgrave, Trees Southern Africa, ed. 3, rev.: 16k of appendix (1988). Type: Zambia, S Province, Bombwe, *Martin* 135/31 (FHO; K, holotype).
    Canthium martinii Dunkley in Bull. Misc. Inform., Kew **1934**: 187 (1934). —F. White, F.F.N.R.: 402 (1962). —Drummond in Kirkia **10**: 276 (1975). —K. Coates Palgrave, Trees Southern Africa, ed. 3, rev.: 883 (1988).

Shrub, slender tree or climber 1–3.5 m tall, with patent or almost patent lateral branches; young stems glabrous or pubescent; bark mid- to light grey. Leaves deciduous, restricted to apex of lateral branches or sometimes clustered on very reduced lateral spurs; blades 3–6 × 1.2–3 cm, elliptic, obtuse or sometimes subacuminate at apex, acute at base, membranous, glabrous to sparsely pubescent on both surfaces; lateral nerves in 3–4 main pairs; tertiary nerves obscure to apparent, rather coarsely reticulate; domatia pocket-like, ciliate; petioles 3–8 mm long, glabrous to pubescent; stipules 1–2 mm long, triangular. Flowers 4-merous, borne in 10–40-flowered sessile umbels; pedicels 5–8 mm long, glabrous to pubescent. Calyx tube 1 mm long, glabrous to densely pubescent; limb 1.75 mm long, repand, ciliate. Corolla white; tube 2.75–3 mm long, somewhat pubescent inside but lacking a well defined ring of deflexed hairs; lobes 2.5–3 × 1.25–1.5 mm, oblong, obtuse. Anthers reflexed. Style 6.5–8 mm long; pollen presenter 1 × 0.75 mm long, scarcely ribbed; disk glabrous. Fruit 7 × 10 × 3 mm, broadly oblong in outline, scarcely indented at apex, black when mature. Pyrene 7 × 4 × 3 mm, rugulose.

Zambia. E: Great East Road, between Hofmeyr turn-off and Kachalola, fl. 12.xii.1958, *Robson* 911 (BM; K; LISC; PRE). S: Kalomo, fr. 16.ii.1965, *Fanshawe* 9097 (K). **Zimbabwe**. N: Hurungwe Distr., Mana Pools, Nyamacheru Pan, sterile 2.v.1970, *P.Guy* 785 (SRGH). **Malawi**. S: Liwonde National Park, Likwenu River area, fl. 7.xi.1988, *J.Phiri* 185 (K).

Not known outside the Flora Zambesiaca area. In low altitude riverine thicket and wooded grassland on ant hills, in *Brachystegia–Bauhinia* woodland and *Baikiaea* forest; 500–650 m.

Sterile sheets of a taxon that may prove to be *P. martinii* have been collected from Zimbabwe C, E, S and Mozambique MS but from the present information formal recognition is considered unwise.

8. **Psydrax locuples** (K. Schum.) Bridson in Kew Bull. **40**: 708 (1985). —von Breitenbach in J. Dendrol. **5**: 88, fig. on p. 90 (1985). —K. Coates Palgrave, Trees Southern Africa, ed. 3, rev.: 16k of appendix (1988). —Pooley, Trees of Natal, Zululand & Transkei: 476, figs. (1993). Type: Mozambique, Delagoa Bay, *Monteiro* 52 (B†, holotype; K).
    Plectronia locuples K. Schum. in Bot. Jahrb. Syst. **28**: 75 (1899). —Schinz in Mém. Herb. Boissier, No. 10: 68 (1900).
    Canthium locuples (K. Schum.) Codd in Kirkia **1**: 108 (1961), pro parte, excl. *P. fragrantissima* K. Schum. —Cavaco in Portugaliae Acta Biol., Sér. B, Sist. **11**: 219 (1972). — Ross, Fl. Natal: 335 (1973). —Palmer & Pitman, Trees Southern Africa **3**: 2103, fig. on 2104 (1973), pro parte. —Gibson, Wild Fls. Natal: 7, pl. 102 (1975). —Moll, Trees of Natal: 259 (1981). —K. Coates Palgrave, Trees Southern Africa, ed. 3, rev.: 883 & page 16k of appendix (1988).

Shrub 2.75–5 m tall; very young stems puberulous or pubescent; bark fawn-coloured or almost white. Leaf blades 2–4.7 × 0.9–2.7 cm, elliptic or rarely oblong-elliptic or broadly elliptic, acute to obtuse at apex, acute at base, coriaceous, shiny on upper surface, glabrous; lateral nerves in 2–3 main pairs; tertiary nerves obscure; domatia present as glabrous or ciliate pockets; petioles 1–3 mm long, glabrous or puberulous; stipules 2–3 mm long, truncate to triangular at their base with a linear decurrent lobe, often puberulous outside. Flowers 4-merous, borne in 6–30-flowered compact pedunculate corymbs; peduncles 3–9 mm long, puberulous;

pedicels 2–6 mm long, puberulous; bracteoles inconspicuous. Calyx tube 1 mm
long, broader than long, glabrous; limb 0.5 mm long, irregularly dentate, ciliate.
Corolla white or greenish-white; tube 1.5–2.5 mm long, pubescent at throat, deflexed
hairs not restricted to a definite ring; lobes 2–2.5 × 1.5 mm, oblong, acute. Anthers
reflexed. Style 4.5–7 mm long; pollen presenter 1 × 0.5 mm long; disk glabrous.
Fruit 4.5–6.5 × 6–9 mm, broadly oblong in outline, not or scarcely indented at apex.
Pyrenes 4.5–5.5 × 3–4.5 × 2.5 mm, ± ellipsoid with ventral face flattened, rugulose.

Mozambique. GI: Gaza Prov., Macia, S. Martinho do Bilene, fl. 8.x.1958, *Barbosa* 8337 (K).
M: between Costa do Sol and Marracuene, Muntanhane, young fr. 13.xi.1960, *Balsinhas* 247
(BM; COI; K; PRE).
    Also in South Africa (KwaZulu-Natal). On coastal plains or dunes, in woodlands and dense
scrub, often on sandy soil or on sandstone outcrops; 0–200 m.
    One fruiting specimen from Xai-Xai (*de Koning* 7801) has the leaves narrower than typical
for this species. A closely related species (*P. faulkneri* Bridson) from Tanzania and Kenya has
juvenile leaves which are small, resembling those of *Buxus sempervirens* L., and quite unlike the
adult leaves. It is possible that this phenomenon may also occur in *P. locuples*, but field
observations and correlated specimens are still required. One gathering from Mozambique,
Inhaca Island (*Mauve & Verdoorn* 86 (PRE)) and several sheets from KwaZulu-Natal could well
represent juvenile foliage forms of this species.

9. **Psydrax obovata** (Klotzsch ex Eckl. & Zeyh.) Bridson in Kew Bull. **40**: 708, fig. 2B & map 2
    (1985). —von Breitenbach in J. Dendrol. **5**: 88, figs. (1985). —K. Coates Palgrave, Trees
    Southern Africa, ed. 3, rev.: 16k of appendix (1988). —Pooley, Trees of Natal, Zululand &
    Transkei: 476, figs. (1993), excl. *P. fragrantissima* in syn. Type from South Africa.
        Canthium obovatum Klotzsch ex Eckl. & Zeyh., Enum. Pl. Afr.: 361 (1837). —Sonder in
    F.C. **3**: 16 (1865). —J.G. Garcia in Mem. Junta Invest. Ultramar **6** (sér. 2): 38 (1959)
    [Contrib. Conhec. Fl. Moçamb. IV (1959)]. —Ross, Fl. Natal: 335 (1973). —Palmer &
    Pitman, Trees Southern Africa **3**: 2101, figs. on 2101–2103 (1973). —Moll, Trees of
    Natal: 260 & 274 (1981). —K. Coates Palgrave, Trees Southern Africa, ed. 3, rev.: 884
    (1988), pro parte.

Shrub or tree 2–15 m tall; young branches square in section, glabrous or
puberulous; bark whitish-grey. Leaf blades 2–7 × 1.2–3.5 cm, obovate to narrowly
obovate or round, or elliptic, rounded or occasionally obtuse at apex, acute or
somewhat attenuated at base, coriaceous, shiny above, glabrous, the margins usually
recurved; lateral nerves in 1–4 main pairs; tertiary nerves obscure or apparent;
domatia pit-like or pocket-like, glabrous or sometimes pubescent; petioles 1–5 mm
long, glabrous or puberulous; stipules 2–4 mm long, truncate at their base with a
linear decurrent lobe, sometimes puberulous outside. Flowers 4–5-merous, borne
in 2–70-flowered pedunculate corymbs; peduncles 2–25 mm long, glabrescent to
puberulous; pedicels 0.5–9 mm long, glabrous to glabrescent; bracteoles
inconspicuous. Calyx tube 1–1.5 mm long, glabrous; limb 0.5–0.75 mm long,
repand. Corolla cream or white; tube 2–3.25 mm long, with a ring of deflexed hairs
at the throat; lobes 2.75–4.5 × 1.5–2 mm, obtuse to acute. Anthers reflexed. Style
5.5–8 mm long; pollen presenter 1.5–2 mm long; disk glabrous. Fruit 5–9 × 7–8
mm, longer than wide, ± square in lateral view or wider than long, scarcely
indented.   Pyrenes 5–7 × 3–4 mm, ellipsoid to flattened broadly ellipsoid,
moderately to strongly rugulose.

Leaves obovate or sometimes narrowly obovate or round, with 1–3 main pairs of arching lateral
    nerves apparent, and with tertiary nerves usually obscure; domatia glabrous; inflorescence
    often large, 8–70-flowered; peduncle 2–25 mm long; fruit longer than wide or slightly wider
    than long · subsp. *obovata*
Leaves mostly elliptic, usually with 3–4 main pairs of scarcely arching lateral nerves, and with
    tertiary nerves often apparent beneath; domatia pubescent; inflorescence smaller, 2–25-
    flowered; peduncle 2–10 mm long; fruit distinctly wider than long · · · · · · subsp. *elliptica*

Subsp. **obovata**
        Canthium pyrifolium Klotzsch ex Eckl. & Zeyh., Enum. Pl. Afr.: 361 (1837). Type from
    South Africa.
        Mitrastigma lucidum Harv. in London J. Bot. **1**: 20 (1842); Thes. Cap. **1**: 14, pl. 22 (1859).
    Type from South Africa.

Phallaria lucida (Harv.) Hochst. in Flora **25**: 238 (1842).
Canthium obovatum var. *pyrifolium* (Klotzsch ex Eckl. & Zeyh.) Sond. in F.C. **3**: 16 (1865).
Plectronia obovata (Klotzsch ex Eckl. & Zeyh.) Sim, For. Fl. Col. Cape Good Hope: 241, pl. 90 (1907).

Mozambique. GI: Gaza Prov., 30 km west of Limpopo R. mouth, bud & fr. 1.iii.1928, *Earthy* 48 (PRE). M: Inhaca Island, fl. 7.iii.1958, *Mogg* 31563 (BM; K).
Also in South Africa (KwaZulu-Natal and Eastern Cape Prov.). Coastal and inland forest and forest margins, also in woodland and dune scrub, on basic or acid soils; 0–20 m.

Subsp. **elliptica** Bridson in Kew Bull. **40**: 711, map 2 (1985). —von Breitenbach in J. Dendrol. **5**: 90 (1985). —K. Coates Palgrave, Trees Southern Africa, ed. 3, rev.: 16k of appendix (1988). Type from South Africa (North Prov. and Mpumalanga).
Canthium obovatum sensu van Wyk, Trees Kruger Nat. Park **2**: 557 (1974). —sensu Drummond in Kirkia **10**: 276 (1975). —sensu K. Coates Palgrave, Trees Southern Africa, ed. 3, rev.: 884 (1988), pro parte, non sensu stricto.

Zimbabwe. E: Mutare Distr., near Watsomba Rd./Stapleford Rd. junction, fl. 27.i.1979, *Müller* 3590 (K; PRE; SRGH). **Mozambique**. MS: Tsetserra, SE slopes, beneath villa of Carvalho, sterile 7.vi.1971, *Müller & Gordon* 1824 (LISC; SRGH).
Also in South Africa (North Prov. and Mpumalanga). Mixed evergreen rain forest on submontane slopes and granite outcrop vegetation; 1600–1980 m.
Ngoni 1520 from southern Zimbabwe, Gonarezhou Game Reserve, near Shabani Fly Gate, may possibly belong here but the leaves (15 × 9 mm) are much smaller than is typical for this subspecies. Additional collections are needed to clarify this point.

10. **Psydrax fragrantissima** (K. Schum.) Bridson in Kew Bull. **40**: 711 (1985). —K. Coates Palgrave, Trees Southern Africa, ed. 3, rev.: 16k of appendix (1988). Type: Mozambique, Maputo (Lourenço Marques), *Schlechter* 11635 (B†, holotype; BR; COI; K; L; LE; WAG).
Plectronia fragrantissima K. Schum. in Bot. Jahrb. Syst. **28**: 75 (1899). —Schinz in Mém. Herb. Boissier, No. 10: 68 (1900).
Canthium locuples sensu Codd in Kirkia **1**: 108 (1961), pro parte, non K. Schum.
Canthium fragrantissimum (K. Schum.) Cavaco in Portugaliae Acta Biol., Sér. B, Sist. **11**: 220 (1972).
Psydrax obovata sensu Pooley, Trees of Natal, Zululand & Transkei: 476, figs. (1993), pro parte.

Shrub or small tree 2.5–6 m tall; young branches square in section, sparsely puberulous to puberulous; bark pale grey. Leaf blades 1.3–3 × 0.9–2.1 cm, obovate, rounded at apex, acute to somewhat attenuate at base, coriaceous, shiny on upper surface, glabrous, margins recurved; lateral nerves in 2–3 main pairs but sometimes only 1 pair clearly apparent; tertiary nerves obscure; domatia pit-like, glabrous or sometimes absent; petioles 2(3) mm long, puberulous; stipules 1–2.5 mm long, truncate at their base with a linear decurrent lobe, puberulous outside. Flowers 4-merous, borne in 10–35-flowered pedunculate corymbs; peduncles 8–22 mm long, puberulous or sparsely pubescent; pedicels 1–4 mm long, glabrescent to sparsely pubescent; bracteoles inconspicuous. Calyx tube 1 mm long, glabrous to glabrescent; limb 0.25 mm long, repand or somewhat dentate. Corolla cream-coloured; tube 1.5–2.5 mm long, with a ring of deflexed hairs at the throat; lobes 2.5–3 × 1–1.25 mm, oblong, acute. Anthers reflexed. Style c. 6 mm long; pollen presenter 0.75–1.25 mm long; disk glabrous. Immature fruit 6.5 × 6 mm, ± oblong in lateral view.

Mozambique. M: Marracuene to Maputo (Lourenço Marques), fr. i.iv.1983, *Groenendijk & de Koning* 292 (K; LMU).
Also in northern KwaZulu-Natal. Principally with *Afzelia quanzensis*, *Syzygium*, *Mimusops caffra* and *Dialium schlechteri*; 0–50 m.
This species is very close to *Psydrax obovata*, differing in the smaller size of the leaves and flowers. More gatherings may indicate that varietal rank within *Psydrax obovata* would be more appropriate, or even that it cannot be kept separate at any level. However, as mature fruit have not yet been seen, the present status is maintained for the time being.

11. **Psydrax mutimushii** Bridson in Kew Bull. **40**: 713, fig. 5 A–F (1985). Type: Zambia, Mwinilunga Distr., Matonchi Farm Road, *Mutimushi* 3490 (K, holotype).
Canthium captum sensu F. White, F.F.N.R.: 404 (1962), non Bullock.

Shrub or tree 2.5–9 m tall, glabrous; young branches terete or sometimes square in section; bark dark greyish-brown. Leaf blades 4–12 × 2–6 cm, elliptic or more often broadly elliptic, sometimes ovate, acuminate at apex, acute and sometimes attenuate at base, coriaceous, distinctly shiny on upper surface, moderately shiny beneath, somewhat discolorous (when dry); midrib somewhat prominent on both surfaces, drying pale; lateral nerves 5(6) in main pairs, apparent but not conspicuous; tertiary nerves obscure; domatia absent, or present as glabrous pits; petioles 4–8 mm long; stipules 5–7 mm long, triangular, somewhat keeled. Flowers 4–5-merous, borne in shortly pedunculate 20–60-flowered subumbellate cymes; peduncle 2–7 mm long; pedicels 3–8 mm long, sparsely pubescent to pubescent; bracts and bracteoles inconspicuous. Calyx tube 1.25 mm long, glabrous; limb reduced to a dentate rim, ciliate. Corolla white; tube 3–3.5 mm long, with a ring of deflexed hairs at the throat; lobes 3–3.25 × 1.25–1.5 mm long, oblong, acute and thickened at the apex. Anthers reflexed. Style 6–8.5 mm long; pollen presenter 1.5 mm long; disk glabrous. Immature fruit 8 × 5.5 mm; pedicels accrescent, up to 10 mm long.

Subsp. **mutimushii**

Zambia. B: Zambezi Distr., fl. & young fr. vii.1933, *Trapnell* 1296 (K). N: Kawambwa Distr., Mukabi P.F.A., fl. bud 11.vi.1962, *Lawton* 903 (FHO; K). W: Mwinilunga Distr., Mwanamitowa River, young fr. 22.viii.1930, *Milne-Redhead* 943 (BR; K).

Also known from Zaire (Dem. Rep. Congo) (Shaba Prov.) and Angola. In dry evergreen *Cryptosepalum* forest and thickets (mavunda) on Kalahari Sand, dry evergreen thicket (mateshi), and in miombo and mixed deciduous woodlands, also in swamp forest (mushitu) and riverine thicket vegetation; 1050–1675 m.

Subsp. *wagemansii* Bridson occurs in Kinshasa Province of Zaire (Dem. Rep. Congo), and has shorter stipules and smaller fruit.

12. **Psydrax schimperiana** (A. Rich.) Bridson in Kew Bull. **40**: 714 (1985). —Bridson & Troupin in Fl. Rwanda **3**: notes nomenclaturales, corrigenda for page 150, fig. 44.3 (1985). — Bridson in F.T.E.A., Rubiaceae: 901 (1991). Type from Ethiopia.

    Phallaria schimperi Hochst. in Flora **24**, Intell. 1(2): 27 (1841), nomen nudum.

    Canthium schimperianum A. Rich., Tent. Fl. Abyss. **1**: 350 (1847). —Hiern in F.T.A. **3**: 135 (1877). —S. Moore in J. Linn. Soc., Bot. **37**: 161 (1905). —Robyns in Notizbl. Bot. Gart. Berlin-Dahlem **10**: 617 (1929). —Bullock in Bull. Misc. Inform., Kew **1932**: 385 (1932), pro parte. —Brenan, Check-list For. Trees Shrubs Tang. Terr.: 485 (1949), pro parte. —Dale & Greenway, Kenya Trees & Shrubs: 432, fig. 83 (1961). —Bridson & Troupin in Fl. Pl. Lign. Rwanda: 550, fig. 184.2 (1982); in Fl. Rwanda **3**: 150, fig. 44.3 (1985).

    Plectronia schimperiana (A. Rich.) Vatke in Linnaea **40**: 195 (1876). —K. Schumann in Engler & Prantl, Nat. Pflanzenfam. IV, **4**: 92, fig. 33 E–F (1891). —Engler, Hochgebirgsfl. Afrika: 399 (1892).

    Plectronia nitens sensu K. Schum. in Engler, Pflanzenw. Ost-Afrikas **C**: 385 (1895), quoad *Holst* 8868, non Hiern.

    Plectronia lamprophylla sensu K. Schum. in Bot. Jahrb. Syst. **34**: 335 (1904), pro parte, quoad *P. nitens* sensu auctt., non K. Schum.

    Canthium myrtifolium S. Moore in J. Bot. **45**: 266 (1907). Type from Uganda.

    Plectronia angiensis De Wild., Pl. Bequaert. **3**: 176 (1925). Type from Zaire (Dem. Rep. Congo).

    Canthium euryoides sensu Bullock in Bull. Misc. Inform., Kew **1932**: 384 (1932), pro parte (all specimens cited except the type of *C. nitens* Hiern), non Bullock ex Hutch. & Dalz.

Shrub or tree 2–10 m tall; young branches square in section, often slightly winged on the angles, usually glabrous or rarely puberulous; bark fawn-coloured or greyish. Leaf blades 3–10.5 × 1.3–5 cm, narrowly to broadly elliptic or less often obovate, acute or more often gradually acuminate at apex, acute or occasionally obtuse at base, coriaceous, very shiny on upper surface, glabrous; lateral nerves in (4)5 main pairs; tertiary nerves obscure, or if apparent then coarsely reticulate; domatia absent or inconspicuous; petioles 2–3 mm long; stipules 3–7(10) mm long, triangular at base, terminating in a linear keeled decurrent lobe. Flowers 4–5-merous, borne in sessile to subsessile, 7–50-flowered umbel-like cymes; peduncle if present not exceeding 3 mm long; pedicels 2–7 mm long, pubescent; bracteoles inconspicuous. Calyx tube 1–2 mm long, glabrous or puberulous; limb reduced to a shortly toothed rim. Corolla whitish; tube 2–2.5 mm long, with a ring of deflexed hairs above the mid-point inside; lobes 3–3.5 × 1.5 mm, oblong-lanceolate, acute. Anthers reflexed, narrowly oblong. Style 4–6 mm long; pollen presenter 1.25–1.5 mm long; disk glabrous. Fruit 5–6.5 × 6–7.5 mm, scarcely indented or slightly prominent at the apex, black when mature. Pyrene 6 × 4–4.5 mm, almost semi-circular in outline, with a lateral groove beneath the point of attachment, rugulose to bullate.

Subsp. **schimperiana** —Bridson in F.T.E.A., Rubiaceae: 902 (1991). —Beentje, Kenya Trees, Shrubs Lianas: 539 (1994).

Pedicels 4–7 mm long; wing on young stems moderately apparent.

Zambia. W: Kitwe, young fr. 27.ii.1955, *Fanshawe* 2099 (K). **Malawi.** N: Livingstonia, Manchewe Falls, fl. bud 16.xii.1982, *Dowsett-Lemaire* 543 (K). S: Mt. Mulanje, well down the Likhubula (Likabula) Valley, fl. 25.xi.1988, *Chapman* 8966A (K; MO).

Also in Ethiopia, ? Somalia, Rwanda, Burundi, Uganda, Kenya, Tanzania, Zaire (Dem. Rep. Congo) and the Yemen. In evergreen rain forest, riverine forest and swamp forest (mushitu), and in *Brachystegia* woodland bordering evergreen forest; 860–1100 m.

This species has not been recorded from southern Tanzania or southern Zaire (Dem. Rep. Congo) and the discontinuity of distribution raises some slight doubt that the few specimens recorded from the Flora Zambesiaca area are correctly placed. However, no significant morphological differences have yet been found between these and common northern element. The West African specimens of *P. schimperiana* are placed in subsp. *occidentalis* Bridson, which differs in having shorter pedicels (2–4 mm long) and more prominent wings on the stem angles.

13. **Psydrax kaessneri** (S. Moore) Bridson in Kew Bull. **40**: 719 (1985); in F.T.E.A., Rubiaceae: 904 (1991). —Beentje, Kenya Trees, Shrubs Lianas: 538 (1994). Type from Kenya.
    Canthium kaessneri S. Moore in J. Bot. **43**: 351 (1905). —Bullock in Bull. Misc. Inform., Kew **1932**: 382, fig. 3 (1932). —Brenan, Check-list For. Trees Shrubs Tang. Terr.: 485 (1949). —Dale & Greenway, Kenya Trees & Shrubs: 492 (1961).
    Plectronia longistaminea K. Schum. & K. Krause in Bot. Jahrb. Syst. **39**: 542 (1907). Type is same specimen as above.

Shrub to 3 m tall, often scandent, usually with short perpendicular lateral branches, glabrous; bark pale grey, often with tissue-like flakes peeling off the young stems. Leaves not restricted to apices of branches; blades 3–8 × 1.5–6 cm, broadly elliptic to round or sometimes obovate, rounded or sometimes obtuse at apex, acute or occasionally obtuse at base, subcoriaceous to coriaceous, moderately shiny on upper surface; lateral nerves in 3–4 main pairs; tertiary nerves coarsely reticulate, prominent on both surfaces; domatia present as small punctured blisters; petioles 2–4 mm long, rather stout; stipules 4–5 mm long with a short truncate to triangular base bearing a linear keeled lobe. Flowers 5-merous, borne in subsessile 5–30-flowered, subumbelliform cymes; peduncles up to 2 mm long; pedicels 2–8 mm long, glabrous; bracteoles inconspicuous. Calyx tube 1.25 mm long, glabrous; limb reduced to a dentate rim, ciliate. Corolla whitish; tube 4–5 mm long, with a ring of deflexed hairs above the mid-point inside; lobes 6 × 2.5 mm, oblong, acute or obtuse. Anthers reflexed. Style 9 mm long; pollen presenter 2–2.5 mm long; disk glabrous. Fruit 7–9 × 9–11 mm, prominent at the apex, black; pedicels accrescent. Pyrene almost semicircular in outline, acute at apex, rounded at base, pale fawn-coloured, rugulose.

Mozambique. N: Messalo Rio (M'salu), fl. ii.1912, *C.E.F. Allen* 119 (K).
Also in Kenya and Tanzania. Riverine vegetation.

14. **Psydrax moggii** Bridson in Kew Bull. **40**: 719, figs. 2 D & 7 C–L (1985). —K. Coates Palgrave, Trees Southern Africa, ed. 3, rev.: 16k of appendix (1988). Type: Mozambique, Inhambane, Ponta Barra Falsa, *Mogg* 28962 (K, holotype; LISC).

Semi-scandent shrub or tree 2–5 m tall, sometimes said to be deciduous, usually with short horizontal (or slightly reflexed when young) lateral branches, glabrous; bark pale grey, often with tissue-like flakes peeling off younger stems. Leaves often restricted to apices of branches; blades 2.7–6.5 × 1 × 4.5 cm, narrowly elliptic to round or obovate, rounded or sometimes obtuse at apex, acute or tending to be attenuate at the base, frequently recurved, eventually coriaceous, very shiny on upper surface;

lateral nerves in 3–4 main pairs; tertiary nerves moderately coarsely reticulate, prominent on both surfaces; domatia present as small glabrous pockets; petioles 4–7 mm long; stipules 1.5–3 mm long, shortly truncate-triangular at base, bearing a linear somewhat keeled lobe.   Flowers 5-merous, borne in subsessile 10–20-flowered subumbelliform cymes; peduncles 1–3 mm long; pedicels 4–12 mm long, glabrous; bracteoles inconspicuous. Calyx tube 1–1.25 mm long, glabrous, limb reduced to a repand limb, c. 0.5 mm long.   Corolla tube 3–3.5 mm long, with a ring of deflexed hairs at the throat; lobes 3–4 × 1–1.5 mm, oblong, acute and thickened at apex. Anthers reflexed. Style 7.5–8 mm long; pollen presenter 1.5 mm long; disk glabrous. Fruit 7.5–8 × 11–12 mm, prominent at the apex; pedicels accrescent.   Pyrene 7 × 5 mm, with a crest at the apex and rounded at the base, fawn-coloured, rugulose.

Mozambique. N: Angoche (António Enes), fr. 17.x.1965, *Gomes e Sousa* 4869 (K; PRE). MS: Sofala Prov., Gorongosa National Park, Sangarassa Forest, fr. i.1972, *Tinley* 2332 (K; PRE). GI: Ponta Barra Falsa, Pomene River, fl. 21.xi.1958, *Mogg* 28962 (K; LISC). M: Inhaca Island, west Coast, fr. 7.iii.1958, *Mogg* 31569 (K; LMU).
Not known outside the Flora Zambesiaca area.   In estuarine riverine swamp forest and mangrove forest, also coastal forest on dunes and low altitude riverine forest, sometimes on termite mounds, often on sandy soils; 0–360 m.

15. **Psydrax micans** (Bullock) Bridson in Kew Bull. **40**: 721 (1985). —K. Coates Palgrave, Trees Southern Africa, ed. 3, rev.: 16k of appendix (1988). —Bridson in F.T.E.A., Rubiaceae: 905 (1991). Type from Tanzania.
    Plectronia lamprophylla K. Schum. in Bot. Jahrb. Syst. **34**: 335 (1904), non *Canthium lamprophyllum* F. Muell. Type as above.
    Canthium micans Bullock in Bull. Misc. Inform., Kew **1932**: 382 (1932), pro parte. — Brenan, Check-list For. Trees Shrubs Tang. Terr.: 484 (1949).

Shrub or small tree 3–4 m tall, or sometimes a liana up to 10 m tall (spines on old stems fide *Mwasumbi* 11629, East Africa), glabrous; bark greyish.   Leaves not restricted to apices of branches, very often ternate; blades 3.5–7.5 × 1.3–4 cm, elliptic or less often narrowly elliptic, acute or obtuse at apex, acute at base, coriaceous, distinctly shiny on upper surface; lateral nerves in 3–4 main pairs; tertiary nerves moderately coarsely reticulate, prominent on both surfaces; domatia present as small blisters with a pin-prick-like puncture; petioles 3–4 mm long; stipules c. 3 mm long, with a shallow truncate base and linear lobe.   Flowers 5-merous borne in subsessile, 2–10-flowered subumbelliform cymes; peduncles not exceeding 2 mm long; pedicels 6–16 (25 fide *Bullock*) mm long, glabrous; bracteoles inconspicuous. Calyx tube 1.5 mm long, glabrous; limb reduced to a dentate rim, ciliate.   Corolla greenish-white; tube 3–4 mm long, with a ring of deflexed hairs above the mid-point inside; lobes 3 × 1.25–1.5 mm, oblong-lanceolate, acute.   Style 7–8 mm long; pollen presenter 1.25 mm; disk glabrous.   Immature fruit 7 × 6 mm; pedicels accrescent, up to 22 mm long.

Mozambique. N: Nampula, Mossuril, Floresta de Cruce, very young fr. 18.ii.1984, *de Koning et al.* 9684 (K; LMU). Z: Maganja da Costa, Gobene Forest, 35 km from Vila de Maganja, fl. bud 10.i.1968, *Torre & Correia* 17016 (LISC). MS: 12 km SE of Sengo, young fr. 15.xii.1971, *Müller & Pope* 2048 (K; LISC; PRE; SRGH).
Also in Tanzania. Low altitude coastal plain in moist forest on high watertable sands, riverine or swamp forest, also in *Androstachys johnsonii* closed forest in *Uapaca* woodland; 20–100 m.
The record of this species given for Zambia by Bullock is erroneous.   It was based on a specimen (*Fries* 1852, determined by Krause as *Plectronia lamprophylla* K. Schum.) which in fact came from Zaire (Dem. Rep. Congo) (Kivu Prov.) and is probably *P. schimperiana*.

Subgen. PHALLARIA

Subgen. PHALLARIA (Schumach. & Thonn.) Bridson in Kew Bull. **40**: 722 (1985).
    Phallaria Schumach. & Thonn., Beskr. Guin. Pl.: 112 (1827). —Hepper, Isert & Thonning Herbarium: 178 (1976).
    Canthium sect. *Pleurogaster* DC., Prodr. **4**: 475 (1830).

Lianas, or at least scandent shrubs; corolla lobes acuminate-apiculate, corolla tube inside with a ring of deflexed hairs below mid-point; pollen presenter widened at the base; fruit didymous, flattened. *Species* 16.

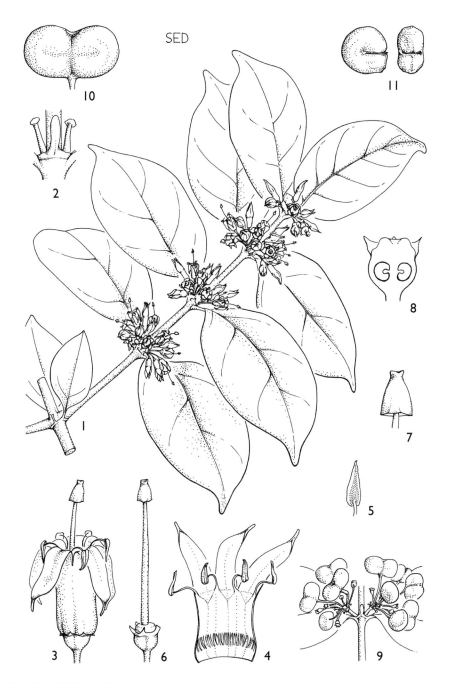

SED

Tab. 71. PSYDRAX KRAUSSIOIDES. 1, flowering branch (× ²⁄₃), from *Burtt* 6580; 2, stipule (× 2), from *Richards* 1156; 3, flower (× 4); 4, part of corolla opened out (× 4); 5, anther, dorsal view (× 8); 6, calyx with style and pollen presenter (× 4); 7, pollen presenter (× 8); 8, longitudinal section through ovary (× 8), 3–8 from *Burtt* 6580; 9, fruiting node (× ²⁄₃); 10, fruit (× 2); 11, pyrene, 2 views (× 2), 9–11 from *Pawek* 2901. Drawn by Sally Dawson. From F.T.E.A.

16. **Psydrax kraussioides** (Hiern) Bridson in Kew Bull. **40**: 723 (1985). —K. Coates Palgrave, Trees Southern Africa, ed. 3, rev.: 16k of appendix (1988). —Bridson in F.T.E.A., Rubiaceae: 907, fig. 161 (1991). TABS. **47**/B4 & **71**. Type from Angola.

Canthium kraussioides Hiern, Cat. Afr. Pl. Welw. **1**: 473 (1898); in F.W.T.A. ed. 1, **2**: 113 (1931).

Plectronia kraussioides (Hiern) K. Schum. in Just's Bot. Jahresber., 26th year, part 1: 393 (1900).

Plectronia pulchra K. Schum. ex De Wild., Études Fl. Katanga [Ann. Mus. Congo, Sér. IV, Bot.] **1**: 229 (1903). —Krause in R.E. Fries, Wiss. Ergebn. Schwed. Rhod.-Kongo-Exped. (Erganzungsheft): 14 (1921). Type from Zaire (Dem. Rep. Congo).

Plectronia malacocarpa K. Schum. & K. Krause in Bot. Jahrb. Syst. **39**: 540 (1907). Type from Tanzania.

Canthium egregium Bullock in Bull. Misc. Inform., Kew **1932**: 348 (1932). —Brenan, Check-list For. Trees Shrubs Tang. Terr.: 485 (1949). Type from Tanzania.

Canthium malacocarpum (K. Schum. & K. Krause) Bullock in Bull. Misc. Inform., Kew **1932**: 348 (1932). —Brenan, Check-list For. Trees Shrubs Tang. Terr.: 485 (1949).

Canthium henriquesianum sensu G. Taylor in Exell, Cat. S. Tomé: 210 (1944), pro parte. —J.G. Garcia in Mem. Junta Invest. Ultramar **6** (sér. 2): 30 (1959) [Contrib. Conhec. Fl. Moçamb. IV (1959)]. —sensu Heppei in F.W.T.A. ed. 2, **2**: 181 (1963). —sensu Drummond in Kirkia **10**: 276 (1975), non (K. Schum.) G. Taylor.

Canthium anomocarpum sensu F. White, F.F.N.R.: 402 (1962), non DC.

Scandent shrub or climber 1.5–7 m tall, with short perpendicular or somewhat backwardly curved lateral branches, glabrous; young branches round to square in cross section, occasionally furrowed; bark fawn-coloured. Leaf blades 4–14 × 1.5–6.5 cm, elliptic to broadly elliptic or sometimes oblong-elliptic, acute to acuminate or sometimes abruptly acuminate at apex, acute or rounded at base, subcoriaceous, moderately shiny to shiny on upper surface; lateral nerves in (4)5–8 main pairs, often impressed above; tertiary nerves completely obscure; domatia absent, or present as unerupted or erupted blisters; petioles 5–9 mm long; stipules 4–7 mm long, truncate at the base and with a linear to narrowly oblong decurrent lobe with plane face perpendicular to the stem. Flowers 5-merous, borne in 2–15-flowered sessile to shortly pedunculate umbellate cymes; peduncles 0–4 mm long; pedicels 5–8 mm long in flowering stage, glabrous; bracteoles glabrous. Calyx tube 1 mm long, glabrous; limb reduced to a dentate ciliate rim. Corolla white or cream; tube 3.75–6 mm long, with a ring of deflexed hairs below the mid-point inside; lobes 3.5–6 × 1.25–2.5 mm, triangular-lanceolate, with a tapering apiculum 0.75–1.75 mm long, sparsely pubescent towards base on upper surface. Anthers reflexed. Style 7.5–10 mm long; pollen presenter 1–1.75 mm long, ± cylindrical, widening towards the base. Disk glabrous. Fruit 7–7.5 × 14–15 mm, almost bispherical, greyish-green, sometimes galled; pedicels accrescent. Pyrene 7 mm wide, semicircular to almost circular in outline, with a shallow groove from the point of attachment to the centre, cartilaginous, scarcely bullate.

Zambia. N: Kasama Distr., Mungwi, fl. 4.ix.1960, *E.A. Robinson* 3805 (K; M). W: Chingola, fr. 27.viii.1954, *Fanshawe* 1504 (K). W: Mwinilunga, fr. 4.x.1955, *Holmes* H1235 (K). C: Chiwefwe, fl. 1.v.1957, *Fanshawe* 3254 (K). **Zimbabwe**. E: Mutare Distr., Stapleford, Nyamkwarara (Nyamakarara) Valley, fl. & fr. 4.xi.1967, *Mavi* 473 (K; LISC; SRGH). **Malawi**. N: Mzimba Distr., Mzuzu, Marymount, fl. & fr. 11.x.1969, *Pawek* 2901 (K). C: Lilongwe Distr., Dzalanyama Forest Reserve, Chaulongwe Falls, fl. 28.iii.1970, *Brummitt* 9479 (K). S: Mulanje Distr., foot of Mt. Mulanje, in Likhubula (Likabula Valley), fl. 21.i.1989, *J.D. & E.G. Chapman* 9483 (K; MO). **Mozambique**. Z: Gurué, Rio Malema, Marope, c. 22 km from Cidade de Gurué, fr. & fl. bud 1.viii.1979, *de Koning* 7506 (BR; LMU). MS: Barûê, Serra de Choa, andados 12 km de Catandica (Vila Gouveia) para o Posto administrativo Choa, fl. & fr. 26.v.1971, *Torre & Correia* 18674 (COI; K; LISC; LMU; PRE).

Also in West Tropical Africa, Burundi, Tanzania, Zaire (Dem. Rep. Congo) and Angola. In riverine evergreen forest and evergreen rainforest by streams, swamp forest (mushitu), riverine thicket vegetation and *Brachystegia* woodland fringing forests; 450–1300 m.

54. KEETIA E. Phillips

Keetia E. Phillips, Gen. S. African Fl. Pl. [Mem. Bot. Survey S. Afr. No. 10]: 587 (1926); in Bothalia **2**: 369 (1927). —Bridson in Kew Bull. **41**: 965–994 (1986).
Canthium sensu Sond. in F.C. **3**: 16 (1865), pro parte. —Hiern in F.T.A. **3**: 132–146 (1877), pro parte. —Bullock in Bull. Misc. Inform., Kew **1932**: 360–373 (1932) pro parte. —Hepper in F.W.T.A. ed. 2, **2**: 181–185 (1963), pro parte, non Lam.

Climbers or scandent shrubs; stems glabrous or frequently pubescent. Leaves paired, petiolate, not restricted to new growth at apex of branches; blades chartaceous or occasionally coriaceous; leaves subtending lateral branches often smaller and broader than main leaves; stipules interpetiolar, lanceolate to ovate or triangular at the base and acuminate to linear above, never keeled, never with white silky hairs inside. Flowers 4–6-merous, borne in pedunculate usually distinctly branched cymes; bracts and bracteoles often conspicuous. Calyx tube ellipsoid to ovoid; limb equalling or sometimes exceeding the tube in length, repand to dentate or less often lobed. Corolla white, cream or yellow; tube cylindrical, usually ± equal to lobes, sometimes shorter or rarely longer, typically with a ring of deflexed hairs inside; lobes reflexed, often thickened at apex but never apiculate. Stamens set at the throat of the corolla; filaments moderately well-developed; anthers fully or partly exserted but usually never reflexed, narrowly ovate or oblong. Style long slender, ± twice as long as the corolla tube; pollen presenter cylindrical, distinctly longer than wide, hollow to below the apex, bifid at apex when mature. Disk puberulous to pubescent or infrequently glabrous. Ovary 2-locular, each loculus containing 1 ovule attached to upper third of the septum. Fruit a 2-seeded drupe, slightly to strongly bilobed, somewhat laterally flattened, slightly to strongly indented at apex; pyrenes woody or less often cartilaginous, usually ± ovoid with the ventral face flattened, somewhat bullate; point of attachment on the ventral face above centre or near apex; lid-like area completely or incompletely defined, either lying along ventral face above point of attachment, or across the apex, provided with a central crest, eventually dehiscent. Seeds ± ovoid, shaped at apex according to position of lid-like area in pyrene, convoluted; endosperm streaked with tanniniferous areas (resembling a ruminate endosperm, except that the testa is never vaginated) occasionally with tannin granules ± evenly dispersed or less often absent; testa thin, very finely reticulate; embryo straight with radicle erect and small cotyledons lying parallel to ventral face of the seed.

A genus of about 40 species, confined to tropical and southern Africa, 4 in the Flora Zambesiaca area.

1. Tertiary venation of leaves finely reticulate; young stems sparsely pubescent to pubescent; pyrene with lid-like area lying across the apex, TAB. **72**/9 · 2
– Tertiary venation of leaves coarsely reticulate; young stems glabrous to sparsely pubescent; pyrene with lid-like area lying along the ventral face, TAB. **47**/D1 · · · · · · · · · · · · · · · · 3
2. Stipules lanceolate to ovate, gradually acuminate; tertiary venation of leaves evenly reticulate; leaf blade glabrescent to densely pubescent beneath · · · · · · · · · · · 1. *gueinzii*
– Stipules triangular at base, narrowing rather abruptly to a linear-acuminate lobe; tertiary venation scalariform (with those veins perpendicular to midrib more pronounced); leaf blade glabrous save for the nerves beneath, or occasionally pubescent · · · · · · · 2. *venosa*
3. Domatia present on leaf blades as pubescent tufts; pedicels sparsely to densely pubescent; style 5–8 mm long; pollen presenter 0.5–1.25 mm long; pyrenes 8–11 mm long · · · · · · · ·
· 3. *zanzibarica*
– Domatia present as glabrous or ciliate pits; pedicels glabrous to sparsely pubescent; style 8–10 mm long; pollen presenter 1–1.75 mm long; pyrenes 12–15 mm long · · · · 4. *foetida*

1. **Keetia gueinzii** (Sond.) Bridson in Kew Bull. **41**: 970, fig. 1 A–C (1986). —K. Coates Palgrave, Trees Southern Africa, ed. 3, rev.: 16k of appendix (1988). —Bridson in F.T.E.A., Rubiaceae: 911, figs. 131/16, 25; 132/25 & 162 (1991). —Pooley, Trees of Natal, Zululand & Transkei: 478, figs. (1993). —Beentje, Kenya Trees, Shrubs Lianas: 517 (1994). TABS. **47**/B3 & **72**. Type from South Africa.
Canthium gueinzii Sond. in Linnaea **23**: 54 (1850); in F.C. **3**: 16 (1865). —S. Moore in J. Linn. Soc., Bot. **40**: 89 (1911). —Bullock in Hooker's Icon. Pl. ser. 5, **2**: t. 3170 (1932); in Bull. Misc. Inform., Kew **1932**: 368 (1932). —Brenan, Check-list For. Trees Shrubs Tang. Terr.: 487 (1949); in Mem. N.Y. Bot. Gard. **8**: 452 (1954). —Dale & Greenway, Kenya Trees & Shrubs: 428 (1961). —F. White, F.F.N.R.: 403, fig. 68, I (1962). —Palmer & Pitman, Trees Southern Africa **3**: 2093 (1973). —Ross, Fl. Natal: 335 (1973). —Drummond in Kirkia **10**: 276 (1975). —Gibson, Wild Fls. Natal: 6, pl. 102 (1975). —Compton, Fl. Swaziland [J. S. African Bot., Suppl. 11]: 580 (1976). —Moll, Trees of Natal: 263 & 247 (1981). —A.E. Gonçalves in Garcia de Orta, Sér. Bot. **5**: 187 (1982). —Bridson & Troupin in Fl. Pl. Lign. Rwanda: 548, fig. 183.1 (1982); in Fl. Rwanda **3**: 148, fig. 45.1 (1985). —K. Coates Palgrave, Trees Southern Africa, ed. 3, rev.: 881 (1988).

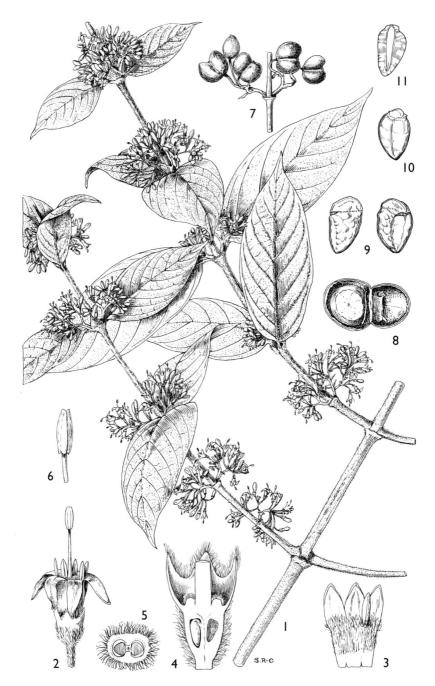

Tab. 72. KEETIA GUEINZII. 1, flowering branch (× ²⁄₃); 2, flower (× 4); 3, part of corolla opened out (× 4); 4, longitudinal section through calyx and ovary (× 12); 5, transverse section through ovary (× 12); 6, pollen presenter (× 8); 7, infructescences (× ²⁄₃); 8, transverse section through fruit (× 2), 1–8 collector unknown; 9, pyrene, 2 views (× 2); 10, seed (× 2); 11, longitudinal section through seed (× 2), 9–11 from *White* 3196. Drawn by Stella Ross-Craig, with 9–11 by Sally Dawson. From F.W.T.A.

Canthium hispidum sensu Hiern in F.T.A. **3**: 140 (1877), pro parte, quoad *Soyaux* 196; Cat. Afr. Pl. Welw. **1**: 476 (1898). —sensu Bullock in Bull. Misc. Inform., Kew **1932**: 369 (1932). —sensu Brenan, Check-list For. Trees Shrubs Tang. Terr.: 486 (1949). —sensu Dale & Greenway, Kenya Trees & Shrubs: 429 (1961), non Benth.

Plectronia hispida sensu K. Schum. in Engler, Pflanzenw. Ost-Afrikas **C**: 386 (1895), pro parte, non (Benth.) K. Schum.

Plectronia gueinzii (Sond.) Sim, For. Fl. Col. Cape Good Hope: 241 (1907). —Eyles in Trans. Roy. Soc. South Africa **5**: 493 (1916). —Bews, Fl. Natal & Zululand: 198 (1921).

Plectronia charadrophila K. Krause in Bot. Jahrb. Syst. **57**: 36 (1920). Type from Tanzania.

Plectronia subcordatifolia De Wild., Pl. Bequaert. **3**: 199 (1925). Type from Zaire (Dem. Rep. Congo).

Keetia transvaalensis E. Phillips in Bothalia **2**: 239 (1927), pro parte. Type from South Africa.

Canthium scabrosum Bullock in Bull. Misc. Inform., Kew **1932**: 367 (1932). —Brenan, Check-list For. Trees Shrubs Tang. Terr.: 486 (1949). Type from Uganda.

Canthium charadrophilum (K. Krause) Bullock in Bull. Misc. Inform., Kew **1932**: 369 (1932).

Canthium sylvaticum sensu Bullock in Bull. Misc. Inform., Kew **1932**: 369 (1932), pro parte. —sensu Brenan, Check-list For. Trees Shrubs Tang. Terr.: 486 (1949), pro parte, non Hiern.

Keetia gueinzii von Breitenbach in J. Dendrol. **5**: 90, fig. on p. 91 (1985), nom invalid.

Scandent shrub or liana, 3–25 m tall; young branches sparsely to densely covered with crisped or spreading golden to rust-coloured hairs. Leaf blades 5.5–13.5 × 3.5–6 cm, oblong-lanceolate to ovate, acuminate at apex, rounded, truncate or more frequently subcordate to cordate at base, often bullate, usually drying brown, glabrous to sparsely pubescent on upper surface, glabrescent to densely pubescent beneath; lateral nerves in 6–9 main pairs; tertiary venation finely reticulate; domatia present as tufts of hair; petioles 3–7 mm long sparsely to densely covered with crisped or patent hairs; stipules 9–13 mm long, up to 6 mm wide at base, lanceolate to ovate, gradually acuminate, pubescent outside; leaves subtending lateral branches smaller, orbicular. Flowers 5-merous, borne in 20–50-flowered pedunculate cymes; peduncles 5–15 mm long, sparsely to densely pubescent; pedicels 5–7 mm long, pubescent to densely pubescent; bracteoles 3–6 mm long, linear-lanceolate to lanceolate. Calyx tube 1 mm long, densely covered with straight or crisped hairs or occasionally glabrescent; limb 1.25–1.5 mm long, divided into teeth for one third to half its length, glabrescent to sparsely pubescent, usually ciliate. Corolla creamy-white; tube 2.25–4 mm long, with a ring of deflexed hairs just below the throat inside; lobes 2.5–4 × 1.25–2.25 mm, oblong-lanceolate to ovate, acute and thickened at apex. Anthers fully exserted but seldom reflexed. Style 5–10 mm long, glabrous; pollen presenter 1.25–2.25 mm long; disk pubescent. Fruit 7–9 × 11–14 mm, broadly oblong in outline, slightly indented at apex, glabrous to glabrescent, black when mature. Pyrene 9–11 × 6–7 mm, obovoid with ventral face flattened or hemispherical; lid-like area lying across the apex with a central crest. Seed with endosperm streaked with granules.

Zambia. N: Mbala Distr., Chinakila, path to Loye Flats, fl. 10.i.1965, *Richards* 19454 (BR; K). W: Ndola, fl. & fr. 1.x.1954, *Fanshawe* 1586 (BR; K). C: Serenje Distr., Kundalila Falls, fl. 13.x.1963, *E.A. Robinson* 5700 (K). **Zimbabwe**. E: Mutare Distr., Vumba Mts., Nimbus Farm, fl. 15.x.1959, *Chase* 7019 (BR; K; LISC; PRE; SRGH). **Malawi**. N: Chitipa Distr., Mafinga Hills (Mts.), middle slopes below Namitawa summit, fl. 3.iii.1982, *Brummitt, Polhill & Banda* 16285 (K). C: Dedza Distr., Chongoni Forest, dambo below staff house at school, fr. 28.ix.1962, *Adlard* 491 (K; LISC; SRGH). S: Mulanje Distr., Chapaluka (Chapanuka) Stream, fl. 4.xi.1972, *Leach* 14939 (BR; K; PRE; SRGH). **Mozambique**. N: Ribáuè, Mepáluè, fl. 9.xii.1967, *Torre & Correia* 16430 (LISC). Z: Milange, Serra Tumbine, fl. 19.i.1966, *Correia* 473 (LISC). T: between Vila Mouzinho and Zóbuè, fr. 19.vii.1949, *Barbosa & Carvalho* 3680 (LISC). MS: Manica, Rotanda between Mussapa River and frontier at Tandara (?Mutambara in Zimbabwe), fl. 19.xi.1965, *Torre & Correia* 13143 (LISC).

Also in Cameroon, Central African Republic, Sudan, Ethiopia, Rwanda, Burundi, Uganda, Kenya, Tanzania, Zaire (Dem. Rep. Congo), Angola and South Africa (North Prov., Mpumalanga and Eastern Cape Prov.). Locally frequent at forest margins, and in most types of woodland, also in thickets; 800–2200 m.

2. **Keetia venosa** (Oliv.) Bridson in Kew Bull. **41**: 974 (1986). —K. Coates Palgrave, Trees Southern Africa, ed. 3, rev.: 16k of appendix (1988). —Bridson in F.T.E.A., Rubiaceae: 914 (1991). —Beentje, Kenya Trees, Shrubs Lianas: 518 (1994). Type from Uganda.

Plectronia venosa Oliv. in Trans. Linn. Soc., London **29**: 85, t. 49 (1873). —K. Schumann in Engler, Pflanzenw. Ost-Afrikas **C**: 386 (1895). —De Wildeman, Pl. Bequaert. **3**: 200 (1925). *Canthium venosum* (Oliv.) Hiern in F.T.A. **3**: 144 (1877). —Bullock in Bull. Misc. Inform., Kew **1932**: 371, fig. 1 (1932). —Brenan, Check-list For. Trees Shrubs Tang. Terr.: 487 (1949). —J.G. Garcia in Mem. Junta Invest. Ultramar **6** (sér. 2): 29 (1959) [Contrib. Conhec. Fl. Moçamb. IV (1959)]. —Dale & Greenway, Kenya Trees & Shrubs: 433 (1961). —F. White, F.F.N.R.: 403 (1962). —Hepper in F.W.T.A. ed. 2, **2**: 184 (1963), pro parte. — Drummond in Kirkia **10**: 276 (1975). —K. Coates Palgrave, Trees Southern Africa, ed. 3, rev.: 887 (1988). —Bridson & Troupin in Fl. Pl. Lign. Rwanda: 552, fig. 184.3 (1982); in Fl. Rwanda **3**: 150, fig. 45.3 (1985).
   Canthium venosum var. *pubescens* Hiern in F.T.A. **3**: 144 (1877). Type from Nigeria.
   Canthium barteri Hiern in F.T.A. **3**: 143 (1877). Type from Nigeria.
   Plectronia cuspido-stipulata K. Schum. ex Engler in Abh. Preuss. Akad. Wiss.: 53 (1894), nomen, based on *Holst* 2426.
   Plectronia hispida sensu K. Schum. in Engler, Pflanzenw. Ost-Afrikas **C**: 386 (1895), pro parte, quoad syn. *P. cuspido-stipulata*, non (Benth.) K. Schum.
   Canthium sylvaticum Hiern, Cat. Afr. Pl. Welw. **1**: 477 (1898). —Bullock in Bull. Misc. Inform., Kew **1932**: 369 (1932), pro parte. —Brenan, Check-list For. Trees Shrubs Tang. Terr.: 487 (1949), pro parte. Type from Angola.
   Plectronia sylvatica (Hiern) K. Schum. in Just's Bot. Jahresber., 26th year, part 1: 393 (1900).
   Plectronia barteri (Hiern) De Wild. in Ann. Mus. Congo, Sér. II, Bot. **1**, 2: 33 (1900).
   Plectronia stipulata De Wild. in Pl. Nov. Herb. Horti. Then. **1**: 171, t. 38 (1905). Type: Mozambique, Morrumbala Forest, *Luja* 430 (BR, holotype).
   Plectronia myriantha K. Krause in Mildbraed, Wiss. Ergebn. Deutsch. Zentr.-Afrika Exped., Bot., part 4: 327 (1911), non Schltr. & K. Krause 1908. Type from Zaire (Dem. Rep. Congo)/Uganda.
   Plectronia vanderysti De Wild. in Bull. Jard. Bot. État **5**: 31 (1915). Types from Zaire (Dem. Rep. Congo).
   Plectronia dundusanensis De Wild., Pl. Bequaert. **3**: 183 (1925). Types from Zaire (Dem. Rep. Congo).
   Plectronia reygaerti De Wild., Pl. Bequaert. **3**: 195 (1925). Types from Zaire (Dem. Rep. Congo).
   Keetia transvaalensis sensu E. Phillips in Bothalia **2**: 369 (1927), pro parte, quoad *Schlechter* 12290, non sensu stricto.
   Canthium gueinzii sensu Hepper in F.W.T.A. ed. 2, **2**: 184 (1963), pro parte.
   Canthium zanzibaricum sensu Hepper in F.W.T.A. ed. 2, **2**: 184 (1963), pro parte.
   Canthium dundusanense (De Wild.) Evrard in Bull. Jard. Bot. État **37**: 457 (1967).
   Canthium sp. C of Bridson & Troupin in Fl. Pl. Lign. Rwanda: 554, fig. 185.2 (1982).

Scandent shrub or climber 2–7 m tall; young branches sparsely to densely covered with rusty-coloured hairs. Leaf blades 4.5–14 × 2.5–6(8) cm, oblong-elliptic or narrowly to broadly elliptic rarely round, acuminate at apex, obtuse to rounded or rarely subcordate at base, glabrous or rarely glabrescent on upper surface, glabrous save for the sparsely to densely pubescent nerves, or occasionally sparsely pubescent beneath; lateral nerves in 5–9 main pairs; tertiary venation finely reticulate, always with element perpendicular to midrib the more conspicuous giving a scalariform appearance and frequently either raised or impressed above; domatia present as inconspicuous tufts of hair; petioles 5–15 mm long, pubescent; stipules 5–14(17) mm long, up to 5 mm wide at base, triangular at base, usually rather abruptly narrowing to a linear or acuminate apex, pubescent outside; leaves subtending lateral branches smaller, ± orbicular. Flowers 4–5(6)-merous, borne in 20–70-flowered pedunculate cymes; peduncles 5–17 mm long, pubescent or rarely glabrous; pedicels 2–5 mm long, pubescent or rarely glabrous; bracteoles 2.5–6 mm long, linear-lanceolate. Calyx tube 0.75–1 mm long, glabrous or pubescent; limb 0.75–1.25 mm long, divided into teeth usually for less than half its length, ciliate. Corolla creamy-white; tube 2.25–3 mm long, with a ring of deflexed hairs just below the throat inside; lobes 1.5–2.25 × 1–1.75 mm, lanceolate to ovate, acute. Anthers fully exserted but seldom reflexing. Style 4–6.5 mm long, glabrous; pollen presenter 0.75–1.25 mm long; disk pubescent. Fruit 8–11 × 11–15 mm, slightly indented at apex, glabrous, black when mature. Pyrene 6–7 × 5–6 mm, hemispherical to suborbicular, with lid-like area lying across the apex with a small crest. Seed with endosperm streaked with granules.

Zambia. B: Kabompo Distr., 59 km WSW of Kabompo on Zambezi (Balovale) road, fl. bud 24.iii.1961, *Drummond & Rutherford-Smith* 7305 (K; LISC; PRE; SRGH). N: Chinsali Distr., near Great North Road, 48 km north of Mpika, fr. 28.xi.1952, *White* 3763 (FHO; K). W: Ndola, fr. 8.xii.1953, *Fanshawe* 556 (K). C: Kamaila Forest Reserve, 40 km north of Lusaka, fr. 12.ix.1963,

Angus 3735 (FHO; K); Luangwa Valley, Mutinondo Stream, Muchinga Escarpment, corolla fallen 23.viii.1966, *Astle* 4975 (K). S: Pemba, fl. 1933, *Trapnell* 1303 (K). **Zimbabwe.** E: Chirinda Forest, fl. & fr. 18.x.1947, *Wild* 2026 (K; SRGH). **Malawi.** N: Nkhata Bay Distr., border of Chombe Tea Estate near Mweza Village, fl. 13.iv.1960, *Adlard* 337 (K; LISC; SRGH). S: Mulanje Distr., Likhubula River, above Likhubula House, west side of Mt. Mulanje, fl. 6.v.1980, *Blackmore & Brummitt* 1467 (K). **Mozambique.** N: between Mocimboa da Praia and Diaca, c. 20 km from Mocimboa da Praia, fl. bud 25.iii.1961, *Gomes e Sousa* 4666 (COI; K; PRE). Z: Milange, fl. 24.ii.1943, *Torre* 4831 (LISC). MS: Sofala Prov., Inhansato, region of Inhaminga, fl. 13.iii.1955, *Gomes e Sousa* 4286 (COI; K). M: Maputo, corolla fallen 1908, *Sim* 20860 (PRE).

Also in West Tropical Africa, Cameroon, Central African Republic, Sudan, Rwanda, Burundi, Uganda, Kenya, Tanzania, Zaire (Dem. Rep. Congo) and Angola. In evergreen forest understorey, swamp forest (mushitu) and riverine vegetation, in mixed deciduous woodlands on Kalahari Sand and in miombo woodlands often on termitaria, on sandy soils; 50–1525 m.

3. **Keetia zanzibarica** (Klotzsch) Bridson in Kew Bull. **41**: 979, fig. 1, D–F (1986); in F.T.E.A., Rubiaceae: 917 (1991). —Beentje, Kenya Trees, Shrubs Lianas: 518 (1994). TAB. **47**/D1 & D2. Type from Zanzibar Island.

    Canthium zanzibaricum Klotzsch in Peters, Naturw. Reise Mossamb. **6** part 1: 291 (1861). —Hiern in F.T.A. **3**: 138 (1877), pro parte. —Bullock in Bull. Misc. Inform., Kew **1932**: 373 (1932), pro parte. —Brenan, Check-list For. Trees Shrubs Tang. Terr.: 488 (1949). —J.G. Garcia in Mem. Junta Invest. Ultramar **6** (sér. 2): 31 (1959) [Contrib. Conhec. Fl. Moçamb. IV (1959)]. —Dale & Greenway, Kenya Trees & Shrubs: 433 (1961).

    Plectronia zanzibarica (Klotzsch) Vatke in Oesterr. Bot. Z. **25**: 231 (1875). —Engler in Abh. Preuss. Akad. Wiss.: 26 (1894).

    Canthium zanzibaricum var. *glabristylum* sensu Hiern in F.T.A. **3**: 139 (1877), pro parte.

Scandent shrubs, small trees or lianas; young stems glabrous to sparsely pubescent or very occasionally pubescent; bark pale greyish. Leaf blade 5–15 × 2–7.5 cm, narrowly elliptic to round or sometimes ovate to broadly ovate, acuminate or less often obtuse to acute and frequently apiculate at the apex, obtuse rounded truncate or rarely subcordate at the base, usually chartaceous or occasionally almost subcoriaceous, dull or sometimes shiny on upper surface, entirely glabrous with the nerves pubescent beneath or glabrescent to pubescent beneath; lateral nerves in 7–9(10) main pairs; tertiary venation coarsely reticulate; domatia usually present as hairy tufts; petiole 5–15 mm, glabrous to pubescent; stipules 4–12 mm long, with a triangular base topped by a subulate to linear lobe; leaves subtending lateral branches smaller but scarcely differing in shape. Flowers sweetly scented, 4–5-merous, borne in 30–60-flowered pedunculate cymes; peduncles 5–15 mm long, glabrescent to sparsely pubescent; pedicels 2–8 mm long, glabrescent to densely pubescent; bracteoles up to 2 mm long. Calyx tube 0.75–1 mm long, glabrous to densely pubescent; limb dentate 0.5–1 mm long, glabrous except for the ciliate teeth or sometimes pubescent. Corolla white; tube 2–3.25 mm long, with a ring of deflexed hairs above the mid-point inside; lobes 1.5–2.75 × 1–1.5 mm, oblong-lanceolate, acute to obtuse. Anthers fully exserted, erect. Style 5–8 mm long, glabrous or pubescent; pollen presenter 0.5–1.25 mm long; disk pubescent. Fruit 8–13 × 12–17 mm, oblong or obcordate in outline. Pyrene 8–11 × 7.5–8 mm, broadly ellipsoid with ventral face flattened; area above point of attachment on ventral plane circular with a central crest.

Subsp. **zanzibarica** —Bridson in Kew Bull. **41**: 979 (1986); in F.T.E.A., Rubiaceae: 918 (1991).

Calyx tube glabrous, or sometimes hairy towards the base. Style distinctly pubescent, with a few hairs or rarely glabrous; pollen presenter (0.75)1–1.25 mm long. Leaf blades ovate to broadly ovate, sometimes round to broadly elliptic, rounded to truncate, rarely subcordate or occasionally obtuse at the base, glabrous or with the petioles and nerves beneath very sparsely hairy, always dull on upper surface.

Mozambique. N: Nampula, fl. 14.iv.1937, *Torre* 1363 (COI; LISC). Z: Lugela–Mocuba Distr., Namagoa Estate, fl. & fr. vi.ix.1944, *Faulkner* in PRE 46 (BR; COI; K; PRE). MS: right bank of Luaua (Luabo) R., fr. 17.v.1858, *Kirk* 36 (K).

Also in coastal regions of Kenya and Tanzania. In riverine forest; c. 60–120 m.

Some specimens from Mozambique (N) at altitudes over 450 m are ± intermediate between subspecies *zanzibarica* and *cornelioides*, e.g. *Gomes e Sousa* 4131A (K; LISC; PRE) from Mutuáli, between rivers Nalume and Nualo, near Catholic Mission, fl. 26.ix.1953.

Subsp. **cornelioides** (De Wild.) Bridson in Kew Bull. **41**: 979 (1986); in F.T.E.A., Rubiaceae: 918 (1991). Type from Zaire (Dem. Rep. Congo) (Shaba Prov.).

> *Canthium zanzibaricum* var. *glabristylum* Hiern in F.T.A. **3**: 139 (1877), pro parte. Type: Malawi, west shore of Lake Malawi, near entrance of Dwangwa (Roangwa) R., *Kirk* s.n. (K, lectotype).
>
> *Plectronia cornelioides* De Wild., Études Fl. Katanga [Ann. Mus. Congo, Sér. IV, Bot.] **1**: 159 (1903).
>
> *Plectronia hispida* var. *glabrescens* K. Krause in Wiss. Ergebn. Schwed. Rhod.-Kongo-Exped. (Erganzungsheft): 15 (1921), nomen, based on *Fries* 556 & 556a.
>
> *Keetia transvaalensis* sensu E. Phillips in Bothalia **2**: 369 (1927), pro parte, quoad *Borle* 293, non sensu stricto.
>
> *Canthium zanzibaricum* sensu Brenan in Mem. N.Y. Bot. Gard. **8**: 452 (1954). —F. White, F.F.N.R.: 403 (1962), non sensu stricto.

Calyx tube sparsely to densely pubescent or occasionally only pubescent towards the base. Style glabrous, very rarely with a few hairs; pollen presenter 0.75–1(1.25) mm long. Leaf blades oblong-ovate or occasionally elliptic or ovate, obtuse to rounded or infrequently truncate at the base, glabrous or glabrescent to pubescent beneath, dull to somewhat shiny on upper surface.

Zambia. B: near Senanga, fl. 30.vii.1952, *Codd* 7258 (BR; EA; K; PRE). N: Tambatamba, Lufila River, fl. 13.viii.1938, *Greenway & Trapnell* 5602 (EA; K). W: Kitwe, Kafue R., fl. 18.viii.1967, *Mutimushi* 2021 (K). C: Mkushi, fr. 2.v.1957, *Fanshawe* 3275 (K). E: Lundazi, fl. 20.x.1967, *Mutimushi* 2222 (K). S: Namwala, on banks of Kafue River at Iteshi Teshi (Meshi Teshi), fl. 24.ix.1963, *van Rensburg* 2473 (K). **Malawi**. N: Rumphi Distr., c. 3 km east of Rumphi Gorge, fr. 27.vii.1975, *Pawek* 9905 (K). C: Nkhota Kota Distr., Chia area, fl. 5.ix.1946, *Brass* 17540 (K; PRE). S: Zomba and vicinity, young fr. xii.1896, *Whyte* s.n. (K). **Mozambique**. Z: Milange, fl. 10.ix.1941, *Torre* 3385 (LISC).

Also in Zaire (Dem. Rep. Congo) (Shaba Prov.). Frequent on river banks, lake shores and islands, often in seasonally inundated localities, in evergreen riverine forest margins and in riverine vegetation where it can become locally common forming thickets, also in swamp forest (mushitu) and *Brachystegia* woodland; 480–1170 m.

In my opinion Bullock's inclusion of the Angolan *Canthium gracilis* Hiern and *C. tenuiflorum* Hiern in his synonymy for *Canthium zanzibaricum* was erroneous.

4. **Keetia foetida** (Hiern) Bridson in Kew Bull. **41**: 981 (1986). Type: Malawi, Mpatamanga (Muata Manga), *Kirk* s.n. (K, holotype).

> *Canthium foetidum* Hiern in F.T.A. **3**: 142 (1877).
>
> *Plectronia foetida* (Hiern) K. Schum. in Engler, Pflanzenw. Ost-Afrikas **C**: 386 (1895).
>
> *Canthium tophami* Bullock & Dunkley in Bull. Misc. Inform., Kew **1937**: 420 (1937). Type: Malawi, Zomba Distr., Namitawa, *Clements* 561 (K, holotype).

Climber, or semi-climber, with a fluted trunk; young stems glabrous. Leaf blades 8.5–15.5 × 3–7.5 cm, oblong-elliptic to broadly oblong-elliptic, acute or more often gradually acuminate at apex, sometimes apiculate, rounded and sometimes unequal at base, usually subcoriaceous, somewhat shiny on upper surface, glabrous; lateral nerves in 7(8) main pairs; tertiary venation coarsely reticulate, not very conspicuous; domatia present as glabrous or ciliate pits, sometimes conspicuous; petiole 9–15 mm long, glabrous; stipules 5–12 mm long, with a broadly triangular base topped by a linear lobe; leaves subtending lateral branches all fallen in specimens seen. Flowers with an unpleasant odour, 5-merous, borne in many-flowered pedunculate cymes; peduncles 7–13 mm long, glabrous to sparsely pubescent; pedicels 5–12 mm long, glabrous to sparsely pubescent; bracteoles up to 1.5 mm long. Calyx tube c. 1 mm long, glabrous to glabrescent; limb 0.5–0.75 mm long, dentate, ciliate. Corolla tube 3.5–4.5 mm long with a ring of deflexed hairs at mid-point inside; lobes 2–3 × 1.25–1.5 mm, oblong, lanceolate to ovate, acute. Anthers fully exserted, erect. Style 8–10 mm long, glabrous; pollen presenter 1–1.75 mm long; disk pubescent. Fruit 12–15 × 16–18 mm, scarcely laterally compressed, indented at apex, glabrous. Pyrene 12–15 × 11 × 10 mm, broadly ellipsoid with ventral face flattened; lid-like area on ventral face well defined, circular with a prominent central crest.

Malawi. S: foot of Chiradzulu Mt., SW side near Lisau Forestry Depot, fr. 14.iv.1970, *Brummitt & Banda* 9830 (K).

Not known outside the Flora Zambesiaca area. In evergreen rainforest; 990–1250 m.

This species is very close to *K. zanzibarica* and can be distinguished chiefly by its glabrous to sparsely pubescent inflorescence branches and rather thicker leaves with pit-like domatia. Also the fruits are larger than in typical *K. zanzibarica* subsp. *cornelioides*. The flowers of *K. foetida* are invariably recorded as having a markedly unpleasant odour, while those of *K. zanzibarica* are either said to be sweetly scented or their scent is not mentioned. *K. foetida* may be sympatric with *K. zanzibarica* in the Zomba District; modern collections of the latter are however lacking and additional collections with good field notes from this area would be of great interest.

Vanguerieae genus unknown.

Shrub c. 2 m tall; stems at first densely fulvous pubescent, later glabrous and with a flaking epidermis revealing a minutely flaking red-brown underlayer. Leaves paired, 0.8–6.5 × 0.7–4 cm, elliptic, rounded at the apex, cuneate at the base, very discolorous, densely shortly pubescent on upper surface, thickly fulvous-velvety beneath; petiole 2–10 mm long, velvety; stipule base 1.5 mm long with dark subulate appendage c. 3 mm long. Cymes 1–3-flowered, known only in young fruit; peduncle 3 mm long; pedicels 2–6 mm long. Calyx limb poorly developed, probably minutely toothed. Ovary 2(?3)-locular.

Mozambique. Z: Milange, about 4 km from Sabelua towards Môngoè, about 10.8 km into the forest in the direction of Serra do Chiperone, 17.ii.1972, *Correia & Marques* 2701 (BR; LMU).

Rocky hillsides with *Sterculia appendiculata, Heteropyxis, Ficus, Combretum, Terminalia, Pterocarpus, Diplorhynchus* and *Commiphora*; 700 m.

Technically this is a *Rytigynia* but without corollas it is best left doubtful.

INDEX TO BOTANICAL NAMES

371